Food Processing Technologies: Quality and Safety

Food Processing Technologies: Quality and Safety

Edited by Dorothy Green

SYRAWOOD
PUBLISHING HOUSE

New York

Published by Syrawood Publishing House,
750 Third Avenue, 9th Floor,
New York, NY 10017, USA
www.syrawoodpublishinghouse.com

Food Processing Technologies: Quality and Safety
Edited by Dorothy Green

International Standard Book Number: 978-1-68286-574-3 (Hardback)

Cataloging-in-Publication Data

Food processing technologies : quality and safety / edited by Dorothy Green.
    p. cm.
Includes bibliographical references and index.
ISBN 978-1-68286-574-3
1. Food industry and trade. 2. Food industry and trade--Technological innovations.
3. Food--Quality. 4. Food industry and trade--Safety measures. I. Green, Dorothy.
TP370 .F66 2018
664.02--dc23

TABLE OF CONTENTS

PREFACE

The process of converting raw materials into edible items is known as food processing. These processes can be chemical or physical in nature. Some of the popular techniques of food processing include pasteurization, mining, preservation, liquefaction among many others. The objective of this book is to give a general view of the different areas of food processing and its applications. Scientists and students actively engaged in this field will find this book full of crucial and unexplored concepts.

This book unites the global concepts and researches in an organized manner for a comprehensive understanding of the subject. It is a ripe text for all researchers, students, scientists or anyone else who is interested in acquiring a better knowledge of this dynamic field.

I extend my sincere thanks to the contributors for such eloquent research chapters. Finally, I thank my family for being a source of support and help.

Editor

Malting Process Effects on Antioxidant Activity of Finger Millet and Amaranth Grains

Sara Najdi Hejazi[1] & Valérie Orsat[1]

[1] Bioresource Engineering Department, McGill University, Ste-Anne-de-Bellevue, Canada

Correspondence: Sara Najdi Hejazi, Bioresource Engineering Department, McGill University, Qc, H9X 3V9, Canada. Tel: 1-514-992-4192. E-mail: sara.najdihejazi@mail.mcgill.ca

Abstract

Finger millet (Eleusine coracana) and amaranth (Amaranthus caudatus) are two nutritious and gluten-free grains with high contents of phenolic compounds. Phenols are known as the main source of antioxidants, with numerous health benefits. Being rich in phenol makes these grains good choices for the functional food industry. In this study, the effects of malting/germination factors, duration and temperature, on the phenolic content and antioxidant activities of these grains are thoroughly investigated and optimized. Based on a central composite design, the grains were germinated for 24, 36, and 48 hrs at 22, 26, and 30 °C. Both temperature and duration factors are found to be significantly influential on the monitored quantities. While malting of amaranth grains for 48 hrs at 26 °C increased the total phenol content four times, in case of millet, a 25% reduction was observed. Linear correlations between the included phenol content and antioxidant activity in terms of DPPH and ABTS scavenging activities were observed.

Keywords: Amaranth, finger millet, malting/germination process, phenol content, ABTS, DPPH, colorimetric properties

1. Introduction

Even though oxidation reactions are important for cellular metabolism, they may threaten cell normal functions, especially in the case of chain reactions (Banerjee, Sanjay, Chethan, & Malleshi, 2012). When a chain oxidation reaction initiates, excessive amounts of free radicals are produced that may cause serious damage or death to the cell. These excessive free radicals may cause oxidative stress (Hegde, Rajasekaran, & Chandra, 2005). Oxidative stress plays a significant role in many human body malfunctions, most importantly coronary heart disease and cancer (Najdi Hejazi, 2012; Saleh, Zhang, Chen, & Shen, 2013).

Antioxidants are molecules that prevent oxidation of other components by being oxidized themselves (Viswanath, Urooj, & Malleshi, 2009). In fact, antioxidants are known as reducing agents that terminate the chain reactions through removing free radical intermediates and inhibit excessive oxidation reactions. In the food industry, antioxidants act as natural preservatives in the products by preventing degradation of their lipid and protein contents (Venskutonis & Kraujalis, 2013). In cereal grains, polyphenolic compounds are assumed to be the main source of antioxidants (Towo, Svanberg, & Ndossi, 2003). Naturally, polyphenols are responsible in the defense of plants against ultraviolet radiations and pathogens aggression (Banerjee et al., 2012). These compounds are referred to as the secondary metabolites of plants. Epidemiological studies have proven that long term consumption of cereal grains, which contain appreciable amounts of phytochemicals, may significantly reduce oxidative stress in the body and keep the desired balance between oxidants and antioxidants levels (Dykes & Rooney, 2007; Hegde et al., 2005).

Recently, with increasing concerns about worldwide food security, increasing attention is given on the cultivation of drought-resistant crops that can be grown in arid and semi-arid regions (Malleshi & Desikachar, 1986). Among them, millet and amaranth grains are gaining significant appreciation. Having good sensory qualities, short growing season, and appropriate nutrient profiles besides being gluten-free are the main factors for this attention (Belton & Taylor, 2002; Charalampopoulos, Wang, Pandiella, & Webb, 2002).The high phenolic contents of these seeds could represent a good source of antioxidant, particularly in the arid regions, where other commercial crops cannot be grown (de la Rosa et al., 2009; Shobana et al., 2012). Among a vast variety of amaranth grains, three of them are of more importance, Amaranthus caudatusm, Amaranthus cruentus, and Amaranthus hypochondriacus (Venskutonis & Kraujalis, 2013). Millets are categorized into two groups; major millets, which include species that are most widely cultivated, and minor millets (Issoufou, Mahamadou, & Guo-Wei, 2013). While Sorghum and Pearl Millet belongs to the first category, Finger Millet, Foxtail Millet, Little Millet, Barnyard Millet, Proso Millet, and Kodo Millet are mostly recognized as minor millets (Obilana & Manyasa, 2002). In this study, total phenolic content and antioxidant activities of Finger millet (Eleusine coracana) and Amaranth (Caudatus amaranthus) are investigated. In addition, the effect of a malting/germination process on the resulting phenolic content and antioxidant activities is explored.

A study performed on two varieties of Caudatus amaranth, Centenario and Oscar Blanco, demonstrated that their total phenolic contents were 98.7 and 112.9 mg/100 g, respectively, using Gallic acid as the reference (Repo-Carrasco-Valencia, Pena, Kallio, & Salminen, 2009). In addition, the authors reported that antioxidant activities using DPPH method were 410 and 398.1 µMol Trolox/g for Centenario and Oscar Blanco, respectively. The ABTS activities were reported at 827.6 and 670.1 µMol Trolox/g, respectively.

Pasko et al., (2009) studied the phenolic contents and antioxidant activity of two varieties, Aztec and Rawa, of Cruentus amaranth. It has been shown that for the native seeds, average total phenols were 2.95 and 3.0 mg GAE/kg, average DPPH activities were 4.42 and 3.15 mMol Trolox/kg, and average ABTS activities were 12.71 and 11.42 mMol Trolox/kg for Aztec and Rawa seeds, respectively. Besides, in the performed study, effects of germination duration on these quantities were investigated. Grains were germinated at room temperature between four to seven days in daylight or darkness. It has been found that sprout antioxidant activities depend on the growth duration, where maximum values were observed at the 4th day. A significant increase in the DPPH and ABTS scavenging activities and a slight decrease in the total phenolic content were reported. Besides, effects of light during germination on the antioxidant activities were claimed to be significant. For Rawa seed, ABTS was in the range of 112.9 to 151.3 mMol Trolox/kg for the grains that were germinated in daylight and 78.8 to 176.1 for those germinated in darkness. ABTS values for Aztec were in the range of 133.1 to 222.1 and 99.5 to 17.5 mMol Trolox/kg for sprouts grown in daylight and darkness, respectively. Finally, significant linear correlations between ABTS and DPPH radical scavenging activities (R^2=0.87), as well as between total phenol content and ABTS and DPPH (R^2=0.98) were obtained (Paśko et al., 2009). Nevertheless, the present authors believe that this conclusion may not remain accurate if the values of native grains were included. This is due to the fact that the phenol content did not significantly change after germination, while the antioxidant activities remarkably (more than 10 times) increased throughout the germination process.

In another recent study, effect of germination on the phenolic content and DPPH radical scavenging activity was investigated for Cruentus amaranth (Alvarez-Jubete, Wijngaard, Arendt, & Gallagher, 2010). Grains were germinated for 98 hrs at 10 °C. Total phenol content increased from 21.2 up to 82.2 mg GAE/100 g after germination. A slight decrease in the DPPH activities from 28.4 to 27.1 mg Trolox/100 g was observed.

For finger millet, Siwela et al.(2007) reported a low phenolic content for white varieties with values below 0.09 mg GAE/100 mg. Comparably higher amounts were obtained for brown finger millets, ranging from 0.34 to 1.84 mg GAE/100 mg. The results are aligned with findings of Ramachandra et al. (1977), who detected phenolic content of 0.06 to 0.1 mg/100 mg, in Chlorogenic acid equivalent (CGA), for white and 0.34 to 2.44 mg/100 mg in brown varieties. A similar seed color dependency was observed by Chethan and Malleshi (2007), who reported pholyphenol content of finger millet ranging from 0.3 to 0.5 % (GAE) in white, versus 1.2 to 2.3 % in brown seeds. Antioxidant activity of white finger millets were reported 37.5 to 75.9 mM Trolox/Kg, while these values for the dark varieties increased up to 117.1 to 195.4 (Siwela et al., 2007). DPPH activity for finger millet was observed to be 1.73 mg Trolox/g by Sreeramulu et al. (2009). They reported 373.15± 70.07 mg GAE/100 g for the finger millet phenolic content (Sreeramulu et al., 2009).

Towo et al.(2003) investigated the effect of germination process on the total phenolic content of finger millet. Grains were germinated in darkness at 25 °C. Germination duration was not specified in their paper. Total phenol decreased from 4.2 to 3.3 mg/g in Catechin Equivalents (CE). Sripriya et al. (1996) reported a similar decrease in phenolic content during germination of finger millet. In their study, total phenol decreased from 102 to 67 mg CGA/100 g (Sripriya et al., 1996). Again, germination parameters including duration were not specified.

The observed diversity in the phenolic contents and antioxidant properties of amaranth and millet grains in the literature is due to differences in employed extraction approaches and sample preparation. Besides, having different evaluation procedures as well as different reference calibrated curves in expressing the results are other main factors for these discrepancies.

It has been shown that food processing steps may alter the phenolic contents of cereal and pseudo-cereal grains (Kunyanga, Imungi, Okoth, Biesalski, & Vadivel, 2012; Queiroz, Manólio, Capriles, Torres, & Areas, 2009;

Saleh et al., 2013; Shobana et al., 2012). One of the simplest and widely employed pre-treatments is the malting/germination process, where grains are soaked, germinated, dried, and ground to a flour (Najdi Hejazi, Orsat, Azadi, & Kubow, 2015; Swami, Thakor, & Gurav, 2013; Traoré, Mouquet, Icard-Vernière, Traore, & Trèche, 2004). In the present study, effects of malting/germination parameters, duration and temperature, on the total phenol content and antioxidant activities of finger millet and amaranth grains were investigated and optimized.

2. Material and Methods

2.1 Materials

Brown finger millet (Eleusine coracana) was procured from University of Agricultural Sciences (Dharwad, India) and amaranth (Amaranthus caudatus L. (love-lies-bleeding)) seeds were purchased from a local market (Bulk Barn store, Qc., Canada).

Trolox (6-hydroxy-2, 5, 7, 8-tetramethylchroman-2-carboxylic acid), Folin-Ciocalteu reagent, chlorogenic acid, 2,2-diphenyl-1-picrylhydrazyl (DPPH), 2,2'-azino-bis (3-ethylbenzthiazoline-6-sulphonic acid) (ABTS), and potassium persulfate were obtained from Sigma-Aldrich (St. Louis, MO, USA). Methanol was purchased from Fisher Scientific (Fair Lawn, NJ, USA). Ethanol was obtained from Commercial Alcohols (Industrial and Beverage Alcohol Division of Green-field Ethanol Inc., Ontario, Canada). Double distilled water (ddH2O) was prepared using Simplicity TM water purification system (Millipore, USA). All other reagents were of analytical and HPLC grades (Teow et al., 2007).

2.2 Malting/Germination Process

Finger millet and amaranth seeds were cleaned thoroughly with sterile water and surface-air dried by airflow. The seeds were steeped (seed/water ratio of 1:5 (w/v)) overnight, and sprouted at 22, 26, and 30 °C in a B.O.D incubator (Benchmark Incu-Shaker Mini) for 24, 36, and 48 hrs. Germinated seeds were freeze-dried (FreeZone® 2.5 l Freeze Dry System, Labonco Corporation, MO, USA) for a week to reach constant dry weight. Native and germinated seeds were ground in an electric grinder (Bodum 10903, PRC, Intertek, USA) for further assays. Samples were stored in hermetic plastic containers at 4 °C.

2.3 Statistical Analysis

The experimental design used in the present study was a Central Composite Design (CCD) with two independent factors, germination duration and temperature, each at three levels, (24, 36, 48 hours) and (22, 26, 30 °C). All the experiments were performed in triplicate and the data presented as mean ± standard deviations. This design has 4 factorial, 4 axial, and 4 central points. For analyzing the data, JMP software version 11 (SAS Institute Inc., Cary, NC, USA) was used. The linear, quadratic, and combined effects of each factor were investigated using ANOVA analysis and regression models, and expressed as follows;

$$Y = Intercept + \beta_1 \times Time + \beta_2 \times Temp + \beta_{11} \times Time^2 + \beta_{22} \times Temp^2 + \beta_{12} \times Time \times Temp$$

In this equation, β_1 and β_2 represent the regression coefficients of the linear, β_{11} and β_{22} are the coefficients of the quadratic, while β_{12} indicates the interactive or bilinear effects. Time and Temp represent the deviation of independent germination variables, germination duration and temperature with respect to their centroids values, 36 hrs and 26 °C, respectively.

2.4 Dry Matter

Dry matters of flours were determined using AACC Method 44 – 15.02 (AACC, 1999).

2.5 Bioactive Compounds and Antioxidant Activity Determination

To determine the total phenol content and antioxidant activities, the phenolic compounds were initially extracted and their radical scavenging activities using ABTS and DPPH approaches were evaluated as follows;

2.5.1 Phenolic Compounds Extraction

Methanolic crude extracts were obtained from native and malted samples (Makkar, 2003). Briefly, 100 mg of fine flour sample was weighed in 1.5 ml eppendorf and 900 μL of methanol (90%) was added. The mixture was sonicated in a dark cold room for 30 min, and centrifuged at 3000 rpm for 10 min at 4 °C. The supernatant was recovered and the pellet was re-centrifuged with addition of 600 μL methanol (90%) under the same conditions. Finally, the two supernatants were pooled and used for further total phenolic content and antioxidant scavenging activity assays.

2.5.2 Total Phenol Content

Total phenolic content was determined by adapting the Folin-Ciocalteu method (Singleton & Rossi, 1965). Initially, 2 ml ddH$_2$O was added to 100 μL of methanolic sample extract. Subsequently, 200 μL of Folin-Ciocalteu reagent (2N) was added to the mixture with vigorous vortexing. After 30 min of incubation in the dark at room temperature, 1 ml of aqueous sodium carbonate solution (7.5%) was added and vortexed. Absorbance of samples was read at 765nm (Ultraspec1000, Amersham Pharmacia Biotech, NJ, USA) against methanol as a blank after one hour more of incubation at ambient temperature. Using chlorogenic acid as the standard curve, total phenol was expressed in mg CGA/100g db.

2.5.3 DPPH Antioxidant Scavenging Activity

The approach to assess antioxidant activity using the DPPH free radical scavenging assay was adapted from the method of Martinez-Valverda et al. (2002).

To perform this assay, fresh DPPH stock solution (1mM) was made and appropriately diluted with absolute methanol to reach into an absorbance range of 0.5 to 0.9 unit. Briefly, 1.5 ml of the prepared DPPH solution was added to 100 μL of sample extracts, vortexed and incubated at room temperature for 30 min. The absorbance of the resulted solution was read at 517 nm against air as the blank. The free radical scavenging activity was estimated using standard curve of Trolox in different concentrations (0-500 mM) with R^2=0.992. Additional dilution was required if the absorbance was over the linear range of the standard. In this regard, for the finger millet samples, a dilution factor of 10 was required, while no additional dilution was needed for amaranth. The final results were expressed as mg Trolox equivalent per 100g on a dry basis (mg TE/100 g db).

2.5.4 ABTS Antioxidant Scavenging Activity

ABTS free radical scavenging assay was implemented after some modification of the method of Re et al. (1999). Initially, an ABTS (7mM) stock solution and a potassium persulfate (2.45 mM) solution were prepared. The working reagent was obtained by mixing the two stock solutions in equal quantities and left to be incubated in the dark for 12 hrs at room temperature. The procedure resulted into the production of radical cations of ABTS (ABTS ˙). ABTS radical solution was diluted with 95% ethanol to obtain an absorbance of 0.7 ± 0.05 unit at 734 nm.

To assess the ABTS activities, 1.2 ml of the prepared working solution was added to 100 μL of the sample extracts. The absorbance of the mixture was determined within 1 to 3 min at 734 nm against air as a blank. For finger millet, a dilution factor of 10 was used to fit the absorbance in the standard curve. This dilution was not required for amaranth samples. The antioxidant activity was estimated using standard curve of Trolox in different concentrations (0-500 mM with R^2=0.99) and results were reported as mg TE/ 100g db.

2.5.5 Color Determination

Colorimetric properties of native and germinated samples were assessed using CIE L*a*b* (CIELAB) approach (Leon, Mery, Pedreschi, & Leon, 2006) by chromameter (Model- CR300, Konica-Minolta®, USA). The measurement was done by covering the 5 g of flour with the plastic food wrap with gentle spreading. The flour was fitted perfectly the diameter of the chromameter's testing window. The flour was located on the black material to prevent unwanted interference from ambient light and scattering of light from the source.

The values for L* (0 = black, 100 = white), a* (+ values = redness, - values = green), and b* (+ values = yellowness, -values=blueness) were obtained for each germination treatment. Total color difference ($\Delta E^* = \sqrt{(\Delta L^*)^2 + (\Delta a^*)^2 + (\Delta b^*)^2}$) is evaluated versus the control sample (native grain).

3. Results

The obtained results are presented for amaranth and finger millet in the following subsections.

3.1 Amaranth

Total phenol contents were assessed for the selected design combinations as described in section 0 and are presented in Table 1. From the table, it was observed that the germination process significantly increased the phenolic compounds in amaranth sprouts comparing to the initial value (71.55 ± 1.76 mg CGA100 g db) of native grains (GM0). An increase of nearly four times was observed for some of the germination treatments (e.g. GM5 cases). A similar increase was reported by Alvarez-Jubete et al. (2010), where the phenol content of amaranth sprouts showed a four-time increase (from initial 21.2 to final 82.2 mg GAE/100 g) after 96 hrs of germination at 10 °C. DPPH and ABTS scavenging activities for the studied cases are presented in Table 1 as well. Again, similar to the phenol content, an increasing trend in the antioxidant activities throughout germination was observed.

Table 1. Total phenol (mg CGA/100 g db) content and DPPH (mg TE/100 g db), and ABTS (mg TE/100 g db) activities of amaranth grain for the tested germination treatments

Exp. #	Germination factors and their levels		Total Phenol	DPPH	ABTS
	Temperature	Duration	(mg CGA/100 g db)	mg Trolox/100 g db	mg Trolox/100 g db
GM0	Control		71.55±1.76	43.16± 1.10	52.27± 2.13
GM1	22	24	126.80±3.59	56.71± 4.39	90.67± 1.79
GM2	22	36	187.54±14.00	59.84± 2.77	78.18± 3.99
GM3	22	48	220.56±7.47	75.21± 4.19	117.23± 8.51
GM4	26	24	144.01±1.47	70.11± 4.27	91.97± 3.74
GM5.1	26	36	262.52±9.53	108.02± 6.24	129.82± 4.51
GM5.2	26	36	264.68±0.75	108.62± 6.30	118.84± 7.06
GM5.3	26	36	259.82±16.00	109.00± 5.30	121.45± 8.21
GM5.4	26	36	261.75±7.17	110.67± 1.30	118.94± 5.60
GM6	26	48	250.01±5.89	91.89± 4.82	127.63± 8.92
GM7	30	24	118.82±0.67	60.39± 3.15	80.96± 1.28
GM8	30	36	160.84±1.97	69.18± 3.40	90.88± 3.15
GM9	30	48	269.15±0.25	107.74± 2.37	126.07± 9.22

The obtained phenol and antioxidant activities are graphically presented in Figure 1. From this figure, the following conclusions may be drawn. Firstly, germination process increased the bulk phenol content when comparing with the control sample (GM0). Secondly, the phenol content is a strong function of germination duration. Thirdly, having an increase and then decrease in the averaged phenol contents versus temperature confirms its second order dependency to the process temperature. Lastly, DPPH and ABTS activities are following a similar trend as the total phenol content indicating strong correlations between these two variables.

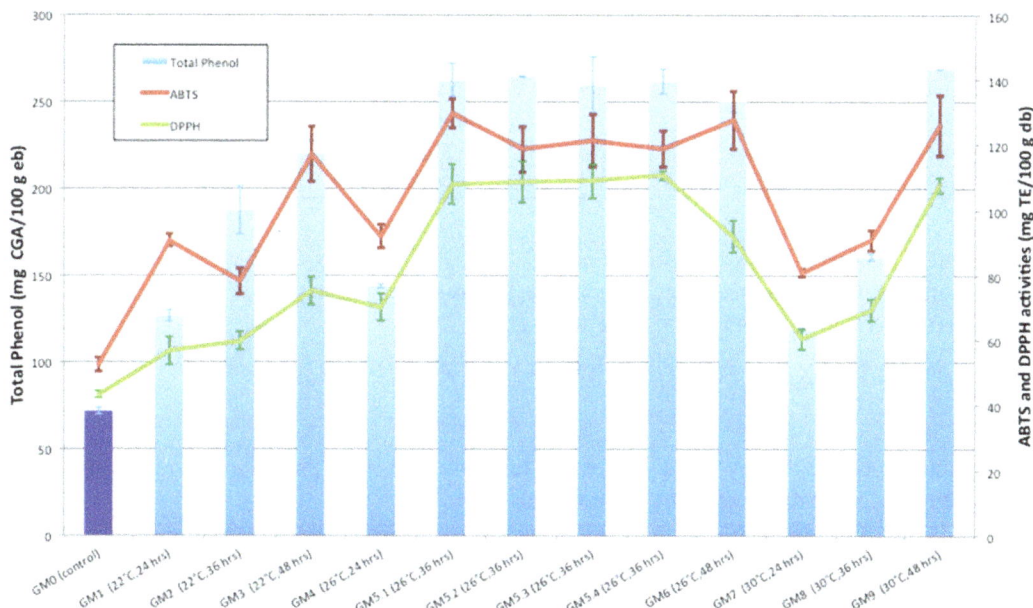

Figure 1. Total Phenol, ABTS, and DPPH activities of amaranth grain for the selected design factor combinations

In Table 2, reported data in the literature for the interested quantities, phenol content, ABTS, and DPPH antioxidant activities, are presented. For comparison, all the data are converted to be presented in mg/100g unit. Besides, data are categorized based on the different employed calibration curves; Gallic acid, Chlorogenic acid, and Trolox equivalents. Excluding a few exemption, our results are aligned with other reported data. Having different phenolic extraction approaches and different reading durations for ABTS and DPPH assays significantly influence the results, which is clearly observed in the reported data.

Table 2. Total phenol content, DPPH, and ABTS antioxidant activities (mg/100 g db) of amaranth grain

Amaranth Variety	Total Phenol (mg/100 g) Gallic acid equivalent	DPPH (mg/100 g) Trolox equivalent	ABTS (mg/100 g) Trolox equivalent
A. Caudatus Centenario (Repo-Carrasco-Valencia et al., 2009)	98.7	10261	20698
A. Caudatus Oscar Blanco (Repo-Carrasco-Valencia et al., 2009)	112.9	9961	16772
A. cruentus (raw) (Alvarez-Jubete et al., 2010)	21.2 ± 2.3	28.4 ± 1.3	-
A. cruentus (germinated)(Alvarez-Jubete et al., 2010)	82.2 ± 4.6	27.1 ± 2.7	-
A. Caudatus (Klimczak, Małecka, & Pachołek, 2002)	39.17	-	-
A. Paniculatus(Klimczak et al., 2002)	56.12	-	-
Amaranth (Mošovská, Mikulášová, Brindzová, Valík, & Mikušova, 2010)	104.1±2.2	-	-
A. cruentus var. Aztec (raw) (Paśko et al., 2009)	295 ±7	110.6 ± 12.5	302.85± 27.53
A. cruentus var. Aztec (sprout) (Paśko et al., 2009)	160 to 300	-	1972 to 4407
A. cruentusvar. Rawa (raw) (Paśko et al., 2009)	300±42	78.8 ±7.5	285.83±30.03
A. cruentusvar. Rawa (sprout) (Paśko et al., 2009)	150 to 250	-	2490 to 5560
A. cruentusvar. Aztec seeds (PAŚKO et al., 2007)	-	110.6 ± 12.0	321.37± 23.03
A. cruentusvar. Rawa seeds (PAŚKO et al., 2007)	-	78.8 ± 6	290.59 ± 16.27
Amaranth (Czerwiński et al., 2004)	14.72 to 14.91	-	-
Amaranth (Chlopicka et al., 2012)	271 ± 1	90.1 ± 8.5	-
Amaranth (Queiroz et al., 2009)	3170	-	-
		Gallic acid equivalent	
Amaranth (Asao & Watanabe, 2010)	51	22600	-
		Percentage basis	
A. hypochondriacus (López, Razzeto, Giménez, & Escudero, 2011)	57.1±1.0	86.93±1.4%	-
	Chlorogenic acid equivalent		
A. cruentus (Kunyanga et al., 2012)	1080	84.67±1.18%	-
		DPPH	
A. cruentus (Ogrodowska et al., 2012)	27.26 to 61.53	436.11 to 604.49	-
	Tannic acid equivalent		
A. cruentus (raw) (Gamel, Linssen, Mesallam, Damir, & Shekib, 2006)	516 to 524	-	-
A. cruentus (germinated) (Gamel et al., 2006)	368 to 420	-	-

Phenolic content data were analyzed using response surface methodology. Table 3 presents a summary of the obtained results for the performed calculations. The proposed response surface based on the selected design factors, X_1 = (temperature – 26) and X_2 = (duration – 36) is,

$$TP = 2.3183\ X_1 + 58.3483\ X_2 - 50.59X_1^2 − 27.77\ X_2^2 + 14.1425\ X_1\ X_2 + 249.7216 \tag{1}$$

Table 3. Summary of the ANOVA analysis of the responses for total phenol content and the corresponding parameter estimates of the different terms for amaranth grain

Term	Estimate	Std Error	t Ratio	Prob>|t|
Intercept	249.7216	12.592	19.83	**<.0001**
Temperature (X_1)	2.3183	11.2632	0.21	0.8437
Duration (X_2)	**58.3483**	11.26312	5.18	**0.0021**
Duration×Temperature (X_1X_2)	14.1425	13.7945	1.03	0.3448
Temperature×Temperature (X_1^2)	**-50.59**	16.8948	-2.99	**0.0242**
Duration×Duration (X_2^2)	-27.77	16.8948	-1.64	0.1513

The analysis states that germination duration and square of germination temperature are significant terms with $p<0.005$ and $p<0.05$, respectively. A positive dependency of the obtained response surface to the duration indicates that the phenol content increases as germination time increases. Furthermore, having a positive coefficient for the quadratic term, X_1^2, means that a maximum should be expected in the constructed response surface.

The obtained response surface (Equation 1) is graphically plotted in Figure 2 (a). As it was pointed out, the surface has a maximum at 26.7 °C and 49.14 hrs. The predicted maximum value is 281.88 mg CGA/100 g db. This means that based on the conducted germination treatments and experiments, if it is desired to maximize the phenolic content and consequently its potential for higher antioxidant activity of selected amaranth grains, they should be germinated approximately for 48 hrs and at 26 °C. This conclusion is based on the performed ANOVA study and constructed response surface. However, based on the obtained raw experimental data (Table 1), the best phenolic content was observed for seeds germinating for 36 hrs at 26 °C. The relative difference is below 5% and may be neglected. The leverage plot (Figure 2 (b)) emphasizes on the appropriate prediction and interpolation of the experimental data by the proposed response surface. Since the confidence curves (dashed-line curves) crossed the horizontal line, the correlation of design factors is significant ($p < 0.05$).

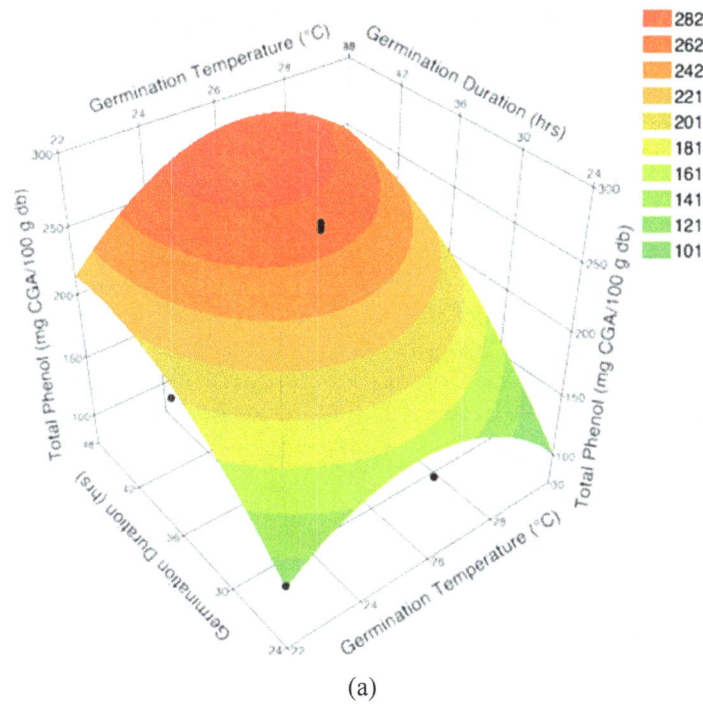

(a)

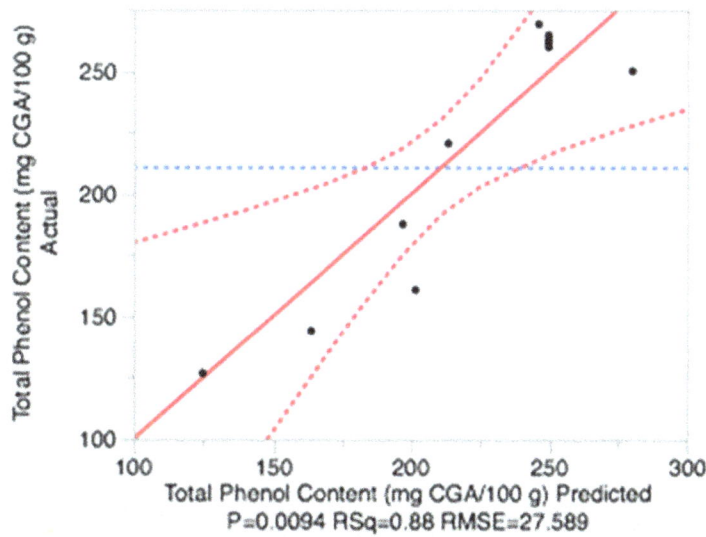

(b)

Figure 2. (a) The response surface for the total phenol content (mg CGA/100 g db) of amaranth grain and (b) the leverage plot based on the ANOVA analysis

Similar procedures were performed for the DPPH and ABTS activities. The response surfaces were constructed and the linear, quadratic, and bi-linear effects of the germination factors were analyzed. The results are presented in Table 4. For all quantities, germination duration (Time) and square of germination temperature (Temp×Temp), the results are significant.

Table 4. Summary of the ANOVA analysis for total phenol, DPPH and ABTS of amaranth grain

	Total Phenol	DPPH	ABTS
Intercept	**249.7216[a]**	**102.6475[a]**	**116.985[a]**
Temp	2.3183	7.5916	1.9716
Time	**58.3483[d]**	**14.605[f]**	**17.8883[c]**
Time×Temp	14.1425	7.2125	4.6375
Temp×Temp	**-50.59[f]**	**-25.2775[f]**	**-21.9[f]**
Time×Time	-27.77	-8.7875	3.37
R^2	0.881	0.806	0.805
R^2adj	0.783	0.645	0.643
RMS	27.59	13.31	11.59

a: $p<0.0001$,   b: $p<0.0005$,   c: $p<0.001$,   d: $p<0.005$,   e: $p<0.01$, f: $p<0.05$.

As it was observed from Figure 1 and Table 4, ABTS and DPPH are showing a similar pattern to total phenol. This suggests the presence of a linear correlation between these quantities. Existence of this correlation has been reported in the literature (Chlopicka, Pasko, Gorinstein, Jedryas, & Zagrodzki, 2012; PAŚKO, BARTOŃ, FOŁTA, & GWIŻDŻ, 2007; Paśko et al., 2009). In figure 3, scattered plots of the phenol, DPPH, and ABTS versus each other are presented. Unlike the data of Pasko et al. (2009), the results of native seeds are also included. Appropriate linear correlations with acceptable R^2 values are observed.

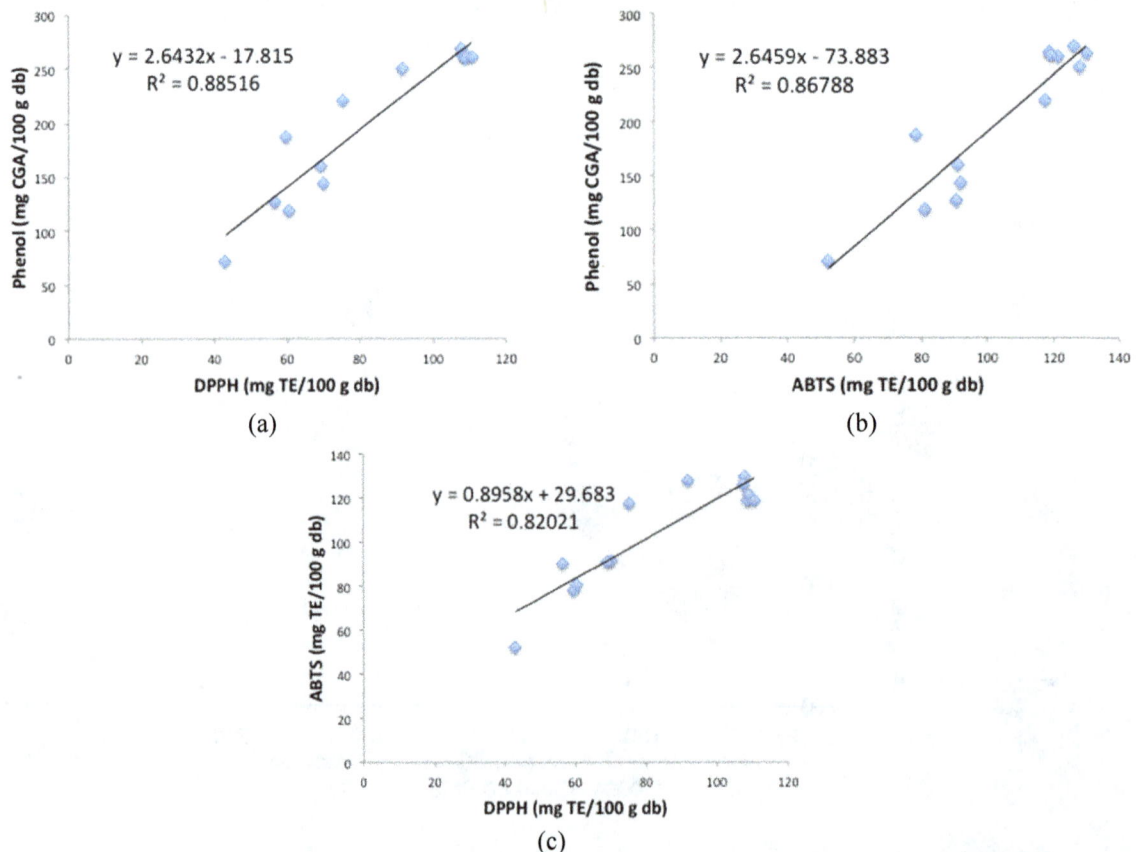

(a)

(b)

(c)

Figure 3. Linear regression curves representing the linear correlations between total phenol content and the antioxidant activities, DPPH and ABTS (a) Phenol vs. DPPH (b) Phenol vs. ABTS, and (c) ABTS vs. DPPH of amaranth seeds

Colorimetric properties of the native and germinated amaranth grains are presented in Table 5. A significant color change (ΔE values) was observed in the amaranth grain during germination, where sprouts' color gradually changed to red. This is clearly observed from the increased a* values shown in the table. The peak of color change occurs in the case of CM6, where grains were germinated for 48 hrs at 26 °C. This is the case in which phenol and antioxidant activities reached their maximums.

Table 5. Colorimetric properties of native and germinated amaranth grain using CIELab approach

Exp. #	Germination factors and their levels		L*	a*	b*	ΔE
	Temperature	Duration				
GM0	Control		86.98±0.53	0.91± 0.08	12.95± 0.17	0.00±0.56
GM1	22	24	89.44±0.31	0.61± 0.05	11.76± 0.12	2.75±0.34
GM2	22	36	88.16± 0.59	1.64± 0.07	11.75± 0.42	1.84±0.73
GM3	22	48	88.68±0.51	1.19± 0.06	12.20± 0.33	1.88±0.62
GM4	26	24	88.38±0.33	0.83± 0.08	12.15± 0.16	1.61±0.37
GM5.1	26	36	86.19±0.45	3.28± 0.06	11.21± 0.11	3.04±0.46
GM5.2	26	36	86.60±0.75	3.19± 0.06	11.05± 0.17	2.99±0.66
GM5.3	26	36	86.04± 0.64	3.32± 0.17	11.27± 0.49	3.08±0.82
GM5.4	26	36	86.54±0.64	3.18± 0.08	11.16± 0.17	2.92± 0.52
GM6	26	48	83.78±0.70	4.34± 0.12	11.63± 0.44	4.87±0.83
GM7	30	24	88.59±0.43	0.54± 0.09	12.15± 0.33	1.84±0.54
GM8	30	36	88.58±0.61	0.97± 0.05	12.21± 0.23	1.77±0.66
GM9	30	48	85.81±0.77	1.99± 0.09	13.75± 0.27	1.78±0.82

3.2 Millet

A similar experimental procedure and analysis were performed for the finger millet grains. The obtained data are numerically presented in table 6 and graphically in Figure 4. A clear observation from these data is that the germination process decreased the phenol content. The highest decrease was observed for the grains germinated at 26 °C for 48 hrs (25% reduction in total phenol content). Besides, germination duration seems to be an important factor, while temperature was not significantly influential.

Table 6. Total phenol (mg CGA/100 g db) content and DPPH (mg TE/100 g db), and ABTS (mg TE/100 g db) activities of finger millet grain for the designed germination treatments

Exp. #	Germination factors and their levels		Total Phenol	DPPH	ABTS
	Temperature	Duration	(mg CGA/100 g db)	(mg TE/100db)	(mg TE/100db)
GM0	Control		627.45±3.93	952.39± 11.30	861.11± 11.10
GM1	22	24	543.29 ±5.87	898.89± 29.28	729.00± 27.29
GM2	22	36	508.60 ± 3.80	829.70± 10.42	643.81± 1.13
GM3	22	48	498.24 ±1.73	812.50± 43.70	657.54± 51.60
GM4	26	24	530.69 ±8.89	794.72 ± 11.43	623.14± 15.69
GM5.1	26	36	480.21 ±11.49	793.22 ± 20.98	587.73± 33.93
GM5.2	26	36	489.85 ±11.59	786.69 ± 14.10	547.57± 25.51
GM5.3	26	36	500.60 ±7.44	808.86± 10.81	621.22± 29.74
GM5.4	26	36	493.76 ±4.99	783.11 ± 92.87	646.71± 26.63
GM6	26	48	466.19 ±10.62	743.11± 24.81	483.11± 21.88
GM7	30	24	515.78±7.99	839.79± 27.30	651.15± 17.53
GM8	30	36	473.59 ±2.78	732.36 ± 41.80	547.09±23.80
GM9	30	48	499.57±0.18	756.54± 58.83	623.05± 25.28

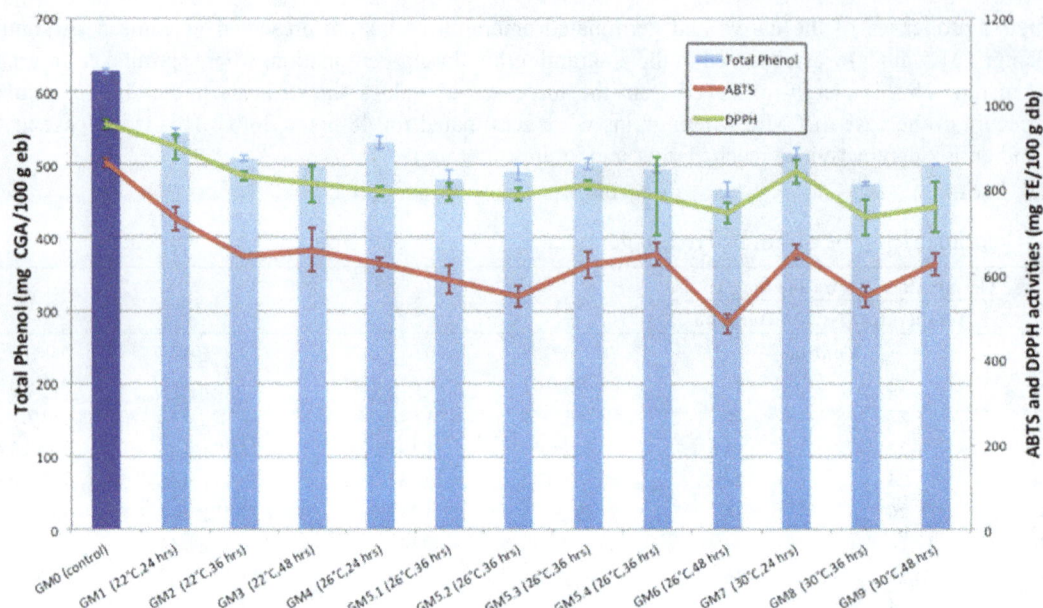

Figure 4. Total Phenol, ABTS, and DPPH activities of finger millet grain for the selected design factor combinations

Table 7. Total phenol content, DPPH, and ABTS antioxidant activities (mg/100 g db) of finger millet grain

Finger millet Variety	Total Phenol (mg/100 g)	DPPH (mg/100 g)	ABTS (mg/100 g)
	Gallic acid equivalent	Trolox equivalent	Trolox equivalent
Brown finger millet (Siwela et al., 2007)	340 to 1840	-	1734 to 4890
White finger millet (Siwela et al., 2007)	<90	-	<1364
Finger millet (Sreeramulu et al., 2009)	373 ±70	173±3	-
Brown color (Chethan & Malleshi, 2007; McDonough, Rooney, & Earp, 1986)	1200 to 2300		-
White color (Chethan & Malleshi, 2007)	300 to 500	-	
			Gallic acid equivalent
Millet (Asao & Watanabe, 2010)	360	-	1770
	Catechin equivalent		
Raw Finger millet (Towo et al., 2003)	420 ±27	-	-
Germinated (Towo et al., 2003)	330 ±11	-	-
	Chlorogenic acid equivalent	Percentage basis	
Finger millet	1050	81.67 ± 2.36%	-
Indian Brown (Shankara, 1991)	60 to 670	-	-
Indian Brown (raw) (Sripriya et al., 1996)	102	-	-
Indian Brown (germinated) (Sripriya et al., 1996)	67	-	-
Indian White (Sripriya et al., 1996)	3.47	-	-
Indian Brown (McDonough et al., 1986)	550 to 590	-	-
Indian White (Geetha, Virupaksha, & Shadaksharaswamy, 1977)	80 to 90	-	-
Indian Brown (Geetha et al., 1977)	370 to 960	-	-
African Brown (Geetha et al., 1977)	540 to 2440	-	-
Finger millet (Raghavendra Rao, Nagasampige, & Ravikiran, 2011)	7200 ±570	-	-
	Tannic acid equivalent		
Indian Brown (Shankara, 1991)	30 to 570	-	-
	Ferulic acid equivalents	Ferulic acid equivalents	
Cooked finger millet (Chandrasekara & Shahidi, 2012)	233±4	314.6± 3.5	-

The reported phenol content and antioxidant activities of finger millet in the literature are summarized in Table 8, all in mg/100 g db with respect to different employed equivalents.

Table 8. Summary of the ANOVA analysis of the responses for total phenol content and the corresponding parameter estimates of the different terms of finger millet grain

Term	Estimate	Std Error	t Ratio	Prob>|t|
Intercept	488.4733	5.6711	86.13	**<.0001**
Temperature	-10.1983	5.0724	-2.01	0.0911
Duration	-20.96	5.0724	-4.13	**0.0061**
Duration×Temperature	7.21	6.2124	1.16	0.2899
Temperature×Temperature	7.885	7.6086	1.04	0.3400
Duration×Duration	15.23	7.6086	2.00	0.0922

It is reported that the phenol content in finger millet strongly depends to its color and its geographical cultivation location (Siwela et al., 2007). Generally, white finger millet has lower phenol contents, while brown finger millet is considerably rich in phenolic compounds. The results of our investigated native Indian brown finger millet are completely in range with other reports (Towo et al., 2003). Beside, a similar decrease in the finger millet total phenol content throughout germination process is pointed out in a few other studies (Towo et al., 2003). Towo et al. reported 21% reduction in total phenol after germination of finger millet (Towo et al., 2003). A 34% reduction was reported by Sripriya et al. in germination of Indian brown finger millet (Sripriya et al., 1996). Abdelrahaman et al. reported a 35–42% decrease in polyphenol content of millet germinated for six days in several studied varieties (Abdelrahaman et al., 2007). In this study, phenolic content data were analyzed for finger millet using response surface methodology. Results are presented in Table 9, where only germination duration is found to be a significant factor. Based on this table the following response surface is derived;

$$TP = -10.1983\ X_1 - 20.96\ X_2 + 7.885\ X_1^2 + 15.23\ X_2^2 + 7.21\ X_1 X_2 + 488.4733 \qquad (2)$$

Table 9. Summary of the ANOVA analysis of the responses for total phenol, DPPH and ABTS of finger millet

	Total Phenol	DPPH	ABTS
Intercept	488.4733[a]	781.3108[a]	581.2379[a]
Temp	-10.1983	**-35.4[c]**	-34.8433
Time	**-20.96[b]**	**-36.875[c]**	-39.9316
Time×Temp	7.21	0.785	10.84
Temp×Temp	7.885	23.0375	53.3513
Time×Time	15.23	10.9225	11.0263
R^2	0.832	0.799	0.604
R^2adj	0.692	0.632	0.274
RMS	12.424	27.530	54.747

a: p<0.0001,   b: p<0.01, c: p<0.05.

For case of finger millet, in contrast to amaranth, a negative correlation between phenol content and germination duration exists. This means that as germination duration increases, phenol content decreases in the sprouts. Furthermore, having positive coefficients for the quadratic terms indicates the existence of a minimum in the proposed response surface. The constructed surface is plotted in Figure 5, and it aligns with its corresponding leverage plot. There is a minimum for this surface at 27.49 °C and 43 hrs with an interpolated phenol content of 480.28 mg CGA/100 g db. In the performed experiments, the minimum obtained value (466.19 mg CGA/100 g db) for phenol content belongs to case GM6 (Table 6), where the grains were germinated for 48 hrs at 26 °C.

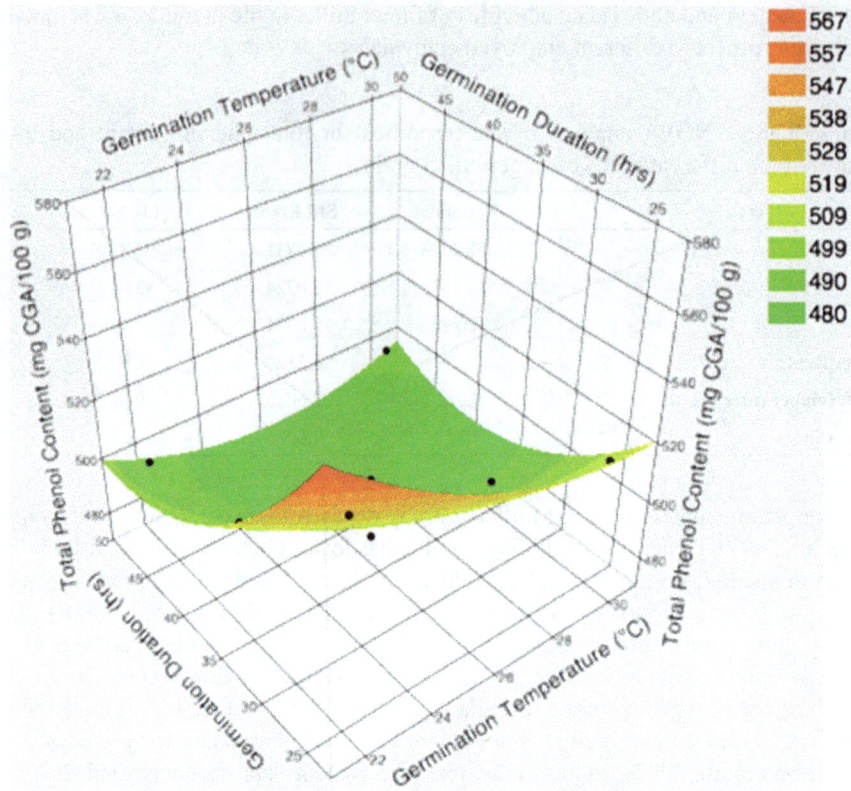

(a)

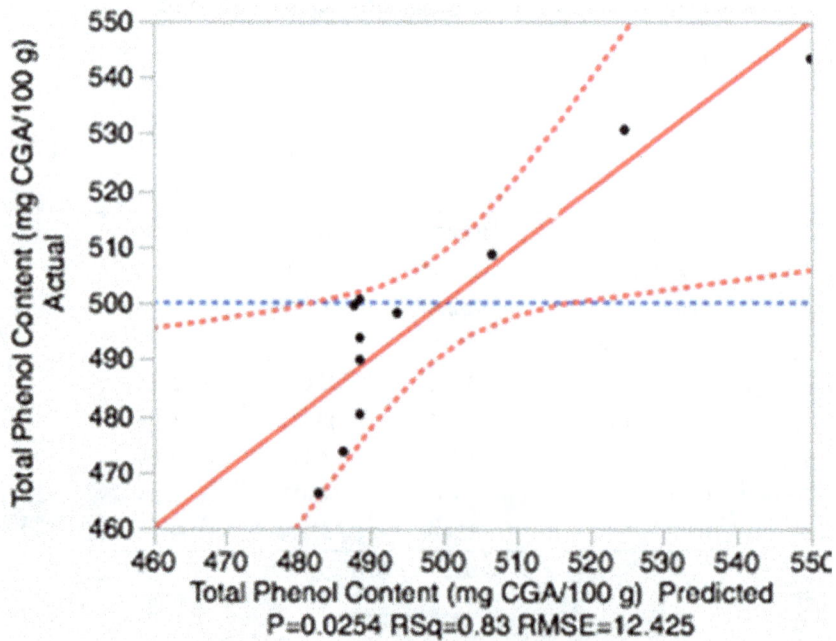

(b)

Figure 5. (a) The response surface for the total phenol content (mg CGA/100 g db) and (b) the leverage plot
based on the ANOVA analysis of finger millet grain

DPPH and ABTS activities are similarly analyzed. Results are presented in Table 10. Unlike for phenol content, temperature seems to be an influential factor for the DPPH activities. In addition, despite the proposed response surface for ABTS activity, none of the factors are significantly influential. This is highlighted in the low obtained R^2 value for the ABTS response surface (Table 9).

Table 10. Colorimetric properties of native and germinated finger millet grain using CIELab approach

Exp. #	Germination factors and their levels		L*	a*	b*	E
	Temperature	Duration				
GM0	Control		78.31±0.68	2.77± 0.10	7.23± 0.12	0.00±0.70
GM1	22	24	81.34±0.50	1.82± 0.04	6.50± 0.13	3.26±0.52
GM2	22	36	82.50± 0.51	1.70± 0.05	6.58± 0.15	4.37±0.53
GM3	22	48	82.14±0.56	1.71± 0.07	6.58± 0.14	4.03 ±0.58
GM4	26	24	82.63±0.82	1.73± 0.03	6.39± 0.11	4.52±0.82
GM5.1	26	36	81.98±0.30	1.69± 0.07	6.74± 0.16	3.86± 0.35
GM5.2	26	36	81.91±0.43	1.64± 0.07	6.50± 0.13	3.84±0.46
GM5.3	26	36	81.72± 0.45	1.71± 0.04	6.46± 0.09	3.66±0.46
GM5.4	26	36	81.66±0.49	1.74± 0.10	6.60± 0.07	3.56± 0.51
GM6	26	48	82.16±0.44	1.76± 0.09	6.67± 0.16	4.02±0.48
GM7	30	24	81.47±0.36	1.86± 0.06	6.48± 0.07	3.37±0.37
GM8	30	36	82.11±0.33	1.64± 0.11	6.46± 0.07	4.04± 0.35
GM9	30	48	82.69±0.37	1.81± 0.05	6.79± 0.06	4.50±0.37

Similar to amaranth, possibility of existence of linear correlations between phenol content and antioxidant activities was investigated. In figure 6, the scattered plots of the phenol content, DPPH, and ABTS activities versus each other are presented for native finger millet and its sprouts. Again, acceptable linear correlations were observed, indicating that antioxidant activities decreased as phenol content decreased.

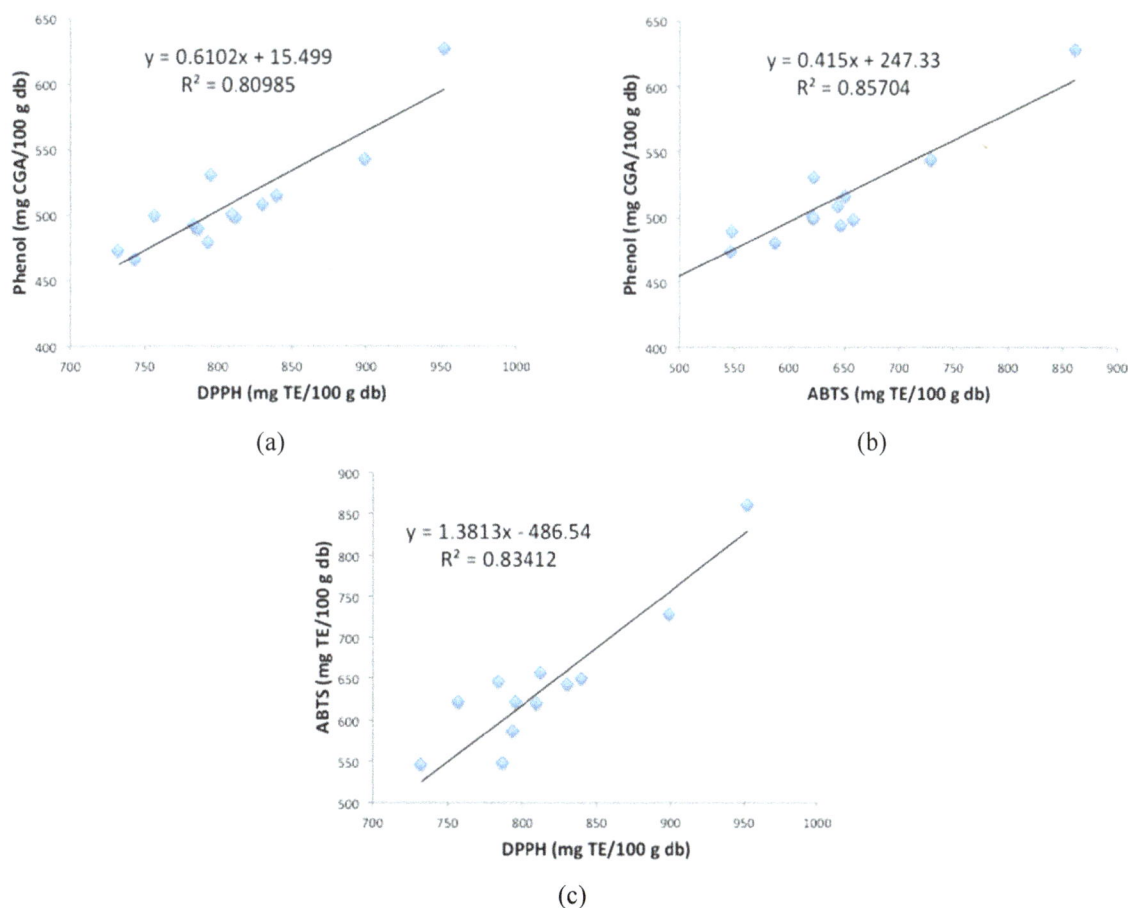

(a)

(b)

(c)

Figure 6. Linear regression curves representing the linear correlations between total phenol content and the antioxidant activities, DPPH and ABTS (a) Phenol vs. DPPH (b) Phenol vs. ABTS, and (c) ABTS vs. DPPH of finger millet grain

Colorimetric values for germinated finger millet grains are presented in table 10, where not a significant pattern was concluded in the color versus germination factors.

4. Conclusion

Finger millet and amaranth grains were demonstrated as two potentially rich sources of phenolic compounds possessing high antioxidant scavenging activities. Malting/germination of these seeds might be an appropriate pre-treatment in the food industry to optimize the phenolic content quantity and quality. In the present study, effects of two important germinating factors, duration and temperature, on the total phenol contents and radical scavenging activities of finger millet and amaranth seeds were investigated. Using a central composite design with three levels for each design factor enabled us to construct a second order response surface for the total phenol content and ABTS and DPPH antioxidant scavenging activities over the interested range of variation (26 °C <germination temperature<30 °C and 24 hrs <germination duration<48 hrs). It was shown that germination significantly (up to four times) increased the phenolic content and antioxidant activities of amaranth seeds. In addition, a significant color change towards pink color was observed in amaranth sprouts, which may increase its consumer acceptability. In case of finger millet, germination slightly decreased the phenol content and DPPH and ABTS activities at most by 25%. Nevertheless, phenolic content of finger millet would still remain appreciably high. For both grains, the extremum point is found to be 26 °C and 48 hrs.

References

AACC. (1999). Approved Methods of the American Association of the Cereal Chemists.

Abdelrahaman, S. M., Elmaki, H. B., Idris, W. H., Hassan, A. B., Babiker, E. E., & El Tinay, A. H. (2007). Antinutritional factor content and hydrochloric acid extractability of minerals in pearl millet cultivars as affected by germination. *International journal of food sciences and nutrition, 58*(1), 6-17. http://dx.doi.org/10.1080/09637480601093236

Alvarez-Jubete, L., Wijngaard, H., Arendt, E., & Gallagher, E. (2010). Polyphenol composition and in vitro antioxidant activity of amaranth, quinoa buckwheat and wheat as affected by sprouting and baking. *Food Chemistry, 119*(2), 770-778. http://dx.doi.org/10.1016/j.foodchem.2009.07.032

Asao, M., & Watanabe, K. (2010). Functional and bioactive properties of quinoa and amaranth. *Food science and technology research, 16*(2), 163-168. http://dx.doi.org/10.3136/fstr.16.163

Banerjee, S., Sanjay, K., Chethan, S., & Malleshi, N. (2012). Finger millet (Eleusine coracana) polyphenols: Investigation of their antioxidant capacity and antimicrobial activity. *African Journal of Food Science, 6*(13), 362-374. http://dx.doi.org/10.5897/AJFS12.031

Belton, P. S., & Taylor, J. R. (2002). *Pseudocereals and less common cereals: grain properties and utilization potential*: Springer.

Chandrasekara, A., & Shahidi, F. (2012). Bioaccessibility and antioxidant potential of millet grain phenolics as affected by simulated in vitro digestion and microbial fermentation. *Journal of Functional Foods, 4*(1), 226-237. http://dx.doi.org/10.1016/j.jff.2011.11.001

Charalampopoulos, D., Wang, R., Pandiella, S., & Webb, C. (2002). Application of cereals and cereal components in functional foods: a review. *International Journal of Food Microbiology, 79*(1), 131-141. http://dx.doi.org/10.1016/S0168-1605(02)00187-3

Chethan, S., & Malleshi, N. (2007). Finger millet polyphenols: Optimization of extraction and the effect of pH on their stability. *Food Chemistry, 105*(2), 862-870. http://dx.doi.org/10.1016/j.foodchem.2007.02.012

Chlopicka, J., Pasko, P., Gorinstein, S., Jedryas, A., & Zagrodzki, P. (2012). Total phenolic and total flavonoid content, antioxidant activity and sensory evaluation of pseudocereal breads. *LWT-Food Science and Technology, 46*(2), 548-555. http://dx.doi.org/10.1016/j.lwt.2011.11.009

Czerwiński, J., Bartnikowska, E., Leontowicz, H., Lange, E., Leontowicz, M., Katrich, E., ... Gorinstein, S. (2004). Oat (Avena sativa L.) and amaranth (Amaranthus hypochondriacus) meals positively affect plasma lipid profile in rats fed cholesterol-containing diets. *The Journal of nutritional biochemistry, 15*(10), 622-629. http://dx.doi.org/10.1016/j.jnutbio.2004.06.002

de la Rosa, A. B., Fomsgaard, I. S., Laursen, B., Mortensen, A. G., Olvera-Martínez, L., Silva-Sánchez, C., ... De León-Rodríguez, A. (2009). Amaranth (Amaranthus hypochondriacus) as an alternative crop for sustainable food production: Phenolic acids and flavonoids with potential impact on its nutraceutical quality. *Journal of Cereal Science, 49*(1), 117-121. http://dx.doi.org/10.1016/j.jcs.2008.07.012

Dykes, L., & Rooney, L. (2007). Phenolic compounds in cereal grains and their health benefits. *Cereal Foods World, 52*(3), 105-111. http://dx.doi.org/10.1094/cfw-52-3-0105

Gamel, T. H., Linssen, J. P., Mesallam, A. S., Damir, A. A., & Shekib, L. A. (2006). Seed treatments affect functional and antinutritional properties of amaranth flours. *Journal of the Science of Food and Agriculture, 86*(7), 1095-1102. http://dx.doi.org/10.1002/jsfa.2463

Geetha, R., Virupaksha, T., & Shadaksharaswamy, M. (1977). Relationship between tannin levels and in vitro protein digestibility in finger millet (Eleusine Coracana). *J Agric Food Chem, 25*, 1101-1104. http://dx.doi.org/10.1021/jf60213a046

Hegde, P. S., Rajasekaran, N. S., & Chandra, T. (2005). Effects of the antioxidant properties of millet species on oxidative stress and glycemic status in alloxan-induced rats. *Nutrition Research, 25*(12), 1109-1120. http://dx.doi.org/10.1016/j.nutres.2005.09.020

Issoufou, A., Mahamadou, E. G., & Guo-Wei, L. (2013). Millets: Nutritional composition, some health benefits and processing—A review. *Emirates Journal of Food and Agriculture, 25*(7). http://dx.doi.org/10.9755/ejfa

Klimczak, I., Małecka, M., & Pachołek, B. (2002). Antioxidant activity of ethanolic extracts of amaranth seeds. *Food/Nahrung, 46*(3), 184-186. http://dx.doi.org/10.1002/1521-3803(20020501)46:3<184::AID-FOOD184 >3.0.CO;2-H

Kunyanga, C. N., Imungi, J. K., Okoth, M. W., Biesalski, H. K., & Vadivel, V. (2012). Total phenolic content, antioxidant and antidiabetic properties of methanolic extract of raw and traditionally processed Kenyan indigenous food ingredients. *LWT-Food Science and Technology, 45*(2), 269-276. http://dx.doi.org/10.1016/j.lwt.2011.08.006

Leon, K., Mery, D., Pedreschi, F., & Leon, J. (2006). Color measurement in L* a* b* units from RGB digital images. *Food Research International, 39*(10), 1084-1091. http://dx.doi.org/10.1016/j.foodres.2006.03.006

López, V. R. L., Razzeto, G. S., Giménez, M. S., & Escudero, N. L. (2011). Antioxidant properties of Amaranthus hypochondriacus seeds and their effect on the liver of alcohol-treated rats. *Plant Foods for Human Nutrition, 66*(2), 157-162. http://dx.doi.org/10.1007/s11130-011-0218-4

Makkar, H. P. (2003). *Quantification of tannins in tree and shrub foliage: a laboratory manual*: Springer Science & Business Media. http://dx.doi.org/10.1007/978-94-017-0273-7

Malleshi, N., & Desikachar, H. (1986). Studies on comparative malting characteristics of some tropical cereals and millets. *Journal of the Institute of Brewing, 92*(2), 174-176. http://dx.doi.org/10.1002/j.2050-0416.1986.tb04393.x

Martínez-Valverde, I., Periago, M. J., Provan, G., & Chesson, A. (2002). Phenolic compounds, lycopene and antioxidant activity in commercial varieties of tomato (Lycopersicum esculentum). *Journal of the Science of Food and Agriculture, 82*(3), 323-330. http://dx.doi.org/10.1002/jsfa.1035

McDonough, C., Rooney, L., & Earp, C. (1986). Structural characteristics of Eleusine coracana (finger millet) using scanning electron and fluorescence microscopy. *Food microstructure (USA)*.

Mošovská, S., Mikulášová, M., Brindzová, L., Valík, Ĺ., & Mikušova, L. (2010). Genotoxic and antimutagenic activities of extracts from pseudocereals in the Salmonella mutagenicity assay. *Food and Chemical Toxicology, 48*(6), 1483-1487. http://dx.doi.org/10.1016/j.fct.2010.03.015

Najdi Hejazi, S. (2012). Impact of neonatal total parenteral nutrition and early glucose-enriched diet on glucose metabolism and physical phenotypes in Guinea Pig.

Najdi Hejazi, S., Orsat, V., Azadi, B., & Kubow, S. (2015). Improvement of the in vitro protein digestibility of amaranth grain through optimization of the malting process. *Journal of Cereal Science, 68*, 59-65. http://dx.doi.org/10.1016/j.jcs.2015.11.007

Obilana, A. B., & Manyasa, E. (2002). Millets *Pseudocereals and less common cereals* (pp. 177-217): Springer. http://dx.doi.org/10.1007/978-3-662-09544-7_6

Ogrodowska, D., Czaplicki, S., Zadernowski, R., Mattila, P., Hellström, J., & Naczk, M. (2012). Phenolic acids in seeds and products obtained from Amaranthus cruentus. *Journal of food and nutrition research*.

PAŚKO, P., BARTOŃ, H., FOŁTA, M., & GWIŻDŻ, J. (2007). Evaluation of antioxidant activity of amaranth (Amaranthus cruentus) grain and by-products (flour, popping, cereal). *Rocz. Państw. Zakl. Hig, 58*, 35-40.

Paśko, P., Bartoń, H., Zagrodzki, P., Gorinstein, S., Fołta, M., & Zachwieja, Z. (2009). Anthocyanins, total polyphenols and antioxidant activity in amaranth and quinoa seeds and sprouts during their growth. *Food Chemistry, 115*(3), 994-998. http://dx.doi.org/10.1016/j.foodchem.2009.01.037

Queiroz, Y., Manólio, S. R., Capriles, V. D., Torres, E., & Areas, J. (2009). [Effect of processing on the antioxidant activity of amaranth grain]. *Arch Latinoam Nutr, 59*(4), 419-424.

Raghavendra Rao, B., Nagasampige, M., & Ravikiran, M. (2011). Evaluation of nutraceutical properties of selected small millets. *Journal of Pharmacy and Bioalternative Science, 3*, 277-279. http://dx.doi.org/10.4103/0975-7406.80775

Ramachandra, G., Virupaksha, T. K., & Shadaksharaswamy, M. (1977). Relation between tannin levels and in vitro protein digestibility in finger millet (Eleusine coracana Gaertn.). *Journal of Agricultural and Food Chemistry, 25*(5), 1101-1104. http://dx.doi.org/10.1021/jf60213a046

Re, R., Pellegrini, N., Proteggente, A., Pannala, A., Yang, M., & Rice-Evans, C. (1999). Antioxidant activity applying an improved ABTS radical cation decolorization assay. *Free radical biology and medicine, 26*(9), 1231-1237. http://dx.doi.org/10.1016/S0891-5849(98)00315-3

Repo-Carrasco-Valencia, R., Pena, J., Kallio, H., & Salminen, S. (2009). Dietary fiber and other functional components in two varieties of crude and extruded kiwicha (Amaranthus caudatus). *Journal of Cereal Science, 49*(2), 219-224. http://dx.doi.org/10.1016/j.jcs.2008.10.003

Saleh, A. S., Zhang, Q., Chen, J., & Shen, Q. (2013). Millet Grains: Nutritional Quality, Processing, and Potential Health Benefits. *Comprehensive Reviews in Food Science and Food Safety, 12*(3), 281-295. http://dx.doi.org/10.1111/1541-4337.12012

Shankara, P. (1991). *Investigations on pre-harvest and post harvest aspects of finger millet.* Ph. D. thesis, University of Mysore, India.

Shobana, S., Krishnaswamy, K., Sudha, V., Malleshi, N., Anjana, R., Palaniappan, L., & Mohan, V. (2012). Finger Millet (Ragi, Eleusine coracana L.): A Review of Its Nutritional Properties, Processing, and Plausible Health Benefits. *Advances in food and nutrition research, 69*, 1-39. http://dx.doi.org/10.1016/B978-0-12-410540-9.00001-6

Singleton, V. L., & Rossi, J. A. (1965). Colorimetry of total phenolics with phosphomolybdic-phosphotungstic acid reagents. *American journal of Enology and Viticulture, 16*(3), 144-158.

Siwela, M., Taylor, J. R., de Milliano, W. A., & Duodu, K. G. (2007). Occurrence and location of tannins in finger millet grain and antioxidant activity of different grain types. *Cereal chemistry, 84*(2), 169-174. http://dx.doi.org/10.1094/CCHEM-84-2-0169

Sreeramulu, D., Reddy, C. V. K., & Raghunath, M. (2009). Antioxidant activity of commonly consumed cereals, millets, pulses and legumes in India. *Indian journal of biochemistry & biophysics, 46*(1), 112.

Sripriya, G., Chandrasekharan, K., Murty, V., & Chandra, T. (1996). ESR spectroscopic studies on free radical quenching action of finger millet (Eleusine coracana). *Food Chemistry, 57*(4), 537-540. http://dx.doi.org/10.1016/S0308-8146(96)00187-2

Swami, S. B., Thakor, N. J., & Gurav, H. S. (2013). Effect of soaking and malting on finger millet (EleusineCoracana) grain. *Agricultural Engineering International: CIGR Journal, 15*(1), 194-200.

Teow, C. C., Truong, V.-D., McFeeters, R. F., Thompson, R. L., Pecota, K. V., & Yencho, G. C. (2007). Antioxidant activities, phenolic and β-carotene contents of sweet potato genotypes with varying flesh colours. *Food Chemistry, 103*(3), 829-838. http://dx.doi.org/10.1016/j.foodchem.2006.09.033

Towo, E. E., Svanberg, U., & Ndossi, G. D. (2003). Effect of grain pre-treatment on different extractable phenolic groups in cereals and legumes commonly consumed in Tanzania. *Journal of the Science of Food and Agriculture, 83*(9), 980-986. http://dx.doi.org/10.1002/jsfa.1435

Traoré, T., Mouquet, C., Icard-Vernière, C., Traore, A., & Trèche, S. (2004). Changes in nutrient composition, phytate and cyanide contents and α-amylase activity during cereal malting in small production units in Ouagadougou (Burkina Faso). *Food Chemistry, 88*(1), 105-114. http://dx.doi.org/10.1016/j.foodchem.2004.01.032

Venskutonis, P. R., & Kraujalis, P. (2013). Nutritional components of amaranth seeds and vegetables: a review on composition, properties, and uses. *Comprehensive Reviews in Food Science and Food Safety, 12*(4), 381-412. http://dx.doi.org/10.1111/1541-4337.12021

Quality Evaluation of Flaxseed for Food Use Specifications

Anuradha Vegi[1], Charlene E. Wolf-Hall[2,3], & Clifford A. Hall III[1]

[1]Department of Plant Sciences, North Dakota State University, North Dakota, USA

[2]Department of Veterinary and Microbiological Sciences, North Dakota State University, North Dakota, USA

[3]Office of the Provost, North Dakota State University, North Dakota, USA

Correspondence: Clifford A. Hall III, Department of Plant Sciences, North Dakota State University, Dept. 7670, PO Box 6050, Fargo, North Dakota, 58108-6050, USA. E-mail: Clifford.Hall@ndsu.edu

Abstract

A Northern Great Plains regional survey of microbiological loads in flaxseed was completed for years 2008 and 2009. Effects of cleaning flaxseed on microbial loads including aerobic plate counts (APCs), mold counts (MCs) yeast counts (YCs), coliform counts (CCs), *Escherichia coli* counts, and Enterobacteriaceae counts (ECs) were determined. Chemical analyses including oil and linolenic acid -ALA indicated that all flaxseed had near normal oil content. This was the first reported survey for flaxseed. The pre-cleaned flaxseed had an average of 5.7 ± 0.1, 4.1 ± 0.2, 4.5 ± 0.2, 3.6 ± 0.1, and 3.0 ± 0.1 log colony forming units (CFU) g^{-1} of APC, CC, EC, YC and MC respectively. All counts were higher than those for cleaned seed. No *E. coli* was detected. The North Dakota-West (ND-W) region flaxseed had higher MC when compared to Canada, ND-North East (ND-NE) and ND-South East (ND-SE) region flaxseed. For APC, the counts were higher in flaxseed from Canada when compared to North Dakota. Cleaning the flaxseed should be considered an important step in reducing the microbial counts and also for maintaining high quality flaxseed.

Keywords: flaxseed, microbial loads, coliforms, linolenic acid

1. Introduction

Flaxseed (*Linum usitatissimum* Linnaeus) is a rich source of alpha-linolenic acid (ALA) comprising up to 55% of the total flaxseed fatty acid content (Chen et al., 1994). Hence, flaxseed's usage as a food ingredient has increased due to the positive results from health studies involving ALA, an omega-3 fatty acid (Morris & Vaisey-Genser, 2003). Evidence suggests that ALA consumption can reduce the risk for coronary artery disease, fatal ischemic heart disease among women, and high blood pressure (Berry & Hirsch, 1986; Hu et al., 1999; Djoussé et al., 2001). Apart from high ALA content, soluble fibre and proteins ($22g\ g^{-1}$ of seed) are also found in flaxseed (Rubilar et al., 2010). Flaxseed is also a good source for lignans that inhibits some types of diabetes as reported by Mueller et al. (2010). In contrast to the health benefits of flaxseed, quality and safety traits of flaxseed for food-use have not been well defined. Flaxseed is commonly consumed raw, and variable and arbitrary standards exist for raw flaxseed. There is a need to understand what normal microbial loads are for raw flaxseed.

The Federal Grain Inspection Service (FGIS) grades flaxseed into US number 1, US number 2 and US Sample grades, based on test weight, heat damaged kernels, total percent of damaged kernels or a combination thereof. Although these grades are useful to obtain seed of a specific purity, grading of the seed varies from year to year depending on the available crop. The FGIS grading system does not provide information on the microbial counts in flaxseed. Microflora such as bacteria, actinomycetes, molds and yeasts are frequently found on cereal grains (Deible & Swanson, 2001; Manthey et al., 2004). Viable, but dormant bacteria can occur in large numbers in optimally stored cereal grains with low level of water activity (ICMSF, 1998). Both pathogenic, such as *Salmonella* spp. and spoilage microflora, such as molds and yeasts, can occur in cereal grains either during pre- or post- harvest (ICMSF, 1998). Flaxseed and flaxseed based products are susceptible to microbial growth during processing and storage steps where water activities increase.

Microbiological standards for flaxseed and flaxseed products have been deemed necessary by food processors. Current standards used for grain and milled products may not be practical (Manthey et al., 2004; Sperber et al., 2007). This is particularly problematic for flaxseed processors who are expected to meet very stringent market specifications for microbial criteria such as coliforms and enterobacteriaceae, which are essentially arbitrary numbers. Also, setting numbers for indicator microbes like coliforms and enterobacteriaceae is not logical as natural, non-fecal, bacteria on raw plant materials will give positive results for these tests (Doyle & Erickson, 2006). To the best of our knowledge, there is a serious lack of data for establishing scientifically sound criteria for raw flaxseed microbiological quality and safety specifications.

The specific objectives of the study were to evaluate microbial, temporal and chemical aspects of flaxseed and especially, to study the effect of region and year on fatty acid contents of whole flaxseeds for food use. Hence, objectives for the study included collecting flaxseed samples over two-harvest periods for determining temporal, geographic region and cleaning effect on flaxseed microbial loads as well as their physical and chemical characteristics.

2. Materials and Methods

2.1 Sample Collection

During 2002-2007 crop years, U.S. Census of Agriculture, reported that North Dakota (ND) produced 94-97% of the nation's flaxseed (NASS, 2009). Hence harvests in North Dakota were targeted for representative flaxseed samples. Flaxseed samples were collected with the help of Northern Crops Institute, Fargo, ND, USA. The pre-cleaned (n=54) flaxseed samples from two years (2008 and 2009), were collected from 4 different regions (Figure 1; Canada, North Dakota (USA) - north east region (ND-NE), south east region (ND-SE), and west region (ND-W)).

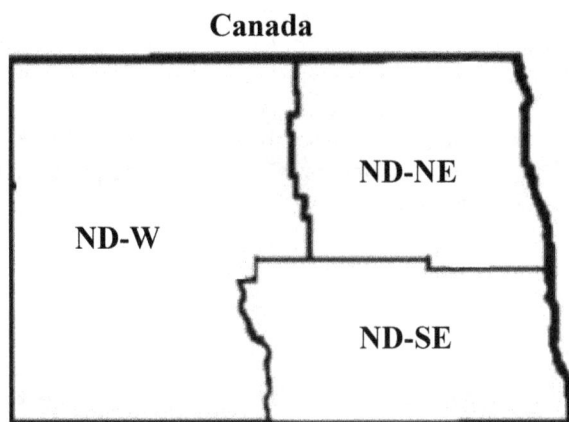

Figure 1. Map of flaxseed collection regions including north east (ND-NE), south east (ND-SE), west (ND-W) regions of North Dakota (USA) and Canada

2.2 Flaxseed Cleaning

Flaxseed cleaning was done following the method described by Manthey et al. (2009). A dockage tester (Carter Day International, Minneapolis, MN, USA) was used to remove weed seeds. The dockage tester was configured with a #25 riddle, #4 top sieve and #2 bottom sieve (no middle sieve was used). The air velocity was set at #3.5. Flaxseed that passed over the #2 sieve was collected. Flaxseed samples were passed through the tester two times.

2.3 Microbial Analyses

Microbial analyses were done on pre-cleaned and cleaned flaxseed to establish microbial loads for populations commonly tested for by the flaxseed industry. These tests included aerobic plate counts (APC), coliform counts (CC), *E. coli* counts, enterobacteriaceae counts (EC), mold counts (MC) and yeast counts (YC). The tests were done by following methods by Manthey et al. (2004), which included use of Petrifilm products (3M Microbiology, St. Paul, MN, USA).

2.4 Physical and Chemical Analyses

The moisture content (%) and test weight (kg m^{-3}, converted from pounds per bushel to kg m^{-3}) of the cleaned flaxseed samples were measured using the Grain Analyzer (GAC 2100, DICKEY-John, Auburn, IL, USA) following the standards set by Grain Inspection, Packers and Stockyards Administration (GIPSA, 2004). The cleaned flaxseed samples (50 g) were ground using a Laboratory Falling Number mill (model 3100, Perten Instruments, Springfield, IL, USA). Crude oil content in ground samples (4 g) was determined using a 16-hr

Soxhlet extraction with hexane following method Ba 3-38 (AOCS, 1998).

Fatty acid composition of the extracted oil was determined by a method described by Lee et al. (2003). The fatty acids were hydrolyzed and determined as methyl esters. Analysis was done on a HP5890 gas chromatograph (Hewlett Packard, Palo Alto, CA, USA) which was fitted with a flame-ionization detector and equipped with a SP2330 fused silica capillary column (30 m × 0.25 mm, i.d., and 0.20 μm film thickness) (Supelco, Bellefonte, CA, USA). The chromatography conditions for 1 μL of sample injected were: column flow rate 1mL min^{-1}, initial column temperature 150 °C (held for 5 min), raised 10 °C min^{-1} to a final temperature of 180 °C, and injector and detector temperatures were held at 200 °C. Individual fatty acids (palmitic acid (PA), stearic acid (SA), oleic acid (OA), linoleic acid (LA), and linolenic acid (ALA)) were confirmed by retention times and quantified against peak area standard plots of known fatty acid concentrations.

2.5 Experimental Design and Statistics

The microbial count data were evaluated for the effect of year, region and type (clean or pre-clean). The effect of region and year was evaluated on physical and chemical characteristics including fatty acid contents of the cleaned flaxseed samples. Analyses of maximum likelihood parameter estimates were performed using the procedure for generalized linear model (PROC GENMOD) of statistical software SAS (Version 9.2; SAS Institute, Inc., Cary, NC, USA) for microbial counts. Analyses of least square estimates were performed using another procedure for general linear model (PROC GLM) of SAS to analyze the physical and chemical data. All statements of statistical significance are based on $P < 0.05$.

3. Results and Discussion

3.1 Microbial Analyses

The temporal effect was evaluated on the microbial counts including APC, CC, EC, MC and YC (Table 1). The effect of geographical region, and type (pre-clean or clean) of flaxseed on microbial counts was also evaluated (Tables 2 and 3). The geographic region where the flaxseed was grown, year and type of flaxseed had a significant ($P < 0.05$) effect on the flaxseed APC and EC values. The pre-cleaned flaxseed APC ranged between 2.1 to 7.1 log CFU g^{-1} (Table 2) for 2008 and 2009 crops, and resulted in an average APC of approximately 5.7 ± 0.1 log CFU g^{-1}. Similarly, the average EC values in pre-cleaned flaxseed were 4.5 ± 0.2 log CFU g^{-1}, and ranged between 0 to 6.4 log CFU g^{-1} (Table 2). The APC and EC were significantly ($P < 0.05$) lower in year 2009 when compared to 2008 (Table 1). Also, the APC were significantly ($P < 0.05$) higher in Canadian samples, when compared to the North Dakota region samples. The ND-NE, ND-SE and ND-W pre-cleaned flaxseed samples were not significantly different in the mean APC values (Table 2). The mean EC were not significantly different among the pre-cleaned flaxseed obtained from different regions. Cleaning the flaxseed significantly reduced ($P < 0.05$) APC and EC values. For cleaned flaxseed, the average APC were 5.0 ± 0.1 log CFU g^{-1} whereas the average EC were 3.7 ± 0.2 log CFU g^{-1} (Table 3). Similar results were observed by Manthey et al. (2004), where cleaning dirty wheat, on average, resulted in approximately a 1-log reduction in CFU g^{-1} for APCs. The pre-cleaned flaxseed contained foreign materials such as shaft and weed seeds, which may have contributed to higher microbial counts.

Table 1. Effect of time on chemical, microbial and physical profile of flaxseed

	Mean Value	
	Year 2008	Year 2009
Aerobic Plate Count (log CFU g^{-1})	5.5[A]	5.1[B]
Coliform Count (log CFU g^{-1})	3.7[A]	3.8[A]
Enterobacteriaceae Count (log CFU g^{-1})	4.3[A]	3.8[B]
Mold Count (log CFU g^{-1})	2.7[A]	2.6[A]
Yeast Count (log CFU g^{-1})	3.2[A]	3.3[A]
Test Weight (kg m^{-3})	629.3[A]	595.9[B]
Moisture (%)	7.7[A]	6.5[B]
Oil (%)	41.6[A]	42.4[A]
Palmitic Acid (%)	5.6[A]	4.7[B]
Stearic Acid (%)	2.3[B]	3.4[A]
Oleic Acid (%)	21.1[A]	19.4[A]
Linoleic Acid (%)	15.8[B]	16.8[A]
Linolenic Acid (%)	55.3[A]	55.6[A]

Different capitalized letters in superscript within the same row indicates mean values are significantly different at α=0.05

Table 2. Microbial counts for pre-cleaned flaxseed obtained from various regions (2008 – 2009)

N	Region	Mean APC ± SE (log CFU g^{-1})	Range for APC (log CFU g^{-1})	Median for APC (log CFU g^{-1})	Mode for APC (log CFU g^{-1})
9	Canada	6.3 ± 0.2^A	5.5 - 7.1	6.2	None
8	ND-NE	5.5 ± 0.1^B	4.9 - 5.9	5.5	5.5
11	ND-SE	5.5 ± 0.2^B	3.8 - 6.2	5.8	None
26	ND-W	5.5 ± 0.2^B	2.1 - 6.3	5.8	5.7
	Average & Range	5.7 ± 0.1	2.1 - 7.1	5.8	5.5

N	Region	Mean CC ± SE (log CFU g^{-1})	Range for CC (log CFU g^{-1})	Median for CC (log CFU g^{-1})	Mode for CC (log CFU g^{-1})
9	Canada	4.6 ± 0.4^A	2.3 - 6.0	4.7	None
8	ND-NE	4.4 ± 0.2^A	3.6 - 5.3	4.5	None
11	ND-SE	4.0 ± 0.4^A	2.2 - 6.0	4.2	None
26	ND-W	3.8 ± 0.3^A	0.0 - 5.8	4.3	1.2
	Average & Range	4.1 ± 0.2	0.0 - 6.0	4.5	4.6

N	Region	Mean EC ± SE (log CFU g^{-1})	Range for EC (log CFU g^{-1})	Median for EC (log CFU g^{-1})	Mode for EC (log CFU g^{-1})
9	Canada	5.4 ± 0.3^A	3.0 - 6.4	5.9	None
8	ND-NE	4.7 ± 0.2^A	3.8 - 5.3	4.8	None
11	ND-SE	4.4 ± 0.4^A	2.2 - 5.7	4.9	5.2
26	ND-W	4.2 ± 0.3^A	0.0 - 6.3	4.8	1.2
	Average & Range	4.5 ± 0.2	0.0 - 6.4	5.0	5.3

N	Region	Mean YC ± SE (log CFU g^{-1})	Range for YC (log CFU g^{-1})	Median for YC (log CFU g^{-1})	Mode for YC (log CFU g^{-1})
9	Canada	3.3 ± 0.2^A	2.6 - 4.3	3.3	None
8	ND-NE	3.3 ± 0.1^A	2.9 - 3.6	3.4	None
11	ND-SE	3.3 ± 0.2^A	2.2 - 4.2	3.6	None
26	ND-W	3.8 ± 0.1^A	2.7 - 5.2	3.7	3.5
	Average & Range	3.6 ± 0.1	2.7 - 5.2	3.6	3.6

N	Region	Mean MC ± SE (log CFU g^{-1})	Range for MC (log CFU g^{-1})	Median for MC (log CFU g^{-1})	Mode for MC (log CFU g^{-1})
9	Canada	2.7 ± 0.2^B	2.1 - 3.5	2.6	None
8	ND-NE	2.7 ± 0.1^B	2.1 - 3.1	2.8	2.9
11	ND-SE	2.6 ± 0.2^B	1.3 - 3.5	2.8	None
26	ND-W	3.5 ± 0.2^A	1.3 - 5.9	3.3	None
	Average & Range	3.0 ± 0.1	1.3 - 5.9	3.0	2.9

Note: ND-NE indicates North Dakota (US) - North East region; ND-SE indicates North Dakota (US) - South East region; ND-W indicates North Dakota - West region; SE indicates Standard Error; APC indicates Aerobic Plate Count; CC indicates Coliform Count; EC indicates Enterobacteriaceae Count; YC indicates Yeast Count; MC indicates Mold Count; Different capitalized letters in superscript within the same column indicates mean values are significantly different at α=0.05

The average CC values of 4.1 ± 0.2 log CFU g^{-1} and 3.4 ± 0.2 log CFU g^{-1} were observed, respectively in the pre-cleaned and cleaned flaxseed samples from 2008 and 2009 (Tables 2 and 3). Similar to the average EC values, there were no significant differences in average CC values among the four flaxseed production regions for both pre-cleaned and cleaned samples (Tables 2-3). Also, no significant differences in CC values were observed between 2008 and 2009 flaxseed samples (Table 1). Regardless of the microorganisms tested, the pre-cleaned flaxseed had significantly (P<0.05) higher microbial numbers. No *E. coli* were detected in the samples in either 2008 or 2009. The *E. coli* assay is a more reliable indicator of fecal contamination, and could serve as an alternative to the fecal coliform assay (Doyle & Erickson, 2006). This reasoning also applies to the EC test.

The average YC and MC for pre-cleaned flaxseed from 2008 and 2009 crop years were 3.6 ± 0.1 log CFU g^{-1} and 3.0 ± 0.1 log CFU g^{-1}, respectively (Table 2). The yeast counts were significantly (P <0.05) lower with a mean value of 3.0 ± 0.1 log CFU g^{-1}, and the mold counts significantly (P <0.05) reduced to 2.3 ± 0.1 log CFU g^{-1} in cleaned flaxseed (Table 3). The 2009 flaxseed yeast and mold counts also were not significantly (P>0.05) different when compared year 2008 (Table 1). Also, when compared to ND-W region MC in pre-cleaned flaxseed were significantly (P <0.05) lower in samples of Canada, ND-NE and ND-SE regions; however, no such difference was observed among different regions for average YC counts in pre-cleaned flaxseed (Table 2). In cleaned flaxseed, the average YC were higher in ND-SE and ND-W region samples were significantly (P<0.05) higher when compared to Canada and ND-NE samples. The MC, however, had higher average values in cleaned flaxseed from Canada and ND-W when compared to ND-NE and ND-SE region samples (Table 3).

Table 3. Microbial counts for cleaned flaxseed obtained from various regions (2008 – 2009)

N	Region	Mean APC ± SE (log CFU g^{-1})	Range for APC (log CFU g^{-1})	Median for APC (log CFU g^{-1})	Mode for APC (log CFU g^{-1})
9	Canada	6.1 ± 0.2^A	5.4 - 7.0	6.0	None
8	ND-NE	4.8 ± 0.3^B	3.8 - 5.6	4.9	5.6
11	ND-SE	5.1 ± 0.3^B	2.7 - 6.0	4.8	None
26	ND-W	4.6 ± 0.2^B	2.3 - 5.9	4.8	None
	Average & Range	5.0 ± 0.1	2.3 - 7.0	5.0	5.6
N	Region	Mean CC ± SE (log CFU g^{-1})	Range for CC (log CFU g^{-1})	Median for CC (log CFU g^{-1})	Mode for CC (log CFU g^{-1})
9	Canada	3.6 ± 0.4^A	1.8 - 5.7	3.4	None
8	ND-NE	3.5 ± 0.3^A	2.6 - 4.9	3.2	None
11	ND-SE	3.6 ± 0.4^A	1.7 - 4.6	4.3	None
26	ND-W	3.2 ± 0.3^A	0.0 - 5.8	3.6	0.0
	Average & Range	3.4 ± 0.2	0.0 - 5.8	3.4	0.0
N	Region	Mean EC ± SE (log CFU g^{-1})	Range for EC (log CFU g^{-1})	Median for EC (log CFU g^{-1})	Mode for EC (log CFU g^{-1})
9	Canada	4.2 ± 0.5^A	2.0 - 5.6	5.2	5.2
8	ND-NE	3.8 ± 0.2^A	2.9 - 5.0	3.5	None
11	ND-SE	4.4 ± 0.3^A	1.9 - 4.3	4.6	5.2
26	ND-W	3.2 ± 0.3^A	0.0 - 5.6	3.2	1.2
	Average & Range	3.7 ± 0.2	0.0 - 5.6	3.8	5.2
N	Region	Mean YC ± SE (log CFU g^{-1})	Range for YC (log CFU g^{-1})	Median for YC (log CFU g^{-1})	Mode for YC (log CFU g^{-1})
9	Canada	2.8 ± 0.1^{BA}	2.3 - 3.4	2.7	None
8	ND-NE	2.4 ± 0.1^B	2.1 - 3.1	2.4	None
11	ND-SE	3.2 ± 0.2^A	2.5 - 3.8	3.0	None
26	ND-W	3.1 ± 0.1^A	1.5 - 4.4	3.1	2.5
	Average & Range	3.0 ± 0.1	1.5 - 4.4	2.9	2.5
N	Region	Mean MC ± SE (log CFU g^{-1})	Range for MC (log CFU g^{-1})	Median for MC (log CFU g^{-1})	Mode for MC (log CFU g^{-1})
9	Canada	2.4 ± 0.1^A	1.9 - 3.2	2.4	2.7
8	ND-NE	1.7 ± 0.1^B	1.3 - 2.0	1.7	1.3
11	ND-SE	2.2 ± 0.1^{BA}	1.6 - 2.9	2.3	None
26	ND-W	2.6 ± 0.2^A	0.0 - 4.6	2.7	3.0
	Average & Range	2.3 ± 0.1	0.0 - 4.6	2.3	2.7

Note: ND-NE indicates North Dakota (US) - North East region; ND-SE indicates North Dakota (US) - South East region; ND-W indicates North Dakota - West region; SE indicates Standard Error; APC indicates Aerobic Plate Count; CC indicates Coliform Count; EC indicates Enterobacteriaceae Count; YC indicates Yeast Count; MC indicates Mold Count; Different capitalized letters in superscript within the same column indicates mean values are significantly different at α=0.05

The cleaned flaxseed microbial data falls in the range reported for cereal grains. Most of the flaxseed samples had APC in the range of $10^2 - 10^5$ (100 – 100,000) CFU g^{-1}. Only a few samples had APC in the $10^6 - 10^7$ (i.e. 1 – 10 million) CFU g^{-1} range. The APC between 10^2 and 10^6 CFU g^{-1} are common for cereal grains. Graves et al. (1967), reported that in Kansas-Nebraska and Pacific Northwest regions, the wheat had total bacterial counts in the range of 15,000 – 660,000 g^{-1} (4.2 – 5.8 log CFU g^{-1}) in wheat. In more modern times, and in clean wheat, the average APC was reported to be 7.2 ± 0.5 log CFU g^{-1},[7] whereas for brown rice, it was around an average of 7.2 ± 0.3 log CFU g^{-1} (Skyrme et al. 1998). Yeast and mold counts are typically in the 10^2 and 10^4 CFU g^{-1} range for cereal grains. Cleaned flaxseed had yeast and mold counts that were 10^3 CFU g^{-1} or lower levels. They were similar to the values for wheat reported by Manthey et al. (2004), where the average mold and yeast counts were

3.5 ± 0.8 log CFU g^{-1}. After cleaning, the CC ranged from none to 6.0 log CFU g^{-1} in flaxseed, whereas for the brown rice the CC were around 2.4 ± 0.6 log CFU g^{-1} according to Skyrme et al. (1998). Furthermore, some flax producing regions did have higher counts than others (Tables 2 and 3).

3.2 Chemical and Physical Analyses

The average test weight was 615.2 ± 6.4 kg m^{-3} for the flaxseed obtained from 2008 and 2009 crop years (Table 4). The test weight for the flaxseed from the 2008 crop year averaged significantly higher (P <0.05) than in year 2009 (Table 1). As a result, the flaxseed obtained from some regions fell below the U.S. No. 2 grade while other samples met the grade No. 1. It should be noted that the additional cleaning may improve the lower test weight samples into the number 2 grade. Of the different regions involved, test weight of the flaxseed obtained from ND-NE and ND-SE were significantly (P <0.05) higher than in Canada and ND-W regions. A similar physical measurement of flaxseed, bulk density, was reported to be in the range of 556-727 kg m^{-3} when the moisture levels of the flaxseed was in the range of 6.1-16.8% (Coskuner & Karababa, 2007).

Table 4. Physical and chemical characteristics of the flaxseed obtained from various regions (2008 – 2009)

N	Region	Mean Test Weight ± SE (kg m^{-3})	Range for Test Weight (kg m^{-3})	Median for Test Weight (kg m^{-3})	Mode for Test Weight (kg m^{-3})
9	Canada	611.3 ± 6.4[B]	574.0 - 635.8	616.5	None
8	ND-NE	646.1 ± 7.7[A]	597.2 - 657.7	655.1	655.1
11	ND-SE	649.9 ± 2.6[A]	631.9 - 667.9	648.6	648.6
26	ND-W	593.3 ± 9.0[B]	491.6 - 661.5	604.9	625.5
	Average & Range	615.2 ± 6.4	491.6 - 667.9	626.8	648.6
N	Region	Mean Moisture ± SE (%)	Range for Moisture (%)	Median for Moisture (%)	Mode for Moisture (%)
9	Canada	8.6 ± 0.1[A]	8.0 - 9.2	8.6	8.3
8	ND-NE	7.1 ± 0.1[B]	6.5 - 7.5	7.2	7.2
11	ND-SE	6.5 ± 0.2[B]	6.0 - 8.0	6.3	6.3
26	ND-W	7.0 ± 0.3[B]	5.8 - 12.5	6.5	6.3
	Average & Range	7.2 ± 0.2	5.8 - 12.5	6.8	6.3
N	Region	Mean Oil ± SE (%)	Range for Oil (%)	Median for Oil (%)	Mode for Oil (%)
9	Canada	43.7 ± 0.6[A]	40.5 - 46.3	43.6	43.6
8	ND-NE	42.0 ± 0.7[A]	39.4 - 44.4	42.4	42.5
11	ND-SE	40.1 ± 0.8[B]	36.1 - 43.5	40.2	37.2
26	ND-W	42.1 ± 0.4[A]	38.0 - 46.5	42.2	None
	Average & Range	41.9 ± 0.3	36.1 - 46.5	42.4	43.6

Note: ND-NE indicates North Dakota (US) - North East region; ND-SE indicates North Dakota (US) - South East region; ND-W indicates North Dakota - West region; SE indicates Standard Error; Different capitalized letters in superscript within the same column indicates mean values are significantly different at α=0.05

Variability in the moisture content was also observed. The highest moisture content was observed in seeds from the 2008 harvest year than in year 2009. The overall flaxseed averaged 7.2 ± 0.2 % moisture, with the highest moisture (8.6 ± 0.1 %) observed in flaxseed from Canada (Table 4). The Canadian samples had significantly higher (P<0.05) moisture when compared to ND-NE, ND-SE and ND-W regions. The production region did not appear to be the determining factor regarding moisture content as flaxseed obtained from ND regions, which were not significantly different from each other except for Canadian samples. Thus, the growth year and level of precipitation at harvest may have resulted in moisture contents being higher in 2008. However, the average moisture content between these two years was 7.2%. Generally, increased moisture contents of flaxseed have been reported to affect some of the physical properties of flaxseed, and in turn may also affect the processing, transport and storage of the flaxseed. Certain physical characteristics such as, one thousand seed weight, bulk density, true density and their relationship to flaxseed moisture content was studied (Coskuner & Karababa, 2007). In that study, when the moisture of flaxseed increased from 6.1 to 16.8%, a linear increase in both one thousand seed weight (4.8 to 5.3 g) and true density (1000 to 1111 kg m^{-3}) was observed; whereas there was a decrease in bulk density from 727 to 556 (Coskuner & Karababa, 2007).

The average oil content of flaxseed from 2009 was not significantly different than the average oil content from the 2008 crop (Table 1). The overall flaxseed oil content ranged from 36.1 to 46.5% and averaged at 41.9 ± 0.3%

(Table 4). The oil contents of the samples are typical of those obtained from previous years typically analyzed in our laboratory. Historical data observed on North Dakota grown flaxseed indicates an oil level of 32-38% (Hettiarachchy et al. 1990). Higher oil content reported in the current study also indicates higher nutritive value of the seed grown in cooler climates. The only significantly (P<0.05) lower oil contents were seen in ND-SE region and rest of the regions were not statistically different in oil content (Table 4). Also, the soxhlet with hexane extracted oil from flaxseed in the current study (range of 36.1 to 46.5%) was very similar in values (45.2% of oil) that were reported by Mueller et al. 2010. Cultivars of the flaxseed were unknown and not analyzed.

The fatty acids (PA, SA, OA and LA) in the flaxseed oil varied among various regions that were tested (Table 5). On average, PA, SA, OA and LA for the flaxseed in the present study were 5.2%, 2.8%, 20.4% and 16.2% respectively. The results were similar to those reported earlier by Bean & Leeson (2002), who observed that the flaxseed had 5.6%, 3.2%, 18.5% and 14.4% respectively of PA, SA, OA and LA. In the present study, Canadian samples had significantly (P<0.05) lower average values of SA (1.7%) and LA (15.2%) when compared to ND regions. ND-SE region had significantly (P<0.05) higher average OA (23.8%) when compared to other regions; whereas ND-NE had significantly (P<0.05) higher PA (6.2%) (Table 5).

Table 5. Fatty acid profile of the flaxseed obtained from various regions (2008 – 2009)

N	Region	Mean PA ± SE (% Oil)	Range for PA (% Oil)	Median for PA (% Oil)	Mode for PA (% Oil)
9	Canada	5.4 ± 0.3^{BA}	4.5 - 6.8	5.2	4.9
8	ND-NE	6.2 ± 0.5^{A}	4.9 - 9.6	6.0	None
11	ND-SE	5.0 ± 0.3^{B}	2.9 - 6.0	5.0	5.2
26	ND-W	5.0 ± 0.2^{B}	3.2 - 7.8	4.8	4.6
	Average & Range	5.2 ± 0.1	2.9 - 9.6	5.0	5.2
N	Region	Mean SA ± SE (% Oil)	Range for SA (% Oil)	Median for SA (% Oil)	Mode for SA (% Oil)
9	Canada	1.7 ± 0.3^{B}	0.6 - 3.1	1.5	1.5
8	ND-NE	2.8 ± 0.3^{A}	1.8 - 4.2	2.8	None
11	ND-SE	3.3 ± 0.2^{A}	1.9 - 4.1	3.6	3.6
26	ND-W	2.9 ± 0.2^{A}	0.0 - 4.1	3.3	None
	Average & Range	2.8 ± 0.1	0.0 - 4.2	3.1	1.5
N	Region	Mean OA ± SE (% Oil)	Range for OA (% Oil)	Median for OA (% Oil)	Mode for OA (% Oil)
9	Canada	18.9 ± 0.9^{B}	15.0 - 24.2	19.1	None
8	ND-NE	20.4 ± 0.5^{B}	18.4 - 22.7	20.3	20.3
11	ND-SE	23.8 ± 1.8^{A}	16.9 - 31.5	20.9	None
26	ND-W	19.4 ± 0.4^{B}	13.4 - 22.8	19.7	None
	Average & Range	20.4 ± 0.5	13.4 - 31.5	19.6	19.6
N	Region	Mean LA ± SE (% Oil)	Range for LA (% Oil)	Median for LA (% Oil)	Mode for LA (% Oil)
9	Canada	15.2 ± 0.4^{B}	13.6 - 16.5	15.6	None
8	ND-NE	16.5 ± 0.4^{A}	14.6 - 17.8	16.8	None
11	ND-SE	16.6 ± 0.3^{A}	14.4 - 18.1	16.7	None
26	ND-W	16.4 ± 0.3^{BA}	12.5 - 20.1	16.4	None
	Average & Range	16.2 ± 0.2	12.5 - 20.1	16.4	16.4
N	Region	Mean ALA ± SE (% Oil)	Range for ALA (% Oil)	Median for ALA (% Oil)	Mode for ALA (% Oil)
9	Canada	58.8 ± 0.8^{A}	53.7 - 61.9	59.1	59.1
8	ND-NE	54.2 ± 0.6^{BC}	50.9 - 56.5	54.2	None
11	ND-SE	51.4 ± 2.0^{C}	43.2 - 59.7	55.3	None
26	ND-W	56.3 ± 0.7^{BA}	50.9 - 65.9	56.5	56.9
	Average & Range	55.4 ± 0.6	43.2 - 65.9	56.1	59.1

Note: ND-NE indicates North Dakota (US) - North East region; ND-SE indicates North Dakota (US) - South East region; ND-W indicates North Dakota - West region; SE indicates Standard Error; SA indicates Stearic Acid; OA indicates Oleic Acid, LA indicates Linoleic Acid; ALA indicates Linolenic Acid; Different capitalized letters in superscript within the same column indicates mean values are significantly different at α=0.05

The ALA composition in the oil appears to be high in both 2008 and 2009 compared to previous existing data (Table 1). The overall average ALA content among all samples from 2008 and 2009 (i.e. two-year average) was $55.4 \pm 0.6\%$ (Table 5). Typical ALA values from flaxseed grown from the early part of the decade were 50-53%. The higher values recently might be explained by improved production practices or simply the effect of growing region. In general, the ALA content of flaxseed from the Northern regions appears to be the highest. The flaxseed from western Canada in 2004 had ALA content as high as 61.9% (DeClercq, 2005), and in year 2009, it was reported to be about 58% (Barthet, 2010). Flaxseed samples obtained from Canada in the present study had similar values with the highest average ALA content of 58.8%, followed by ND-W region (56.3%), and the lowest values were found in ND-SE region (51.4%). This observation supports the basic observations that cooler climates cause plants to produce higher levels of unsaturated fatty acids. Also, the high ALA content could have a positive impact on overall nutritive value and general health benefits for humans ranging from reducing blood cholesterol to reducing platelet adhesivenesss (Loria, 1993; Cunnane & Thompson, 1995; Guan et al., 1998; Oomah, 2001).

The growing season of 2009 was North Dakota's 36th coolest summer and the 21st driest (since 1895), thus the lower moisture contents of flaxseed grown in 2009 may be related to the drier harvest period (Akyuz & Mullins, 2009). The growing season of 2008 was slightly warmer than 2009 in North Dakota, and ranked as the 46th coldest and 46th wettest (since 1895) (Akyuz & Mullins, 2008). Rains during the harvest period may have contributed to the slightly higher moisture levels observed in the 2008 samples. Overall, the basis for the elevated oil and ALA contents in 2009 may be indicative of the cooled growing season and thus, differences between the two years in this study.

4. Conclusion

The pre-cleaned flaxseed had an average of 5.7 ± 0.1, 4.1 ± 0.2, 4.5 ± 0.2, 3.6 ± 0.1, and 3.0 ± 0.1 log colony forming units (CFU) g^{-1} of APC, CC, EC, YC and MC respectively. For cleaned seed, the microbial count averages were 5.0 ± 0.1, 3.4 ± 0.2, 3.7 ± 0.2, 3.0 ± 0.1, and 2.3 ± 0.1 log CFU g^{-1} of APC, CC, EC, YC and MC respectively. The cleaning of flaxseed was beneficial in reducing the microbial counts and should be considered an important step in maintaining high quality flaxseed. Only ND-W region had higher counts than other regions tested. For APC however, the counts were higher in flaxseed from Canada when compared to North Dakota. Although, the flaxseed samples obtained in 2008 and 2009 represent a typical year with regards to oil (average of $41.9 \pm 0.3\%$) and moisture (average of 7.2 ± 0.2 %) contents. However, the high ALA contents ($55.4 \pm 0.6\%$) observed were unique for these two crop years. Although this survey evaluated a limited number of samples (54), the oil and fatty acid profiles indicate that the cooler growing seasons of 2008 and 2009 may have contributed to the higher ALA values. Overall, cool growing regions seem to have a positive effect on the overall ALA (chemical nutrients) content in the flaxseed. Also, cleaning of flaxseed can help in decreasing the overall microbial load on the seed.

Acknowledgments

The project was supported by the USDA Cooperative State Research, Education and Extension Service, special research grant number 2008-34328-19146 through the Midwest Advanced Food Manufacturing Alliance (MAFMA) program.

References

Akyuz, A., & Mullins, B. A. (2008). *2008 Growing season weather summary for North Dakota.* Retrieved from https://www.ndsu.edu/fileadmin/ndsco/ndsco/growing_season/2008.pdf

Akyuz, A., & Mullins, B. A. (2009). *2009 Growing season weather summary for North Dakota.* Retrieved from https://www.ndsu.edu/fileadmin/ndsco/ndsco/growing_season/2009.pdf

AOCS (American Oil Chemist Society). (1998). Method Ba3-38. In *Official Methods of Analysis,* Urbana, IL: American Oil Chemist Society.

Barthet, V. J. (2010). Quality of western Canadian flaxseed 2009. *Canadian Grain Commission,* 1-14.

Bean, L. D., & Leeson, S. (2002). Fatty acid profiles of 23 samples of flaxseed collected from commercial feed mills in Ontario in 2001. *The Journal of Applied Poultry Research, 11,* 209-211. https://doi.org/10.1093/japr/11.2.209

Berry, E. M., & Hirsch, J. (1986). Does dietary linolenic acid influence blood pressure? *The American Journal of Clinical Nutrition, 44,* 336-340.

Chen, Z. Y., Ratnayake, W. M. N., & Cunnane, S. C. (1994). Oxidative stability of flaxseed lipids during baking.

Journal of the American Oil Chemists' Society, 71, 629-632. https://doi.org/10.1007/BF02540591

Coskuner, Y., & Karababa, E. (2007). Some physical properties of flaxseed (*Linum usitatissimum* L.). *Journal of Food Engineering, 78*(3), 1067-1073. https://doi.org/10.1016/j.jfoodeng.2005.12.017

Cunnane, S., & Thompson, L. U (Eds.). (1995). *Flaxseed in human nutrition.* Champaign, IL: AOCS Press.

DeClercq, D. R. (2005). Quality of western Canadian flaxseed 2004. *Canadian Grain Commission,* 1-14.

Deible, K. E., & Swanson, K. M. J. (2001). Cereal and cereal products. In F. P. Downes, & K. Ito (Eds), *Compendium of Methods for the Microbiological Examination of Foods* (pp. 549-553). Washington, DC: American Public Health Association. https://doi.org/10.2105/9780875531755ch55

Djoussé, L., Pankow, J. S., Eckfeldt, J. H., Folsom, A. R., Hopkins, P. N., Province, M. A., Hong, Y., & Ellison, R.C. (2001). Relation between dietary linolenic acid and coronary artery disease in the National Heart, Lung, and Blood Institute family heart study. *The American Journal of Clinical Nutrition, 74,* 612-619.

Doyle, M. P., & Erickson, M. C. (2006). Closing the door on the fecal coliform assay. *Microbe, 1,* 162-163. https://doi.org/10.1128/microbe.1.162.1

GIPSA (Grain Inspection, Packers and Stockyards Administration, Federal Grain Inspection Service). (2004). General information. In *Grain Inspection Handbook* (pp. 1-42). Washington, DC: U.S. Department of Agriculture.

Graves, R. R., Rogers, R. F., Lyons, Jr. A. J., & Hesseltine, C. W. (1967). Bacterial and actinomycete flora of Kansas-Nebraska and Pacific Northwest wheat and wheat flour. *Cereal Chemistry, 44,* 288-299.

Hettiarachchy, N., Hareland, G., Ostenson, A., & Balder-Shank, G. (1990). Composition of eleven flaxseed varieties grown in North Dakota. *Proceedings of 53rd Flax Institute of the United States,* 36-40.

Hu, F. B., Stampfer, M. J., Manson, J. E., Rimm, E. B., Wolk, A., Colditz, G. A., Hennekens, C. H., & Willett, W.C. (1999). Dietary intake of alpha-linolenic acid and risk of fatal ischemic heart disease among women. *The American Journal of Clinical Nutrition, 69*(5), 890-897.

ICMSF (International Commission on Microbiological Specifications for Foods). (1998). Cereal and cereal products. In *Microorganisms in Foods 6: Microbial Ecology of Food commodities* (pp. 313-355). New York, NY: Blackie Academic and Professional.

Lee, R. E., Manthey, F. A., & Hall, C. A. (2003). Effects of boiling, refrigerating, and microwave heating on cooked quality and stability of lipids in macaroni containing ground flaxseed. *Cereal Chemistry, 80*(5), 570-574. https://doi.org/10.1094/CCHEM.2003.80.5.570

Manthey, F. A., Schorno, A. L., & Hall, C.A. (2009). Effect of immature and off-colored seeds on the lipid quality of milled flaxseed. *Journal of Food Lipids, 16,* 407-420. https://doi.org/10.1111/j.1745-4522.2009.01155.x

Manthey, F. A., Wolf-Hall, C. E., Yalla, S., Vijayakumar, C., & Carlson, D. (2004). Microbial loads, mycotoxins, and quality of durum wheat from the 2001 harvest of the northern plains region of the United States. *Journal of Food Protection, 67,* 772-780.

Morris, D., & Vaisey-Genser, M. (2003). Availablity and labeling of flaxseed food products and supplements. In L. U. Thompson, & S. C. Cunnane (Eds.), *Flaxseed in Human Nutrition* (pp. 404-422). Champaign, IL: AOCS Press. https://doi.org/10.1201/9781439831915.ch22

Mueller, K., Eisner, P., Yoshie-Stark, Y., Nakada, R., & Kirchhoff, E. 2010. Functional properties and chemical composition of fractionated brown and yellow linseed meal *(Linum usitatissimum). Journal of Food Engineering, 98*(4), 453-460. https://doi.org/10.1016/j.jfoodeng.2010.01.028

NASS (National Agricultural Statistics Service). (2009). 2007 Census of agriculture, United States, summary and state data. Retrieved from https://www.agcensus.usda.gov/Publications/2007/

Oomah, B. D. (2001). Flaxseed as a functional food source. *Journal of the Science of Food and Agriculture, 81*(9), 889-894. https://doi.org/10.1002/jsfa.898

Rubilar, M., Gutierrez, C., Verdugo, C., Shene, C. & Sineiro, J. (2010). Flaxseed as a source of functional ingredients. *Journal of Soil Science and Plant Nutrition, 10*(3), 373-377. https://doi.org/10.4067/S0718-95162010000100010

Skyrme, D. S., Marks, B. P., Johnson, M. G., & Siebenmorgen, T. J. (1998). Distribution of total aerobic and coliform bacterial counts among rice kernel components. *Journal of Food Science, 63*(1), 154-156.

https://doi.org/10.1111/j.1365-2621.1998.tb15698.x

Sperber, W. H., & The North American Miller's Association Microbiology Working Group. (2007). Role of microbiological guidelines in the production and commercial use of milled cereal grains: a practical approach for the 21st century. *Journal of Food Protection, 70*(4), 1041-1053.

3

Mince from Tilapia-Backbone: Effects of Washing and Cryoprotectant Addition during Frozen Storage

Wendy L. Lizárraga-Mata[1], Celia O. García-Sifuentes[1], Susana M. Scheuren-Acevedo[1], María E. Lugo-Sánchez[1], Libertad Zamorano-García[1], Juan C. Ramirez-Suárez[1], & Marcel Martinez-Porchas[1]

[1]Centro de Investigación en Alimentación y Desarrollo, A. C. Carretera a la Victoria, Km 0.6. Hermosillo, Sonora, C.P. 83304, México

Correspondence: Celia Olivia García-Sifuentes, Centro de Investigación en Alimentación y Desarrollo, A. C. Carretera a la Victoria, Km 0.6. Hermosillo, Sonora, C.P. 83304, México. E-mail: sifuentes@ciad.mx

Abstract

Mince obtained from tilapia (*Oreochromis niloticus*) (backbone) was evaluated; the effect of washing and the addition of a commercial cryoprotectant on the quality of the mince obtained were also assessed. Physicochemical, microbiological and sensorial analyses were carried out at 0, 15, 30, 45, 60, 90, 120 and 180 days of frozen storage (-20 °C). During washing treatment 91% of lipids was removed from the mince (p<0.05). The proximal composition was stable during the storage time (p>0.05). Parameters such as L* and "θ" increased while a*, b*, chroma and TBARS decreased due to the washing treatment (p<0.05). The cryoprotectant effect resulted in a decrease of L*, "θ" (Hue) and TBARS (p<0.05). The addition of the cryoprotectant caused a significant decrease of L*, "θ" and TBARS. During the storage period, the proximate composition was stable and the microbial load remained below the official limits. The panelists detected changes on the odor, color and texture in the mince evaluated. Results suggest that the washing treatment improved the stability of the mince compared to the addition of cryoprotectant.

Keywords: Cryoprotectant, quality, mince, backbone, *Oreochromis niloticus*, washing, frozen storage

1. Introduction

Nile tilapia, *Oreochromis niloticus*, produces approximately 60-70% of by-products (skeletons, meat remains, head, skin, bones, scales, and viscera) during the filleting process (Clement & Lovell, 1994, Abdel-Moemin, 2015). Some of these by-products had been evaluated by developed countries to produce sausages, hamburgers, breaded and other products with fish mince as the main ingredient (Borderías & Sánchez, 2011; FAO, 2014).

Currently, some of the fish species used for mince production are the Alaska pollock (*Theragra chalcogramma*), Pacific hake (*Merluccius productus*), American hake (*Merluccius bilinearis*) and Argentinian hake (*Merluccius bilinearis*), rose salmon, cod, catfish, tilapia, mackerel and herring (Alasalvar et al, 2011). However, by-products generated by filleting teleost fish are usually accompanied with blood, pigments, impurities, lipids, enzymes and sarcoplasmic proteins. These components compromise the stability, organoleptic characteristics (taste, texture, appearance, color) and nutritional properties of the mince (Leelapongwattana et al, 2005; Majumdar et al, 2012; Oliveira et al, 2012). Therefore, washing treatments have to be performed. Asgharzadeh, et al (2010), showed that the washing had an important positive effect on carp mince quality, the content of expressible moisture, TVB-N, FFA and thiobarbituric acid reactive substances were decreased. The washing removed fat, pigments, blood, enzymes and water soluble undesirable compounds in carp mince.

Cryoprotectants are compounds widely used to preserve the quality of biological tissues from freezing damage including fishery products (Parvathy et al, 2014; Santya & Krushna, 2011). These products are required to be nontoxic, cheap and low-molecular weight compounds; herein extracellular agents (do not penetrate cell membranes) improve the osmotic imbalance that occurs during freezing storage, e.g., sucrose, trehalose, and dextrose.

Therefore, the aim of this study was to evaluate the physical and chemical changes of mince obtained from tilapia-skeleton (backbone) as a result of washing and addition of commercial cryoprotectant during frozen storage.

2. Materials and Methods

2.1 Mince Source and Preparation

Tilapia backbones (*Oreochromis niloticus*) were obtained from a local producer who harvested and filleted tilapia in Novillo dam, placed in coolers with ice (0-1°C) and transported to the laboratory at Research Center of Food and Development in a period no longer than 24h after filleting. Upon receipt, backbones were washed with pressurized potable cold water, packed in plastic bags, distributed layered into a cooler (ice-backbones-ice) and stored in a 0-2°C chamber for processing. Backbones previously washed and cooled were processed in mechanical deboning equipment (Bibun (NDX103) Bibun Corp. Fakuyama, Japan). The mince obtained was divided and prepared for lots (treatments) as follow:

1) Unwashed mince without cryoprotectant (M). Represent the mince obtained directly from the deboning machine.

2) Unwashed mince with cryoprotectant (MC). Represent the mince obtained directly from the deboning machine with the addition of cryoprotectant mixture, which consisted in 0.1% of CAFODOS® (based in citric acid, sodium citrate and hydrogen peroxide) and 1.0% of ALTESA® (based in sodium citrate), both diluted in cold drinking water (1°C) and manually incorporated to the mince.

3) Washed mince without cryoprotectant (MW). Represent the mince obtained from the deboning machine and subjected to two washing steps. Washing was carry out with cold water (0 to 4°C) in 1:3 (mince/water) proportion, stirred for 2 min and finally a rest period of 5 min (Oliveira et al, 2012). During each wash, the floating lipids were manually removed using a strainer. After two washes, the mince was recovered by removing the water with a screw dryer (Bibun, Model SR1000 Bidun Corp., Fakuyama, Japan).

4) Washed mince with cryoprotectant (MWC). Represent the washed mince with the addition of cryoprotectant mixture described previously.

Finally, treatments were packed into vacuum bags in portions of 1 kg and 4 cm thick (Prime Source®), and placed in high density polyethylene shrimp trays (28X21X5cm) for freezing. The samples were frozen and stored in a freezer chamber at -20 °C for 6 months. Physicochemical, microbiological and sensory evaluations were evaluated on days 0, 15, 30, 45, 60, 90, 120 and 180.

2.2 Physicochemical Analysis

Proximate analysis of the samples was done in triplicate and was performed by following the A.O.A.C. recommendations (2000). Moisture was assessed by oven drying at 103°C, while, ash were evaluated at 550°C using an electrical muffle furnace (methods 950.46 and 938.08). The crude fat was determined by the Goldfish method (920.39) and the amount of crude protein and non-protein nitrogen was determined by the Micro-Kjeldahl method (A.O.A.C. 960.52). To calculate the protein content the nitrogen factor 6.25 was used.

pH was monitored by using a potentiometer, following the specifications of Woyewoda et al (1986). Total volatile base nitrogen (TVB-N) was determined by the method of magnesium oxide (Woyewoda et al, 1986) and expressed as mg of nitrogen per 100 g of sample.

To evaluate drip loss, two mince blocks (3.5x2.5 cm) were placed into Ziploc bags and the initial weight was recorded; then they were thawed (CODEX STAN 165-1989) and the water drained during thawing was removed; final weigh was recorded again. Results were reported as percent of weight loss.

Change of color was evaluated by using a Konica Minolta CR-400 (Konica Minolta Sensing, Inc., Tokyo, Japan) colorimeter. Color coordinates were considered to measure the degree of L* (lightness), a* (redness or greenness) and b* (yellowness-blueness). The values of L*, a* and b* were used to calculate other color indices such as the hue angle (Θ), chroma and the total color difference (ΔE).

The peroxide value was evaluated in lipids extracted from M and MW, while thiobarbituric acid reactive substances (TBARS) were evaluated directly from samples of M, MW, MC and MWC. Peroxide Value and TBARS were evaluated following the methods described by Woyewoda, Shaw, Ke & Burns (1986).

2.3 Microbiological Analysis

The microbiological analyses were carried out according to the Mexican Official Standards (NOM) and the FDA´s Bacteriological Analytical Manual. Total aerobic plates, incubated at 35 ± 2 °C and 24h (NOM-092-SSA1-1994) and total and fecal coliforms (NOM-112-SSA1-1994) were evaluated. Results were expressed as colony forming units per gram (CFU/g) and most probable number per gram of mince (MPN/g) respectively.

Bacterial pathogens such as *S. aureus* were evaluated on Baird Parker agar (Difco) at 35-37 °C, *Salmonella* spp on Xylose Lysine Desoxycholate (XLD) agar (Difco) at 35-37 °C (NOM-114-SSA1-1994 and BAM 2014). *Listeria monocytogenes* enrichment was performed in UVM Modified Listeria Enrichment Broth (Difco), followed by plating on Oxford (MOX) agar (Difco) incubated at 30°C/18-24 h (NOM-143-SSA1-1995 and BAM 2011). Pathogen evaluation was performed at the beginning and the end of the storage time.

2.4 Sensory Analysis

For sensory evaluation, the sample was prepared by following the method described in CODEX STAN 165-1989, which consists of placing mince blocks (inside a plastic bag) in a water bath at 21 ± 1.5 °C until thawing. Thereafter, the sensory evaluation was conducted by a group of 8 semi-trained panelists. Panelists (males, females, free of allergies and 40-50 years old) received six training sessions to permit the assessment of this study and recommended NMX-F-529-2004 standard was taken as a reference. Furthermore, the panelists have five years of experience in sensory evaluation of fishery products. Changes (odor, color and texture) were evaluated through the storage by using a 16-point scale (NMX-F-529-2004), where: 0= no change (characteristic attributes of the sample) and 16= extreme changes (changes due to contamination or decomposition).

2.5 Statistical Analysis

A complete randomized design was used with a 2 X 2 X 8 factorial arrangement, where the factors were X1 (washing treatment), X2 (cryoprotectant treatment) with two levels (with and without) and X3 Storage time (180 days) under frozen conditions with eight sampling dates. Data were analyzed by an analysis of variance for general linear models (GLM). When a significant effect of factors was found, the Tukey-Kramer comparison test was performed with a 5% significance level using the NCSS statistical software (2007).

3. Results and Discussion

3.1 Physicochemical

The moisture values obtained in this study are in accordance with the moisture range previously reported for tilapia muscle mince 70-80% (Biscalchin-Grÿchek et al , 2003). MC moisture was higher than M because the cryoprotectant added to the mince was diluted in water. There was a 16% increase (average) in moisture content attributable to the washing treatment (Table 1) and 1.9% due to the cryoprotectant effect ($p < 0.05$).

The increase in moisture content after washing (MW and MWC) could be associated to the removal of sarcoplasmic proteins. This change could promote the increment of myofibrillar proteins and finally as a result of these changes, the water retention capacity was increased (Biscalchin-Grÿchek et al , 2003). However, the moisture content M (71.3%) was lower than that reported in other mince species; for instance, 77.1% was documented in mackerel muscle mince (Eymard et al, 2009), 78.7% in muscle mince carp (Majumdar et al, 2012), and 81.3 and 78.3% in tilapia mince backbones (Oliveira et al, 2012; Kirschnik et al, 2013).

No significant difference ($P > 0.05$) was detected in the tilapia mince moisture content during the 6 month of frozen storage. According to Biscalchin-Grÿchek et al (2003) tilapia is considered as high protein source (15-20%); however, after the washing process the content of crude protein decreased ($p < 0.05$) 25.8% (calculated under higher water content). If the content of protein in MW is adjusted to the moisture of M (71.3%) the content of protein in MW decrease as describe previously. The wash-cryoprotectant interaction was not significant for protein content ($p > 0.05$), but was affected by storage time ($p < 0.05$). The decrease in the protein content reported in the present study was 13% lower than that reported for tilapia mince (Oliveira et al, 2012) and 12 % lower than carp muscle mince (Yongsawatdigul et al, 2013).

The decrease in protein content could be associated to the removal of sarcoplasmic- as well as myofibrillar-proteins, but the later in less proportion (Eymard et al, 2009; Yongsawatdigul et al, 2013).

The percentage of lipids removed as a result of the washing treatment was 28% higher than that reported for other Tilapia-skeleton mince (Oliveira et al, 2012) and 11 % than carp muscle mince (Yongsawatdigul et al, 2013) but, lower (31-37 %) than tilapia muscle mince (Biscalchin-Grÿschek et al, 2003).

The washing process resulted in the removing of 91% of lipids ($p < 0.05$) for MW and MWC, while the cryoprotectant addition resulted in 0.8% ($p < 0.05$). The wash-cryoprotectant interaction was significant (Table 1). Significant differences for lipid content were detected comparing washed vs non-washed minces ($p < 0.05$), but no differences were registered between MW and MWC ($p > 0.05$).

Regarding ash content (Table 1) the registered value in M was 0.7±0.2 which is comparable to previous reports for tilapia-skeleton mince (0.9-1.1%) (Oliveira et al, 2012; Kirschnik et al, 2013). However, 70% of the ash content ($p < 0.05$) was removed through the washing process, which is considerably higher than the removal achieved (29-40%) by other authors for Tilapia-skeleton mince (Oliveira et al, 2012; Kirschnik et al, 2013). The addition of cryoprotectant in the mince caused an ash level enhancement of 10% (p < 0.05); possibly due to the presence of citric acid and sodium citrate in the cryoprotectant used. Hoke et al. (2000) observed increases of 4% in the ash content in muscle mince with the addition of sodium citrate. Wash-cryoprotectant interaction had no effect on the percentage of mince ash (p > 0.05) and no changes were observed during frozen storage (p > 0.05).

Table 1. Chemical composition of tilapia minces during frozen storage.

Treatment	Component (%)	Storage time (days)					
		0	15	30	45	90	180
M	Moisture	[A]71.3±1.6	73.6±1.7	72.3±2.0	73.9±0.7	74.3±2.3	72.4±1.6
	Lipids	[A]15.3±0.7	14.3±0.5	14.4±0.5	14.0±0.7	14.2±0.6	14.8±0.6
	Protein	[A]12.0±0.2[ac]	10.2±0.4[b]	10.5±0.1[bc]	10.4±0.8[bc]	11.3±0.3[c]	11.8±0.5[ac]
	Ash	[A]0.7±0.2	0.6±0.1	0.7±0.1	0.6±0.0	0.6±0.1	0.7±0.0
	NPN	[A]0.1±0.0	0.1±0.0	0.1±0.0	n.d.	n.d.	0.0±0.0
MC	Moisture	[B]76.3±3.5	75.0±1.3	75.1±2.0	75.4±1.9	74.8±1.5	75.7±0.9
	Lipids	[B]12.4±1.4	13.6±1.3	13.3±0.8	13.0±0.7	13.5±1.2	13.1±0.7
	Protein	[B]10.5±1.0[a]	10.2±0.6[a]	9.9±0.1[a]	8.2±0.5[b]	11.2±0.6[a]	10.5±0.8[a]
	Ash	[A]0.6±0.2	0.7±0.1	0.6±0.0	0.7±0.1	0.7±0.0	0.7±0.0
	NPN	[B]0.0±0.0	0.1±0.0	0.1±0.0	n.d.	n.d.	0.0±0.0
MW	Moisture	[C]89.5±0.1	89.6±0.2	89.4±0.6	89.8±0.3	89.4±0.9	89.4±0.5
	Lipids	[B]1.4±0.2	1.2±0.1	1.5±0.2	1.3±0.1	1.4±0.3	1.4±0.1
	Protein	[C]8.9±0.3[ac]	7.6±0.1[b]	7.6±0.4[b]	7.8±0.5[b]	8.7±0.2[a]	9.5±0.4[c]
	Ash	[B]0.2±0.1	0.2±0.0	0.2±0.1	0.2±0.0	0.2±0.0	0.2±0.0
	NPN	[B]0.0±0.0	0.0±0.0	0.0±0.0	n.d.	n.d.	0.0±0.0
MWC	Moisture	[C]90.6±0.7	90.9±0.1	90.9±0.8	90.9±0.8	90.9±0.7	91.2±0.7
	Lipids	[B]1.3±0.1	1.3±0.4	1.2±0.2	1.2±0.1	1.2±0.1	1.1±0.2
	Protein	[C]8.7±0.7[a]	6.5±0.5[b]	6.6±0.3[b]	6.5±0.5[b]	7.9±0.7[a]	7.7±0.8[a]
	Ash	[B]0.2±0.0	0.2±0.1	0.2±0.0	0.2±0.0	0.2±0.0	0.2±0.0
	NPN	[B]0.0±0.0	0.0±0.0	0.0±0.0	n.d.	n.d.	0.0±0.0

Values represent the mean (n=16) ± standard deviation. Means with different capital letters in the same column indicate the effect of treatments and means with different little case letters in the same row were significantly different (p < 0.05). n.d.= no determined. NPN: Non-protein nitrogen. M= unwashed mince and non-cryoprotectant. MC= unwashed mince with cryoprotectant. MW= washed mince non-cryoprotectant. MWC= washed mince with cryoprotectant.

Drip loss was also detected in samples; herein, drip loss ranged from 7.2 to 7.9% in all treatments except for MW which recorded the lowest value (0.9 ± 0.4%). Drip loss of M was 1.2% higher than that reported for tilapia muscle mince and 0.8% lower than other tilapia-skeleton mince (Biscalchin-Grýchek et al., 20 03; Kirschnik et al., 2012).

During the six months of storage changes in drip loss were also detected (Figure 1). Drip loss in M increased by 5% at 15th day, while an increase of 0.9% was detected for MW at the end of the storing period (p < 0.05). No differences were detected (p >0.05) for MW and MWC, probably due to the presence of the cryoprotectant.

No studies regarding color attributes have been reported by the scientific community for tilapia-skeleton mince during frozen storage. Analysis of variance (GLM) suggest that the mince brightness (L*) increased in average 8.6 units (p < 0.05) as a result of the washing effect (Table 2), but decreased 0.7 units due to the cryoprotectant effect. Hoke et al. (2000) reported an increase of 5 units on the L* parameter for catfish-skeleton mince, caused by the removal of blood and pigments during the washing process. Furthermore, hydrogen peroxide could be used in some cases to improve color attributes that are important to define the selling price and consumer acceptance. The cryoprotectant mixture used in this study, contains hydrogen peroxide as part of its formulation; however, L* was no improved; Himonides et al. (1999) used hydrogen peroxide as a bleacher in cod mince and found no significant differences for "L" values.

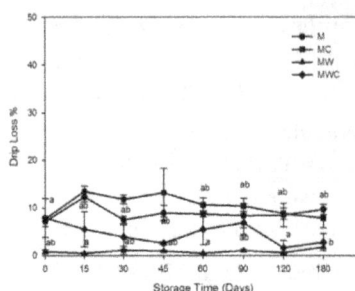

Figure 1. Changes in drip loss of tilapia mince during frozen storage.

Values represent the mean (n=4) ± standard deviation. Means with different letters were significantly different (p < 0.05). M= unwashed mince without cryoprotectant. MC= unwashed mince with cryoprotectant. MW= washed mince without cryoprotectant. MWC= washed mince with cryoprotectant.

Results of this study indicated no significant interaction of washing process and cryoprotectant addition (p >0.05). Results revealed that during storage time, L* value increased by 5, 2.2 and 5.5 units in MW, M and MWC respectively (p < 0.05), at 45th, 60th and 60th days of storing. Herein, Chow et al (2009), concluded that the increase in L* could be caused by the alkaline pH of the cryoprotectant used. However, as described previously our results showed an increment in L* value due to the washing treatment. Similar to M in this study, Pivarnik et al. (2013) observed no changes in L* values of tilapia fillets stored at 5°C during the frozen storage.

For the red-green values (a*, Table 2), a decrease of 9.4 units were detected as a result of washing treatment (p < 0.05). The lower values obtained on a* were probably the associated to the removal of soluble pigments. Myoglobin, hemoglobin, carotenoid and melanin are the main pigments that contribute to fish color (Thiansilakul et al, 2012); herein, Yarnpakdee et al (2012) documented a myoglobin removal of 52% in Tilapia-muscle mince after a washing process.

Table 2. Changes in color of tilapia mince during frozen storage.

Treatments	Color Traits	Storage Time (Days)							
		0	15	30	45	60	90	120	180
M	L*	A40.07±4.0	38.34±2.5	39.76±2.0	38.43±1.6	39.88±2.9	40.21±2.6	39.69±2.9	38.91±3.6
	a*	A8.38±1.1a	9.70±2.2ab	11.14±1.6bc	12.00±1.2c	10.49±1.5bc	10.17±1.3b	9.98±1.6ab	10.21±2.2bc
	b*	A5.65±0.9a	6.07±1.1ac	7.38±0.8b	7.19±0.5b	7.12±0.8b	7.12±0.7b	6.86±0.9bc	7.03±0.7b
MW	L*	C43.56±3.7ac	43.61±6.5ac	49.46±3.6bc	50.64±3.2b	51.62±2.6b	50.39±3.6b	52.03±2.6b	46.16±1.7c
	a*	C1.56±1.0ac	1.47±0.6a	0.74±0.4b	0.62±0.3bc	0.35±0.2ab	0.12±0.2b	0.33±0.3b	0.27±0.2b
	b*	C2.33±1.2ac	3.45±1.4ab	2.63±1.0abc	3.47±1.4ab	3.8±1.2bc	3.56±1.0b	2.82±0.8abc	1.63±0.6ac
MC	L*	B38.96±2.6ab	37.45±2.1a	39.32±2.4ab	38.76±2.7ab	41.19±3.1b	40.16±2.3ab	38.38±2.6ab	38.68±2.6ab
	a*	B10.15±1.9ab	11.66±1.9ab	11.69±2.2ac	11.90±1.9a	11.02±1.8ab	10.53±1.5ab	9.73±1.6bc	9.64±2.0bc
	b*	B6.38±1.0a	6.74±0.7ad	7.67±0.8bc	7.82±0.6bc	8.08±0.6b	7.91±0.4bc	7.15±0.5cd	7.20±0.7cd
MWC	L*	D45.42±4.1ad	45.48±2.5ad	47.67±3.8abc	47.2±3.4abd	48.94±2.4bc	50.88±3.5b	49.31±2.1bc	44.10±2.2d
	a*	D0.54±0.3ab	0.81±0.3a	0.63±0.4ab	0.58±0.3ab	0.51±0.2bc	0.34±0.3c	0.35±0.3bc	0.40±0.2bc
	b*	D1.25±0.8a	1.32±0.8a	2.36±1.1bc	2.24±0.9ab	2.93±1.0b	3.03±1.1b	2.04±0.6ab	1.76±1.0c

Values represent the mean ± standard deviation. Means with different capital letters in the same column indicate the effect of treatments and means with different little case letters in the same row were significantly different (p < 0.05). M = unwashed mince without cryoprotectant. MC = unwashed mince with cryoprotectant. MW = washed mince without cryoprotectant. MWC = washed mince with cryoprotectant.

Regarding the use of the cryoprotectant no significant effect of this practice was detected on the a* value (p > 0.05), neither significant interactions with the washing factor (p<0.05). Finally, M and MW increased 2.8 and 0.8 a* units (p<0.05) at 30th day of storing, while treatments with cryoprotectant (MC and MWC) did not differ during storage and remained stable during the six months of storage.

With regard to yellow-blue parameter (b*, Table 2), results of analysis of variance (GLM) showed a decrease in average of 4.6 units in the washed mince (p < 0.05); similar results were found in previous reports for washed carp-skeleton mince, where b* decreased 4 units (Suvanich et al, 2000). The cryoprotectant treatment had not a significant effect on this parameter (p > 0.05), but the washing-cryoprotectant interaction was significant (p < 0.05). At 30th day of storage, b* increased 1.7, 1.3 and 1.1 units for M, MWC and MC, respectively (p < 0.05). The increase of b* in tilapia mince could promote the browning flesh bloodline development due to the oxidation of heme-proteins and the subsequent reduction of a* values (Pivarnik et al., 2013).

Chroma values decreases after the washing process also (Table 3). Mince red saturation decreased 9.7 units as a result of washing (p < 0.05), but no significant effect was detected by the addition of cryoprotectant (p > 0.05); however, a significant wash-cryoprotectant interaction was recorded (p < 0.05).

The values of hue angle are shown on Table 3, no significant changes were observed for M during the 180 days of storage (p > 0.05). For MC the hue angle increased 4.3 units (60th day), MW increased 8.2 units (15th day) and 7.2 units in MWC after 30th day (p < 0.05), and remained stable for the rest of the storage period. Even though there were a visible color change after washing, ΔE was not affected until 30th day in M, 30th and 180th day in MW, 60th day in MC and 180th day in MWC (Table 3). In consonance with ΔE results, color tones for washed samples end up yellowish, probably associated to b* changes. Unwashed samples (M, MC) were more susceptible to oxidation and discoloration turning brownish-dark pink probably due to presence of oxidized heme-pigments (Suvanish et al, 2000). Chaijan et al (2005) observed a decrease in chroma values of sardine mince due to pigment removal during washing. The color was seen reddish in the M and MC at the beginning of the storage time and became yellow-green at the end of the trial. Moreover, the MW and MWC had a pink-cream color at the beginning of the storage time, and a yellow-cream color at the end of the storage period.

Table 3. Changes in calculated color traits of tilapia mince during frozen storage.

Treatments	Color traits	Storage Time (Days)							
		0	15	30	45	60	90	120	180
M	Θ	[A]34.03±5.2	32.40±3.8	33.76±3.4	31.01±2.4	34.33±3.1	35.21±4.0	34.65±2.3	35.26±6.1
	Chroma	[A]10.14±1.1[ac]	9.29±4.4[a]	13.38±1.6[b]	14.00±1.1[b]	12.69±1.6[b]	12.43±1.2[b]	12.12±1.8[ab]	12.45±2.0[ab]
	ΔE		2.2	2.4	1.6	2.1	0.5	0.61	0.8
MW	Θ	[B]58.45±9.8[a]	66.64±4.8[b]	73.23±6.5[bc]	78.05±8.1[cd]	84.96±1.8[de]	88.02±2.5[e]	83.10±7.7[de]	80.34±6.0[cd]
	Chroma	[C]2.84±1.5[ab]	3.76±1.5[a]	2.75±1.1[ab]	3.55±1.3[a]	3.82±1.2[a]	3.56±1.0[a]	2.86±0.8[ab]	1.67±0.6[ab]
	ΔE		1.1	6.0	1.4	1.1	1.3	1.8	5.99
MC	Θ	[A]32.35±2.8[a]	30.23±2.1[a]	33.6±2.8[ab]	33.60±3.1[ab]	36.62±3.77[bc]	37.20±3.4[bc]	36.71±4.5[bc]	37.24±3.6[c]
	Chroma	[B]12.00±2.0[ab]	10.31±4.4[a]	13.99±2.2[b]	14.26±1.8[b]	13.69±1.67[b]	13.19±1.4[b]	12.11±1.3[bc]	12.06±2.0[b]
	ΔE		2.17	2.09	0.62	2.6	1.1	2.1	0.3
MWC	Θ	[C]64.97±9.3[ab]	55.43±15.1[a]	74.21±10.8[bc]	74.63±10.6[bcd]	78.99±5.2[cd]	84.97±5.6[d]	80.16±7.3[cd]	74.68±9.1[bcd]
	Chroma	[D]1.37±0.8[a]	1.67±0.7[ab]	2.46±1.1[bc]	2.33±1.0[abc]	2.98±1.0[c]	3.06±1.1[c]	2.08±0.6[abc]	1.82±1.0[b]
	ΔE		0.3	2.4	0.49	1.87	1.95	1.86	5.22

Values represent the mean ± standard deviation. Means with different capital letters in the same column indicate the effect of treatments and means with different little case letters in the same row were significantly different ($p < 0.05$). M = unwashed mince without cryoprotectant. MC = unwashed mince with cryoprotectant. MW = washed mince without cryoprotectant. MWC = washed mince with cryoprotectant.

Regarding pH, the results registered at day 0, showed an increase of 0.4 pH units after the washing process ($p < 0.05$), and 0.5 units after the addition of the cryoprotectant ($p < 0.05$); significant wash-cryoprotectant interaction was also detected ($p < 0.05$). Biscalchin-Grýchek et al (2003) observed a gradual increase in pH on washed tilapia-muscle mince, and argued that this pattern can continue if additional wash cycles are performed. Higher pH values in washed mince may occur due to the removal of free fatty acids, free amino acids, lactic acid and other water-soluble acidic substances (Majumdar et al, 2012). The initial values of pH were 7.2, 7.2, 7.4 and 7.7 for M, MW, MC and MWC, respectively. pH decreased ($p < 0.05$) 0.2 units at 45th day in M, 0.4 at 120th day in MC and 0.1 at 90th day in MW; no significant changes were observed for MWC during the storage period.

Significant changes were also detected for TVB-N; a decrease of 61% was associated the washing effect (< 0.05, Figure 2) and <10% to the cryoprotectant addition ($p < 0.05$). No significant washing-cryoprotectant interaction was detected, but a significant effect of storage period was registered ($p < 0.05$). All minces reported TVB-N values lower than the official permitted limits (30 to 35 mg N/100 g). Unlike the present study, previous reports observed an increase of BVT-N content (80%) in frozen tilapia fillets during storage (Emire & Gebremariam, 2009) and 33% after the 60th day of tilapia-sketelon mince freezing storage (Kirschnik et al, 2013).

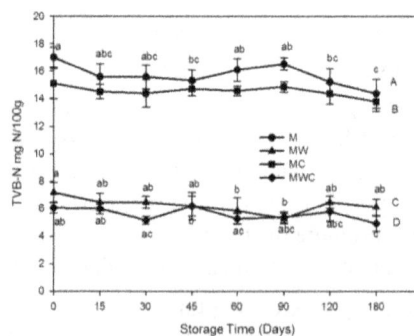

Figure 2. Changes in total volatile bases nitrogen (TVB-N) of tilapia mince during frozen storage.

Values represent the mean (n=6) ± standard deviation. Means with different capital letters indicates the effect of treatments and little case letters indicates the effect of storage time ($p < 0.05$). M= unwashed mince without cryoprotectant. MC= unwashed mince with cryoprotectant. MW= washed mince without cryoprotectant. MWC= washed mince with cryoprotectant.

Peroxide values for washed mince were not determined due to its low lipid content. The PV values at the beginning of storage were 1.1 ± 0.5 meq/kg for M and 0.7 ± 0.4 meq/kg for MC. No effect of the cryoprotectant addition ($p > 0.05$) was observed. Moreover, during frozen storage, no changes were observed for M ($p > 0.05$), but MC increased by 1.0 meq/kg at 60th day of storage ($p < 0.05$) (Data not shown). The PV values were maintained under the acceptable limits (MC= 1.1 ± 0.5 meq/kg and M= 0.7 ± 0.4 meq/kg) compared with normal values suggested for fish 10-20 meq/kg (Sallam, 2007), probably due to the lean characteristic of the mince elaborated in this study.

Thiobarbituric acid reactive substances registered a decrease ($p<0.05$) of 59% and 19% due to the washing and cryoprotectant effect, respectively (Figure 3). Mince washing treatment can prevent lipid oxidation because most of the lipid, hydroperoxides and secondary oxidation compounds are removed, providing greater stability to a food product (Fogaca et al., 2015). Furthermore, Kilinc et al (2009), observed a decrease of 20.5% of TBARS in trout fillets treated with a solution of sodium citrate 2.5%. Citric acid and its salts have been reported to have antioxidant effect when used as a pretreatment to frozen fish storage, acting as synergists, oxygen scavengers and heavy metal chelators (Aubourg et al, 2004; Sanjuás-Rey et al, 2011). Results also indicated that the wash-cryoprotectant interaction was not significant ($p > 0.05$).

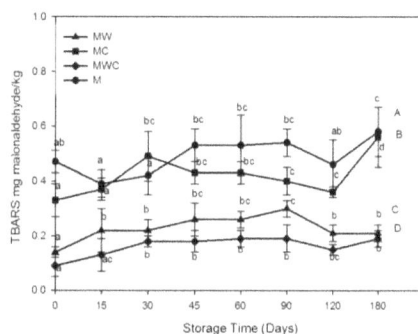

Figure 3. Changes in thiobarbituric acid reactive substances (TBARS) level of tilapia mince during frozen storage.

Values represent the mean (n=6) ± standard deviation. Means with different capital letters indicates the effect of treatments and little case letters indicates the effect of storage time ($p < 0.05$). M= unwashed mince without cryoprotectant. MC= unwashed mince with cryoprotectant. MW= washed mince without cryoprotectant. MWC= washed mince with cryoprotectant.

An increase ($p < 0.05$) of TBARS in mince was observed during storage time. MC and MWC increased 0.2 and 0.1 mg malonaldehyde/kg respectively at 30th day, whereas MW and M increased 0.1 mg of malonaldehyde/kg at 15th and 45th, days respectively. At the end of the storage, all treatments registered TBARS levels below 3 mg of malonaldehyde/kg, which is within the range recommended by Kilinc et al (2009), for human consumption (7-8 mg of malonaldehyde/kg).

3.2 Microbiological Analysis

Microbiological results in unfrozen minces resulted in Mesophiles count of 5.3±0.6, 5.1±0.5, 4.9±0.2 and 4.0±0.1 CFU/g log for M, MW, MC and MCW, respectively. Total coliforms were 15±7, 30±14, 150±32 and 148±61 for M, MW, MC and MCW, respectively. With regards to *S. aureus* all minces registered < 100 CFU/g, whereas *Salmonella* spp and *L. monocytogenes* were absent. Unfrozen and frozen minces showed mesophilic counts and total coliforms below the permitted limits (< 7 CFU/g log, < 400 MPN/g, respectively), suggesting a good quality from the microbiological perspective. The absence of pathogenic microorganisms was also confirmed in accordance with international and the mexican standards (NOM-242-SSA1-2009; CODEX STAN 165-1989; ICMSF, 1986).

Analysis of variance (GLM) showed a reduction in average of 0.6 CFU/g log ($p < 0.05$) in mesophilic count (Table 4) and 0.5 CFU/g log due to the washing and cryoprotectant effect, respectively. During storage, MWC decreased 1.4 CFU/g log ($p < 0.05$), whereas no significant changes were observed in M, MW and MWC treatments ($p > 0.05$). Total mesophiles count demonstrated that after six months of storage, the minces bacterial loads remained under the permitted limits, which can be associated somehow to the washing treatment.

Table 4. Microbiological evaluation of tilapia minces during frozen storage.

Treatments	Microorganism	Storage Time (Days)							
		0	15	30	45	60	90	120	180
M	Mesophiles	[A]5.5±0.6	5.2±0.4	5.2±0.3	4.8±0.1	4.6±1.5	4.1±0.5	4.2±0.6	6.0±0.01
	FC	[A]80.0±17.3	150.0±23.0	126.7±25.1	172.5±80.5	135±51.8	131.5±125.2	130±126.4	240±0.0
MW	Mesophiles	[C]4.4±0.7	4.6±0.1	4.5±0.7	3.8±0.1	3.8±0.1	4.2±0.4	3.6±0.1	4.0±0.6
	FC	[B]30±8.2	6.7±5.9	27.5±27.5	52.5±5.0	30±8.1	11.5±13.3	48.5±51.4	50±49.7
MC	Mesophiles	[B]4.6±0.4[ac]	4.4±0.2[ac]	4.3±0.4[abc]	4.6±0.0[ac]	4.0±0.6[ab]	4.0±0.3[abc]	3.2±0.6[b]	5.1±0.1[c]
	FC	[B]36.7±11.5	17.5±15	37.5±29.9	9±0.0	12±3.5	9±2.3	5.5±1.7	16.8±8.08
MWC	Mesophiles	[D]4±0.1	3.6±0.1	3.8±0.3	3.6±0.1	3.6±0.0	3.8±0.1	4.1±0.7	3.8±0.3
	FC	[B]<3±0	22.5±17.08	15±10	34.5±9.8	<3±2.3	<3±2.3	13±11.5	5.0±2.3

Values represent the mean ± standard deviation. Means with different capital letters in the same column indicate the effect of treatments and means with different little case letters in the same row were significantly different ($p < 0.05$). M = unwashed mince without cryoprotectant. MC = unwashed mince with cryoprotectant. MW = washed mince without cryoprotectant. MWC = washed mince with cryoprotectant. Mesophiles are express as Log CFU/g. FC = Fecal Coliform is express as MPN/g.

The fecal coliform count in mince resulted in a reduction of 60 MPN/g due to the washing effect ($p < 0.05$) and 74 MPN/g due to the cryoprotectant addition (Table 4). The limits recommended in fresh and frozen fish for fecal coliforms should be <230 MPN/g, whereas *E. coli* should be absent (ICMSF, 1986). Lowering the number of sanitary-indicator microorganisms (e.g., coliforms) in food products can be beneficial for assessing effectiveness of safety procedures during processing and handling. Shaviklo & Rafipour (2013) observed a reduction 82 MPN/g of fecal coliform during the washing process of myctophid-mince fish to obtain surimi. In contrast, no effect due to washing process was observed on fecal coliform counts of catfish-skeleton mince (Suvanich et al, 2000).

Regarding cryoprotectant effectiveness to decrease bacterial loads, previous studies reported a decrease of 35 MPN/g fecal coliform in trout fillet due to sodium citrate (2.5%) use (Kilinc et al., 2009). Sanjuás-Rey et al (2012), also observed a decrease of 89 NMP/g fecal coliforms by the addition of an organic acid solution (citric, ascorbic acid and lactic acid) during mackerel storage. In this way, maybe the organic acids and their salts cross the microbial membranes and come into to the cytoplasm. Inside the cytoplasm, acids dissociate and the microorganisms eliminate the anions excess out of the cytoplasm to protect the physiological pH; however, this process causes limiting of growth or microbial die (Kilinc et al., 2009). Therefore, both wash and cryoprotectant treatments evaluated achieved an effective reduction in fecal coliform counts in tilapia-mince backbones; at day 180th all minces registered bacterial loads below the permitted limits of fecal coliforms (400 MPN/g); thus, no significant changes was observed during the storage for fecal coliforms in all minces. Furthermore, 0 NPN/g of *E. coli* was registered for all minces along the storage period. Finally, pathogen evaluation on the tilapia mince at 180th day resulted in *S. aureus* counts < 100 CFU/g and absence of *Salmonella* spp and *L. monocytogenes*.

3.3 Sensory Analysis

Features of tilapia-skeleton mince such as low fat content and low lipid oxidation rates (TBARS), contributed to panelists to detect slight odor changes during storage (Table 5). The sensory panel detected slight odor changes on M after the 120th day of storage ($p < 0.05$). In the case of MW slight changes were detected until the end of storage ($p < 0.05$). In MC and MWC changes in odor occurred after 60th day of storage ($p < 0.05$), suggesting that the cryoprotectant probably promoted a slight but detectable odor by panelists. The reduction in the content of hemoglobin, myoglobin and iron from hemoglobin due to the washing treatment (MW and MWC) increases the stability to lipid oxidation and reduces the formation of volatile (odor producing) compounds (Yarnpakdee et al, 2012). In MWC and MW, the panel detected slight changes on color at 180th day ($p < 0.05$) while in MC and M changes were detected after 60th day of storage ($p < 0.05$). Panelists reported that non-washed treatments (M and MC) had brown color tones during storage while washing treatments (MW and MWC) after 30th day of storage turned to clearer yellow hues, resulting in yellow-cream color at the end of the storage. Color changes detected by the panelists somehow agree with the results evaluated instrumentally and explained in previous sections.

Table 5. Sensory evaluation of tilapia minces during the frozen storage.

Parameter	Storage time (days)	M	MW	MC	MWC
Color	0	0.05 ± 0.08^a	0.13 ± 0.12^a	0.06 ± 0.11^a	0.09 ± 0.14^a
	15	1.44 ± 1.48^{ab}	0.66 ± 0.74^a	2.10 ± 2.34^{abc}	1.48 ± 1.77^{ab}
	30	2.10 ± 1.93^{ab}	1.44 ± 1.36^a	1.81 ± 1.41^{ab}	1.60 ± 1.37^{ab}
	45	2.76 ± 1.57^{abc}	0.76 ± 0.40^a	2.81 ± 2.27^{abc}	1.16 ± 0.94^{ab}
	60	3.57 ± 3.18^{bc}	2.49 ± 1.73^{ac}	4.86 ± 3.16^{bcd}	1.87 ± 1.84^{ab}
	90	2.30 ± 1.47^{abc}	1.33 ± 1.27^a	2.30 ± 2.28^{abc}	1.15 ± 0.9^{ab}
	120	3.76 ± 2.44^{bc}	3.98 ± 2.86^{bc}	5.30 ± 4.26^{cd}	3.35 ± 2.9^{bc}
	180	4.88 ± 2.98^c	5.43 ± 3.38^b	6.73 ± 4.41^d	5.45 ± 3.39^c
Odor	0	0.11 ± 0.13^a	0.13 ± 0.12^a	0.07 ± 0.11^a	0.06 ± 0.09^a
	15	1.74 ± 1.72^{ac}	0.50 ± 0.58^a	1.42 ± 1.25^{ab}	1.61 ± 1.77^{ab}
	30	2.09 ± 1.76^{ab}	1.90 ± 1.82^{ab}	1.91 ± 1.48^{abc}	0.97 ± 0.85^{ac}
	45	2.34 ± 1.99^{ab}	1.36 ± 1.01^{ab}	3.05 ± 1.99^{abc}	2.60 ± 1.28^{ab}
	60	2.58 ± 2.11^{ab}	3.73 ± 3.63^{bc}	4.83 ± 4.15^{bcd}	4.01 ± 3.79^{bd}
	90	2.57 ± 1.44^{ab}	3.67 ± 2.34^{bc}	3.22 ± 1.69^{bcd}	3.27 ± 2.38^{bcd}
	120	4.38 ± 3.36^{ab}	2.72 ± 1.96^{abc}	5.18 ± 4.25^{cd}	3.03 ± 2.39^{bc}
	180	3.90 ± 1.88^{bc}	4.87 ± 3.05^c	6.95 ± 5.18^d	5.90 ± 3.64^d
Texture	0	0.11 ± 0.13^a	0.19 ± 0.12^a	0.07 ± 0.08^a	0.16 ± 0.26^a
	15	1.69 ± 2.23^a	0.28 ± 0.31^a	0.85 ± 1.19^{ab}	0.45 ± 0.44^a
	30	1.60 ± 1.42^a	0.76 ± 0.78^a	0.98 ± 0.55^{ab}	1.23 ± 0.96^{ab}
	45	1.85 ± 1.78^a	0.53 ± 0.53^a	1.69 ± 1.6^{abc}	1.29 ± 1.31^{ab}
	60	2.88 ± 1.87^{ab}	1.76 ± 1.3^{ac}	3.34 ± 3.14^{abc}	3.43 ± 3.41^{bc}
	90	1.72 ± 1.54^a	1.37 ± 1.33^{ac}	1.91 ± 1.99^{abc}	1.78 ± 1.9^{abc}
	120	1.72 ± 1.75^a	2.72 ± 3.4^{bc}	4.05 ± 4.9^{bc}	3.02 ± 3.08^{bc}
	180	5.35 ± 4.38^b	3.87 ± 2.19^b	4.77 ± 4.45^c	4.09 ± 2.31^c

Values represent the mean ± standard deviation. Means with different little case letters in the same row were significantly different (p < 0.05). M = unwashed mince without cryoprotectant. MC = unwashed mince with cryoprotectant. MW = washed mince without cryoprotectant. MWC = washed mince with cryoprotectant.

The panel did not detect changes in the texture of MW (p > 0.05). However, MC registered slight textural changes at 120th day of storage, whereas the M and MWC were at 180th day (p < 0.05). Results showed that M and MC presented a soft texture that offered no resistance when it was pressed; in addition samples presented an evident drip loss, whereas MW and MWC had a firmer texture. The textural changes observed were more related with the treatment (washing) and storage condition (freezing). Previous studies suggested that washed carp mince showed higher water holding capacity than non-washed mince, resulting in a firmer mince but although the mince water holding capacity in both cases decreased during frozen storage, the gel forming ability of carp mince was retained for a minimum of 135 days (Majumdar et al, 2012).

4. Conclusions

The frozen mince resulted to be safe from the microbiological perspective and had a stable physicochemical and sensorial quality for six months of storage, suggesting that the product may have a longer shelf-life, particularly due to the washing treatment compared to the selected concentrations of the commercial cryoprotectant. Minces obtained from tilapia backbone offered good quality and organoleptic characteristics; therefore, have great potential to develop value-added products.

References

A. O. A. C. (2000). *Official Methods of Analysis of AOAC* (Vol II). Washington, D. C., USA: Association of Official Analytical Chemists (Association of official analytical chemists).

Abdel-Moemin, A. R. (2015). Healthy cookies from cooked fish bones. *Food Bioscience, 12,* 114-12. http://dx.doi.org/10.1016/j.fbio.2015.09.003

Alasalvar, C., Shahidi, F., Miyashita, K., & Wanasundra, U. (2011). *Handbook of seafood quality, safety and health applications.* p 576. Wiley-Blackwell. First edition.

Aubourg, S. P., Pérez, F., & Gallardo, J. M. (2004). Studies on rancidity inhibition in frozen horse mackerel (*Trachurus trachurus*) by citric and ascorbic acids. *European Journal of Lipid Science and Technology, 106*(4), 232-240. http://dx.doi.org/10.1002/ejlt.200400937

Asgharzadeh, A., Shabanpour, B., Aubourg, S. P., & Hosseini, H. (2010). Chemical changes in silver carp (*Hypophthalmichthys molitrix*) minced muscle during frozen storage: Effect of a previous washing process. *Grasas y Aceites, 61*(1), 95-101. http://dx.doi.org/10.3989/gya.087109.

Biscalchin-Grÿchek, S. F., Oetterer, M., & Gallo, C. R. (2003). Characterization and frozen storage stability of minced Nile tilapia (*Oreochromis niloticus*) and red tilapia (*Oreochromis* spp.). *Journal of Aquatic Food Product Technology, 12*(3), 57-69. http://dx.doi.org/10.1300/J030v12n03_06

Borderías, A. J., & Sánchez-Alonso I. (2011). First processing steps and the quality of wild and farmed fish. *Journal of Food Science, 76*(1), R1-R5. http://dx.doi.org/10.1111/j.1750-3841.2010.01900.x

Chaijan, M., Benjakul, S., Visessanguan, W., & Fautsman, C. (2005). Changes of pigments and color in sardine (*Sardinella gibbosa*) and mackerel (*Rastrelliger kanagurta*) muscle during ice storage. *Food Chemistry, 93*, 607-617. http://dx.doi.org/10.1016/j.foodchem.2004.10.035

Chow, C., Yang, J., Lee, P., & Ochiai, Y. (2009). Effects of acid and alkaline pretreatment on the discoloration rates of dark muscle and myoglobin extract of skinned tilapia fillet during iced storage. *Fisheries Science. 75*(6), 1481-1488. http://dx.doi.org/10.1007/s12562-009-0168-z

Clement, S., & Lovell, R. T. (1994). Comparison of processing yield and nutrient composition of cultured Nile tilapia (*Oreochromis niloticus*) and channel catfish (*Ictalurus punctatus*). *Aquaculture, 119*, 299-310. http://dx.doi.org/10.1016/0044-8486(94)90184-8

CODEX STAN 165-1989. Norma para bloques de filetes de pescado congelados, carne de pescado picada y mezclas de filetes y carne de pescado picada congelados rápidamente. Codex Alimentarius. Normas Internacionales de los Alimentos.

Emire, S. A., & Gebremariam, M. M. (2009). Influence of frozen period on the proximate composition and microbiological quality of nile tilapia fish (*Oreochromis niloticus*). *Journal of Food Processing and Preservation, 34*, 743-757. http://dx.doi.org/10.1111/j.1745-4549.2009.00392.x

Eymard, S., Baron, C. P., & Jacobcsen, C. (2009). Oxidation of lipid and protein in horse mackerel (*Trachurus trachurus*) mince and washed minces during processing and storage. *Food Chemistry, 114*(1), 57-65. http://dx.doi.org/10.1016/j.foodchem.2008.09.030

FAO. (2014). El estado mundial de la pesca y la acuicultura, oportunidades y desafíos. Organización de las Naciones Unidas para la Alimentación y la Agricultura. pp 192-196. Roma.

Fogaca, F. H., Sant, L. S., Ferreir,a J. A., Giacometti, M., & Carneiro, D. J. (2015). Restructured products from tilapia industry byproducts: The effects of tapioca starch and washing cycles. *Food and Bioproducts Processing, 94*, 482-488. http://dx.doi.org/10.1016/j.fbp.2014.07.003

Himonides, A. T., Taylor, K. D. A., & Knowles, M. J. (1999). The improved whitening of cod and haddock flaps using hydrogen peroxide. *Journal of the Science of Food and Agriculture. 79*(6), 845-850. http://dx.doi.org/10.1002/(SICI)1097-0010(19990501)79:6<845::AID-JSFA297>3.0.CO;2-W

Hoke, M. E., Jahncke, M. L., Silva, J. L., Hearnsberger, J. O., Chamul, R. S., & Suriyaphan, O. (2000). Stability of washed frozen mince from channel catfish frames. *Journal of Food Science, 65*(6), 1083-1086. http://dx.doi.org/10.1111/j.1365-2621.2000.tb09422.x/pdf

ICMSF (International Commission of Microbiological Specification for Food). (1986). *Sampling for microbiological analysis: Principles and specific applications. Microorganisms in food.* International Commission on the Microbiological Specification of Foods. Toronto Press, Toronto, Canada.

Kilinc, B., Cakli, S., Dincer, T., & Tolasa, S. (2009). Microbiological, chemical, sensory, color and textural changes of rainbow trout fillets treated with sodium acetate, sodium lactate, sodium citrate, and stored at 4°C. *Journal of Aquatic Food Product Technology, 18*, 3-17. http://dx.doi.org/10.1080/10498850802580924

Kirschnik, P. G., Trindade, M. A., Gomide, C. A., Gaglianone, M. E., & Macedo, E. M. (2013). Storage stability of nile tilapia meat mechanically separated, washed, added with preservatives and frozen. *Pesquisa Agropecuária Brasileira, 48*(8), 935-942. http://dx.doi.org/10.1590/S0100-204X2013000800018

Leelapongwattana, K., Benjakul, S., Visessanguan, W., & Howell, N. K. (2005). Physicochemical and biochemical changes during frozen storage of mince flesh of lizardfish (*Saurida micropectoralis*). *Food Chemistry, 90*, 141-150. http://dx.doi.org/10.1016/j.foodchem.2004.03.038

Majumdar, R. K., Deb, S., Dhar, B., & Priyadarshini, B. M. (2012). Chemical changes in washed mince of silver carp (*Hypophthalmichthys militrix*) during frozen storage at -20°C with or without cryoprotectants. *Journal of Food Processing and Preservation, 37*(5), 952-961. http://dx.doi.org/10.1111/j.1745-4549.2012.00741.x

Meilgaard, M., Civille, G. V., & Carr, B. T. (1987). *Sensory Evalaution Techniques* (2nd Ed.). CRC Press, INC. Boca Raton, Florida.

NOM-242-SSA1-2009. Productos y servicios. Productos de la pesca frescos, refrigerados, congelados y procesados. Especificaciones sanitarias y métodos de prueba.

Oliveira, P. R. C., Macedo E. M., Kamimura, E. S., & Trindade, M. A. (2012). Evaluation of physicochemical and sensory properties of sausages made with washed and unwashed mince from Nile tilapia by-products. *Journal of Aquatic Food Product Technology, 21*(3), 222-237. http://dx.doi.org/10.1080/10498850.2011.590270

Parvathy, U. Jibina, M. M., & Sajan G. (2014). Effect of cryoprotectants on the frozen storage stability of mince and quality of mince-based products from *Nemipterus japonicus* (Bloch, 1791). *Fishery Technology, 51,* 47-53.

Pivarnik, L. F., Faustman, C., Suman, P. S., Palmer, C., Richard, N. L., P. Christopher Ellis, P. C., & DiLiberti, M. (2013). Quality Assessment of commercially processed carbon monoxide-treated tilapia fillets. *Journal of Food Science, 78*(6), S902-S910. http://dx.doi.org/10.1111/1750-3841.12145

Sallam, K. I. (2007). Antimicrobial and antioxidant effects of sodium acetate, sodium lactate and sodium citrate in refrigerated sliced salmon. *Food Control, 18*(5), 566-575. http://dx.doi.org/10.1016/j.foodcont.2006.02.002.

Satya, S. D., & Krushna, C. D. (2011). Suitability of chitosan as cryoprotectant on croaker fish (*Johnius gangeticus*) surimi during frozen storage. *Journal of Food Science and Technology, 48*(6), 699-705. http://dx.doi.org/10.1007/s13197-010-0197-8.

Sanjuás-Rey, M.,Gallardo, J. M., Barros, J., & Aubourg, P. (2012). Microbial activity inhibition in chilled mackerel (*Scomber scombrus*) by employment of an organic acid-icing system. *Journal of Food Science, 77*(5), 264-269. http://dx.doi.org/10.1111/j.1750-3841.2012.02672.x

Sanjuás-Rey, M., García-Soto, B., Barros-Velázquez, J., Fuertes-Gamundi, J. R., & Aubourg, S. P. (2011). Effect of a two-step natural organic acid treatment on microbial activity and lipid damage during blue whiting (*Micromesistius poutassou*) chilling. *International Journal of Food Science and Technology, 46,* 1021-1030. http://dx.doi.org/10.1111/j.1365-2621.2011.02565.x

Shaviklo, A. R., & Rafipour F. (2013). Surimi and surimi seafood from whole ungutted myctophid mince. *LWT-Food Science and Technology, 54*(2), 463-468. http://dx.doi.org/10.1016/j.lwt.2013.06.019

Suvanich, V., Marshall, D. J., & Jahncke, M.L. (2000). Microbiological and color quality changes of channel catfish frame mince during chilled and frozen storage. *Journal of Food Science, 61*(1), 151-154. http://dx.doi.org/10.1111/j.1365-2621.2000.tb15971.x/pdf

Thiansilakul, Y., Benjakul, S., Park, S. Y., & Richards M. P. (2012). Characteristics of myoglobin and haemoglobin-mediated lipid oxidation in washed mince from bighead carp (*Hypophthalmichthys nobilis*). *Food Chemistry, 132,* 892-900. http://dx.doi.org/10.1016/j.foodchem.2011.11.060

Woyewoda, A. D., Shaw, S. J., Ke, P. J., & Burns B. G. (1986). *Recommended Laboratory Methods for Assessment of Fish Quality.* Canadian Technical Report of Fisheries and Aquatic Sciences No. 1448.

Yarnpakdee, S., Benjakul, S., & Kristinsson, H. G. (2012). Effect of pretreatments on chemical compositions of mince from a Nile tilapia (*Oreochromis niloticus*) and fishy odor development in protein hydrolysate. *International Aquatic Research, 4*(7), 1-16. http://dx.doi.org/10.1186/2008-6970-4-7

Yongsawatdigul, J., Pivisan, S., Wongngam, W., & Benjakul, S. (2013). Gelation characteristics of mince and washed mince from small-scale mud carp and common carp. *Journal of Aquatic Food Product Technology, 22*(5), 460-473. http://dx.doi.org/10.1080/10498850.2012.664251

4

Effects of Amylose-To-Amylopectin Ratios on Binding Capacity of DDGS/Soy-Based Aquafeed Blends

Ferouz Ayadi[1], Kurt A. Rosentrater[2], K. Muthukumarappan[1], & S. Kannadhason[1]

[1]Department of Agricultural and Biosystems Engineering, South Dakota State University, USA

[2]Department of Agricultural and Biosystems Engineering, Iowa State University, USA

Correspondence: Kurt A. Rosentrater, Department of Agricultural and Biosystems Engineering, and Department of Food Science and Human Nutrition, Iowa State University, Ames, IA 50011, USA.
E-mail: karosent@iastate.edu

Abstract

Demands for seafood products are steadily increasing. Alternative protein sources are required to compensate for enormous amounts of fishmeal that is needed for global seafood production. Starch is a food polymer that can be added to fish feed formulations to enhance binding and expanding capabilities of extrudates. Floatability, a key factor for most aqua feeds, can be optimized by the addition of certain starch sources. Six ingredient blends with a similar protein content (~32.5%) containing two starch sources, Hylon VII (containing 70% amylose, 30% amylopectin) or Waxy I (containing 0% amylose, 100% amylopectin), 20% distillers dried grain with solubles (DDGS), and 15, 25, and 35% moisture content were used along with appropriate amounts of soybean meal, menhaden fishmeal, whey, vitamin and mineral mix to investigate nutritionally-balanced feeds for Nile tilapia (*Oreochromis niloticus* L.). The blends were processed using a laboratory single-screw extruder with varying temperature settings (90-90-90°C, 100-120-120°C, and 100-120-140°C), screw speeds (100, 120, and 140 rpm), and length/diameter ratio (3.4, 6.6, 9.2) of the die. Extensive analyses of expansion ratio (ER), unit density (UD), sinking velocity (SV), and pellet durability indices (PDI), water absorption (WAI) and water solubility indices (WSI) were conducted to evaluate the effects of the two starch sources on extrudate binding and floating capacity. By varying process conditions, significant differences (P>0.05) among the blends were detected for all extrudate physical properties. Significantly higher values for ER, UD, and PDI were achieved by using the Waxy I starch source, while values for SV and WAI decreased. For WSI no significant differences were detected. Increasing the moisture content from 15-35% resulted in a significant increase in ER, WAI, and PDI and a significant decrease in UD. WSI showed no clear pattern in changes. The impact of different amylopectin to amylose ratio, temperature and moisture content on extrudate stability, cohesion and physical properties was demonstrated in this study. All formulations yielded viable extrudates while the blends with the amylopectin as the sole source of starch resulted in higher quality extrudates.

Keywords: binding, corn, DDGS, expansion, extrusion, properties, soy, starch

1. Introduction

The world's hunger for seafood products has tremendously exploited and damaged global wild fish resources. Limited supply of seafood can only be regenerated by changing the global attitude towards utilizing nature's food resources and benefiting from aquaculture production. Moreover, aquaculture farming can take pressure of the dependency on wild fishery stocks by using alternative protein sources. In particular, carnivorous species require high amounts of fishmeal and fish oil in their diets. Furthermore, prices for fishmeal are so high that diets often represent 40-70% of aquaculture operating expenses (Thompson et al., 2008). Alternative protein sources are more cost effective and additionally, can support regional and local economies and can reduce environmental impact (Cheng and Hardy, 2004; Ayadi et al., 2009).

Numerous studies have examined alternative protein sources such as products from animal processing waste (e.g. meat and bone meal, poultry-by products, feather meal, etc.) and plant sources (e.g. soybean meal, corn, corn by-products, cottonseed meal, rapeseed meal, etc.). Soybean meal and corn are essential ingredients for fish feed formulations. Additional research is needed to investigate less expensive, more compatible, and sustainable sources that can replace fishmeal. Distillers dried grains with solubles (DDGS), a coproduct of grain fermentation from fuel ethanol or beverage alcohol production is another potential alternative to meet the protein requirements for fish. In contrast to the original grain, it contains about three times the amount of most nutrients due to the fermentation of starch to alcohol (Jacques et al., 2003; Klopfenstein et al., 2008). DDGS does not contain anti-nutritional factors that are commonly found in most plants (Lim et al., 2009). It has less phosphorus than fishmeal and may ultimately reduce the total phosphorus excreted into water and reduces water pollution (Cheng and Hardy, 2004). Changes in domestic energy policies and resulting growth of the fuel ethanol industry will subsequently increase quantities and availabilities of DDGS. DDGS is competitively priced and less expensive than other plant protein sources on a per unit protein basis (Bals et al., 2006; Lim et al., 2009), particularly when compared to soybean meal, which is the most commonly used substitute in aqua feeds.

Traditionally, DDGS has been fed to beef and dairy cattle (Klopfenstein et al., 2008; Schingoethe et al., 2009) and other livestock such as swine (Stein and Shurson, 2009) and poultry (Lumpkins et al., 2004). Since the late 1940s, DDGS has been integrated in fish feed at low inclusion levels (Thompson et al., 2008). In several studies, DDGS has been investigated for species such as Nile tilapia (Wu et al., 1996; Coyle et al., 2004; Lim et al., 2007), channel catfish (Webster et al., 1993; Robinson and Li, 2008; Lim et al., 2009), and rainbow trout (Cheng and Hardy, 2004) where 20-35%, 30-40%, and 22.5%, respectively, could be included without adverse effects on growth performance and weight gain. These studies indicated that quality of feed that contain DDGS as a protein source, plays a key role in fish diets. Several studies demonstrated that processing conditions of fish feed is crucial to extrudate properties from both single and twin screw extrusion, especially when innovative materials are being utilized (Kannadhason et al., 2009a; Rosentrater et al., 2009b). Processing conditions impact pressure and shear forces within the extruder. Moisture content and screw speed have significant effects on extruder throughput, extrudate durability, and color (Chevanan et al., 2008). Feed moisture content, barrel temperature, extruder screw speed, die geometry, protein, and DDGS content significantly affect expansion ratio, sinking velocity, pellet durability index, extrudate color, mass flow rate, pressure at the die, and apparent viscosity (Kannadhason et al., 2009b; Rosentrater et al., 2009a; Chevanan et al., 2007, 2010). These studies ascertained a successful incorporation of up to 40% and 60% DDGS, while DDGS levels between 20-30% are recommended for floating aquaculture feed (Kannadhason et al., 2009a). This is due to the moderately high fiber content of DDGS, which reduces the binding capacity of the blends (Webster et al., 1995). To assure a feed product with adequate floatability and binding capacities, binders such as cellulose or other less expensive starch sources can be added. Extrusion studies combining DDGS with different starch sources showed that cassava and tapioca starch, which have a high amylopectin proportion yielded the highest expansion ratios (Kannadhason et al., 2009a, 2009b; Rosentrater et al., 2009a, 2009b). Expansion, a key factor for floatability, is predominantly impacted by gelatinization of starch, which is governed by moisture content, temperature, pressure and shear forces in the extruder (Lai and Kokini, 1991). As an important food polymer, starch can be added to increase the viscosity, stability, and holding capacity of fat and water in fluids or semi-solid products (Hermansson and Svegmark, 1996). In aqua feeds, starch plays an important role for floatability based on its binding and expanding properties (Webster et al., 1995).

Starch is the reserve carbohydrate of plants and exists as water-insoluble heterogeneous granules with both amorphous and crystalline regions (Keetels et al., 1996a; Yu and Christie, 2005). It consists of the two major polysaccharides, amylose and amylopectin, of which amylopectin commonly makes up 70-80%. The physical and biological properties of amylopectin are affected by a multiple branched structure formed by inter-chain linkage of every 20-25 glucose monomer. On the contrary, amylose content represents 20-30% of the content of most starches and forms longer linear glucose chains that are only lightly branched (Manners, 1989). Amylose and amylopectin have different functionalities and can be altered to desired property by plant breeding. Due to its crystalline order, amylopectin has stabilizing properties and shows higher storage stability whereas amylose tends to form gels and complexes. Waxy corn is a natural mutant without an amylose-producing enzyme that contains 100% amylopectin (Hermansson and Svegmark, 1996). The amylose portion and the branching points of amylopectin form the amorphous region in the starch granules (Keetels et al., 1996; Yu and Christie, 2005). Hylon VII is a commercial available corn hybrid with high amylose content (70%) that is used in the confectionary industry as a gelling and film-forming agent such as for jelly gum and batter coating (NSFI, 2008).

Gelatinization is not only determined by the molecular structure and chemical composition of starch granules, but heat treatment and available water also highly affect the processing of starch. During gelatinization, the granules swell and form a gel as the amylose and the amylopectin solubilize. This process generally occurs in presence of water at temperatures between 60-70°C (Hermansson and Svegmark, 1996). The swelling of the starch granules involves the separation of amylose and amylopectin (i.e., the leaching of amylose out of the granules) (Keetels et al., 1996b) and the destruction of the crystalline structure of the amylopectin (Yu and

Christie, 2005).

Chinnaswamy and Hanna (1988) determined that expansion properties of starches were lowest at 70% and highest at 50% amylose content, when investigating in starch sources with 0, 25, 50, and 70% amylose proportions and were additionally influenced by temperature. Different starch blends expanded best at uniform moisture contents between 13-14%.

Thermal processing of starch leads to degradation (debranching) and is affected by the moisture content. First, during heating at lower temperatures, long chains are cleaved followed by decomposition of the glucose rings. The second scission is affected by the moisture content: the higher the moisture content, the lower the temperature that is required to decompose the glucose ring (Liu et al., 2009).

Extrusion is a continuous process of short time cooking and forming of materials at relatively high pressure, heat, mechanical shear forces, and moisture. This involves order-disorder transition at different temperatures that impact the starch molecules size and shape, which includes starch gelatinization (under excess moisture) and protein denaturation (Lai and Kokini, 1991). Extrusion processing has been widely applied to produce digestible, palatable, durable, water stable, floating feed with high intake for fish.

Whey is the byproduct of the cheese making production. In studies with DDGS-based aquaculture feed, whey improved binding properties of extrudates and increased the pellet durability indices (Chevanan et al., 2009).

This study continued the research previously done by Kannadhason et al. (2009a, 2009b) and Rosentrater et al. (2009a, 2009b) to examine the effect of starch as a binder. Thus, the objectives of this study were: 1) to produce viable extruded feed for juvenile Nile tilapia using DDGS/soy as an alternative protein source in combination with two different starch sources, and 2) to examine the effects of different amylose/amylopectin ratios, various levels of feed moisture content, extruder die temperature, screw speed, and length-to-diameter (L/D) ratio, on the resulting physical properties of the extrudates and on various processing parameters.

2. Materials and Methods

2.1 Feed Blend Preparation

Six ingredient blends suitable for Nile tilapia fish were formulated to a target protein level of 32.5% db using 20% DDGS and 48% soybean meal, and 15% corn [either Hylon VII (containing 70% amylose, 30% amylopectin) or Waxy I (containing 0% amylose, 100% amylopectin)] as the starch source, and three moisture contents (15, 25, and 35%), along with appropriate quantities of soybean meal, menhaden fishmeal, whey, vitamin and mineral mix (Table 1) to prepare nutritionally-balanced diets for Nile tilapia. Hylon VII corn was obtained from National Starch Food Innovation (Bridgewater, NJ), while Waxy No. 1 corn was from Tate and Lyle (Decatur, IL). DDGS was provided by Dakota Ethanol LLC (Wentworth, SD) and soybean meal was from Dakotaland Feeds Inc., LLC (Huron, SD). All were ground with a Wiley Mill (Model 4, Thomas Scientific, Swedesboro, NJ) to a powder with an average particle size of approximately 500 μm. Menhaden fishmeal was purchased from Consumers Supply Distributing Co. (North Sioux City, SD); vitamin C was from DSM Nutritional Products France SAS (Village-Neuf, France); whey, vitamin mix and mineral mix were obtained from Lortscher Agri Service, Inc. (Bern, KS). The ingredients were mixed in a laboratory-scale mixer (N50 mixer, Hobart Co., Troy, OH) for 10 min and adjusted with adequate amounts of water to the target moisture content.

Table 1. Ingredient components (g/100 g, dry basis) in the feed blends used in the study.

	Dry weight of ingredients (g/100g)
DDGS[1]	20
Soybean meal[2]	48
Corn[3]	15
Menhaden fish meal[4]	9
Whey[5]	5
Vitamin/mineral premix[6]	2
Vitamin C mix[7]	1
Total	100

1 Dakota Ethanol Plant (Wentworth, SD)

2 Dakotaland Feeds Inc., LLC (Huron, SD)

3 Hylon VII: National Starch Food Innovation (Bridgewater, NJ); Waxy No. 1: Tate and Lyle (Decatur, IL)

4 Consumers Supply Distribution Co. (North Sioux City, SD)

5 Animal feed dried whey, Midor Ltd (Elroy, WI)

6 Lortscher Agri Service, Inc. (Bern, KS)

7 DSM Nutritional Products France SAS (Village-Neuf, France)

2.2 Extrusion Processing

All blends were processed using a single screw extruder (Brabender Plasti-Corder, Model PL 2000, South Hackensack, NJ) which had a compression ratio of 3:1, with a screw length-to-diameter ratio of 20:1, and a barrel length of 317.5 mm. The center of the die assembly was conical, and tapered from an initial diameter of 6.0 mm to the die diameters of 1.8, 2.6, and 2.8 mm (length-to-diameter ratios of 9.2, 6.6 and 3.4 mm, respectively) at the exiting of the extruder. The screw speed and the barrel temperature were monitored by a computer that was attached to the extruder. The temperature of the feed, transition and die zone were adjusted with external band heaters to three temperature profiles (feed-transition-die sections) in the barrel (90-90-90°C, 100-120-120°C, and 100-120-140°C, respectively). A 7.5 HP (5.5 kW) motor was connected to the extruder to control the screw speed from 0 to 210 rpm (22 rad/s).

The raw blends were funneled into the extruder manually in constant quantities to avoid jamming at the opening to the barrel. The mass flow rate (MFR) was determined by collecting extrudate samples at 30 s intervals during extrusion, and then weighing the collected amount on an electronic balance (PB 5001, Mettler Toledo, Switzerland).

2.3 Extrudate Properties

After processing, the extrudates were dried for 24 h at room temperature (20±1°C) and triplicates (n=3) were then subjected to an extensive physical property analysis of expansion ratio (-), unit density (kg/m^3), water absorption index (-), water solubility index (%), and pellet durability index (%).

Expansion ratio (ER)

Radial expansion ratio has been determined as the ratio of the diameter of the dry extrudate that was measured with a digital calliper (Digimatic calliper, Model No: CD-6"C, Mitutoyo Corp., Tokyo, Japan), to the diameter of the die nozzle (1.8, 2.6, and 2.8 mm, respectively). The results were displayed as the mean of ten (n=10) measurements.

Unit density (UD)

Extrudate at approximately sizes of 25.4 mm were weighed on an analytical balance (Adventurer™, Item No: AR 1140, Ohaus Corp. Pine Brook, NJ), and then measured with a digital calliper (Digimatic calliper, Model No: CD-6"C, Mitutoyo Corp., Tokyo, Japan) to determine their diameter. According to Rosentrater et al. (2005) the unit density (UD, kg/m^3) was calculated as the ratio of the mass M (kg) to the volume V (m^3) of each measured and weighed extrudate sample, assuming a cylindrical shape for each extrudate:

$$UD = \frac{M}{V} \qquad (1)$$

Sinking velocity (SV)

Sinking velocity was determined following Himadri et al. (1993) by measuring the time an extrudate of approximately 25.4 mm length needed to reach the bottom of a 2 L measuring cylinder filled with distilled water. The total distance (0.415 m) required for the time yielded the sinking viscosity (m/s).

Water absorption and water solubility index (WAI and WSI)

The procedure described by Anderson et al. (1969) was used to measure water absorption index (WAI) and water solubility index. These parameters were used to quantify binding capacity of the extrudates. Approximately 2.5 g of finely ground extrudate sample (150 μm) was suspended in 30 mL of distilled water in a tarred centrifuge tube of 50 mL and placed in a laboratory oven (Thelco precision, Jovan Inc., Wincester, VA) at 30°C. At intervals, it was then stirred intermittently for a period of 30 min. Subsequently, the centrifuge tube was centrifuged and the supernatant was decanted into an aluminum dish, which was placed in the oven for 2 h at 135°C (AACC method 44-19, 2000) and then desiccated for 20 min. The dry solids and the gel mass were weighed. WAI (-) was determined as the ratio of gel mass to the original sample mass (2.5 g ground sample). WSI (%) was calculated as the ratio of the mass of the dry solids (recovered from evaporation of the supernatant from the WAI test) to the original sample mass. WAI is the gel weight received per gram of dry sample, whereas WSI quantifies the starch portion that remains in the water phase when exposed to water. The better the binding capacity, the lower the leaching losses to water.

Pellet durability (PDI)

The pellet durability index was quantified according to Method S269.4 (ASAE, 2004). To separate initial fines from each blend, about 100 g of extrudates were manually sieved (U.S.A. standard testing, ASTM E-11 specification, Daigger, Vernon Hills, IL) for about 10 s, and then tumbled in a pellet durability tester (model PDT-110, Seedburo Equipment Company, Chicago, IL) for 10 min. Afterwards, the samples were again sieved for about 10 s, and then weighed on an electronic balance (Explorer Pro, Model: EP4102, Ohaus, Pine Brook, NJ). PDI was calculated as:

$$PDI = \left(\frac{M_a}{M_b}\right) \times 100 \tag{2}$$

where M_a was the mass (g) after tumbling and M_b was the sample mass (g) before tumbling.

2.4 Experimental Design

Six blends were prepared with 20% DDGS, one starch sources (Waxy I or Hylon VII), three levels of moisture content (15, 25, and 35%), with a constant protein content (32.5%). During processing, three temperature profiles were adjusted in the barrel (90-90-90°C, 100-120-120°C, and 100-120-140°C, respectively), and were referred to as temperatures of 100, 120, and 140°C, respectively, at three screw speeds (100, 130, and 160 rpm, respectively), and three levels of die geometry with various nozzle length-to-diameter (L/D) ratios (9.2, 6.6 and 3.4 mm, respectively).

This resulted in 3x3x3x2=162 total treatment combinations (Table 2). Triplicates (n=3) were measured for most physical properties (i.e., dependent variables) for each treatment combination, except for pellet durability index, which was only measured in duplicate.

The data were then analyzed with Proc GLM to determine the main, interaction and treatment combination effects using SAS software (SAS Institute, Cary, NC) with a Type I error rate (α) of 0.05.

Table 2. Experimental design.[*]

Treat-ment	Starch type	Moisture content	Tdie	Screw speed	Die L/D	Treat-ment	Starch type	Moisture content	Tdie	Screw speed	Die L/D	Treat-ment	Starch type	Moisture content	Tdie	Screw speed	Die L/D
		(% db)	(°C)	(rpm)	(-)			(% db)	(°C)	(rpm)	(-)			(% db)	(°C)	(rpm)	(-)
1	Hylon VII	15	100	100	3.4	55		35	100	100	3.4	109		25	100	100	3.4
2					6.6	56					6.6	110					6.6
3					9.2	57					9.2	111					9.2
4				130	3.4	58				130	3.4	112				130	3.4
5					6.6	59					6.6	113					6.6
6					9.2	60					9.2	114					9.2
7				160	3.4	61				160	3.4	115				160	3.4
8					6.6	62					6.6	116					6.6
9					9.2	63					9.2	117					9.2
10			120	100	3.4	64			120	100	3.4	118			120	100	3.4
11					6.6	65					6.6	119					6.6
12					9.2	66					9.2	120					9.2
13				130	3.4	67				130	3.4	121				130	3.4
14					6.6	68					6.6	122					6.6
15					9.2	69					9.2	123					9.2
16				160	3.4	70				160	3.4	124				160	3.4
17					6.6	71					6.6	125					6.6
18					9.2	72					9.2	126					9.2
19			140	100	3.4	73			140	100	3.4	127			140	100	3.4
20					6.6	74					6.6	128					6.6
21					9.2	75					9.2	129					9.2
22				130	3.4	76				130	3.4	130				130	3.4
23					6.6	77					6.6	131					6.6
24					9.2	78					9.2	132					9.2
25				160	3.4	79				160	3.4	133				160	3.4
26					6.6	80					6.6	134					6.6
27					9.2	81					9.2	135					9.2
28		25	100	100	3.4	82	Waxy I	15	100	100	3.4	136		35	100	100	3.4
29					6.6	83					6.6	137					6.6
30					9.2	84					9.2	138					9.2
31				130	3.4	85				130	3.4	139				130	3.4
32					6.6	86					6.6	140					6.6
33					9.2	87					9.2	141					9.2
34				160	3.4	88				160	3.4	142				160	3.4
35					6.6	89					6.6	143					6.6
36					9.2	90					9.2	144					9.2
37			120	100	3.4	91			120	100	3.4	145			120	100	3.4
38					6.6	92					6.6	146					6.6
39					9.2	93					9.2	147					9.2
40				130	3.4	94				130	3.4	148				130	3.4
41					6.6	95					6.6	149					6.6
42					9.2	96					9.2	150					9.2
43				160	3.4	97				160	3.4	151				160	3.4
44					6.6	98					6.6	152					6.6
45					9.2	99					9.2	153					9.2
46			140	100	3.4	100			140	100	3.4	154			140	100	3.4
47					6.6	101					6.6	155					6.6
48					9.2	102					9.2	156					9.2
49				130	3.4	103				130	3.4	157				130	3.4
50					6.6	104					6.6	158					6.6
51					9.2	105					9.2	159					9.2
52				160	3.4	106				160	3.4	160				160	3.4
53					6.6	107					6.6	161					6.6
54					9.2	108					9.2	162					9.2

[*] The experimental design consisted of 2 (starch sources) x 3 (moisture contents) x 3 (die temperatures) x 3 (screw speeds) x 3 (die L/D) = 162 total treatment combinations. Tdie is temperature of the die, die L/D is length-to-diameter ratio of the die.

3. Results and Discussion

Tables 3, 4, and 5 summarize the main treatment effects and interaction effects of the two different amylose/amylopectin ratios, different moisture contents, temperature profiles, screw speeds, and L/D ratios of the die. The different starch sources had significant effects on all extrudate properties at $\alpha=0.05$. Likewise, the moisture contents (15, 25, and 25%) significantly affected all tested parameters, except for WSI, as well as increasing the screw speed from 100 to 130 and 160 rpm resulted in significant differences, except for UD and PDI. Table 3 provides main effects resulting from considering both starches simultaneously; Table 4 also provides the main effects, but type of starch is used as a blocking factor. It can readily be seen that the behaviours of the dependent variables, when considering each starch separately, were similar to the behaviours when the starches were considered together (i.e., Figures 1 and 2).

Table 3. Main effects due to starch source, moisture content, die temperature, screw speed, and die L/D on resulting extrudate physical properties (n=3, $\alpha=0.05$).*

Parameter	Levels	ER (-)	UD (kg/m³)	SV (m/s)	WAI (-)	WSI (-)	PDI (%)
Starch source	Hylon VII	1.17a (0.06)	0.97a (0.08)	0.076a (0.012)	3.04a (0.29)	18.17a (2.56)	92.85a (3.53)
	Waxy I	1.20b (0.09)	0.99b (0.10)	0.068b (0.024)	2.83b (0.30)	18.27a (2.27)	95.98b (1.64)
Moisture content (% db)	15	1.22a (0.06)	1.02a (0.05)	0.081a (0.009)	2.86a (0.29)	18.42a (2.09)	93.98a (4.59)
	25	1.20b (0.08)	0.94b (0.11)	0.065c (0.026)	2.93b (0.30)	17.67b (3.07)	93.17b (2.22)
	35	1.14c (0.07)	0.97c (0.10)	0.069b (0.018)	3.02c (0.31)	18.55a (2.23)	95.54c (2.05)
Temperature of die (°C)	100	1.17a (0.07)	1.04a (0.08)	0.082a (0.009)	2.88a (0.26)	18.12a (2.33)	95.13a (2.64)
	120	1.19b (0.08)	0.98b (0.07)	0.072b (0.015)	2.89a (0.30)	18.61b (2.57)	94.65b (1.81)
	140	1.19b (0.08)	0.92c (0.09)	0.061c (0.026)	3.04b (0.34)	17.93a (2.60)	92.98c (4.53)
Screw speed (rpm)	100	1.17a (0.07)	0.99b (0.10)	0.073a (0.016)	2.94a (0.32)	18.61a (2.83)	93.78a (4.45)
	130	1.19b (0.07)	0.98ab (0.09)	0.074b (0.018)	2.89b (0.30)	18.35b (2.55)	94.58b (2.53)
	160	1.20c (0.09)	0.97a (0.09)	0.067c (0.025)	2.98c (0.30)	17.67c (2.00)	94.67b (1.93)
Die L/D (-)	3.4	1.16a (0.06)	0.97a (0.10)	0.073a (0.022)	2.85a (0.28)	19.55a (2.72)	94.36a (3.52)
	6.6	1.16a (0.06)	0.97a (0.08)	0.077b (0.020)	2.98b (0.30)	17.86b (2.47)	94.23a (2.03)
	9.2	1.24b (0.08)	1.00b (0.10)	0.065c (0.015)	2.96b (0.33)	17.52c (1.93)	-

* Means followed by similar letters within each dependent variable are not significantly different (P>0.05, LSD). Values in parentheses are standard deviation. ER is expansion ratio, UD is unit density, WAI is water absorption index, WSI is water solubility index, PDI is pellet durability index, die L/D is length-to-diameter ratio of the die.

Table 4. Main effects on extrudate physical properties using starch source as a blocking variable (n=3, $\alpha=0.05$).*

Parameter	Levels	Hylon VII ER (-)	Waxy I ER (-)	Hylon VII UD (kg/m³)	Waxy I UD (kg/m³)	Hylon VII SV (m/s)	Waxy I SV (m/s)	Hylon VII WAI (-)	Waxy I WAI (-)	Hylon VII WSI (-)	Waxy I WSI (-)	Hylon VII PDI (%)	Waxy I PDI (%)
Moisture content (% db)	15	1.21ax (0.05)	1.24ay (0.06)	1.02ax (0.05)	1.03ax (0.06)	0.081ax (0.01)	0.080ax (0.01)	2.99ax (0.26)	2.73ay (0.27)	18.59ax (1.93)	18.25ax (2.23)	91.81ax (5.88)	95.95ay (1.13)
	25	1.17bx (0.04)	1.22by (0.09)	0.94cx (0.08)	0.94cx (0.13)	0.075bx (0.01)	0.055by (0.03)	3.05bx (0.27)	2.79by (0.29)	18.03bx (3.12)	17.28bx (2.99)	91.92ax (1.40)	94.80by (2.04)
	35	1.12cx (0.04)	1.15cy (0.08)	0.96bx (0.10)	0.99bx (0.10)	0.071cx (0.01)	0.067cx (0.02)	3.07bx (0.33)	2.97cy (0.28)	17.88bx (2.49)	19.25cy (1.68)	94.36bx (2.00)	96.91cy (0.98)
Temperature die (°C)	100	1.18ax (0.06)	1.17ax (0.08)	1.01ax (0.08)	1.07ay (0.07)	0.083ax (0.01)	0.081ax (0.01)	2.97ax (0.22)	2.78ay (0.26)	18.26ax (2.44)	17.96acx (2.23)	93.75ax (2.66)	96.90ay (1.42)
	120	1.17bx (0.05)	1.22by (0.09)	0.98bx (0.08)	0.98bx (0.07)	0.076bx (0.01)	0.070by (0.02)	3.03bx (0.30)	2.75ay (0.24)	18.46ax (2.55)	18.77bx (2.59)	93.79ax (1.54)	95.81by (1.49)
	140	1.16cx (0.05)	1.22by (0.08)	0.93cx (0.07)	0.91cy (0.10)	0.070cx (0.02)	0.052cy (0.03)	3.11cx (0.32)	2.96by (0.34)	17.79bx (2.84)	18.08acx (2.33)	89.89bx (5.32)	95.29cy (1.63)
Screw speed (rpm)	100	1.16ax (0.05)	1.19ay (0.08)	0.97abcx (0.09)	1.00ax (0.10)	0.075ax (0.01)	0.071ax (0.02)	3.06ax (0.30)	2.82ay (0.30)	18.36ax (3.01)	18.86ax (2.63)	91.70ax (4.94)	96.43ay (1.31)
	130	1.17bx (0.06)	1.21by (0.08)	0.97abx (0.08)	0.99ax (0.10)	0.077bx (0.01)	0.072abx (0.02)	2.94bx (0.28)	2.84ax (0.32)	18.35ax (2.79)	18.34abx (2.24)	96.06bx (2.39)	96.06by (1.74)
	160	1.17bx (0.06)	1.22cy (0.10)	0.98acx (0.08)	0.96bx (0.11)	0.075ax (0.02)	0.060by (0.03)	3.13cx (0.25)	2.84ay (0.28)	17.75bx (1.53)	17.60cx (2.28)	93.89cx (1.81)	95.45cy (1.74)
Die L/D (-)	3.4	1.14ax (0.05)	1.18ay (0.07)	0.96ax (0.09)	0.97ax (0.11)	0.077ax (0.02)	0.070ax (0.02)	2.81ax (0.26)	2.89ax (0.30)	19.46ax (2.89)	19.67ax (2.54)	92.71ax (4.04)	96.01ay (1.77)
	6.6	1.15bx (0.05)	1.17ax (0.06)	0.95ax (0.07)	0.98ay (0.10)	0.083bx (0.01)	0.072by (0.03)	3.14bx (0.27)	2.82by (1.60)	17.59bx (2.50)	18.13bx (2.42)	93.19bx (1.79)	95.81ay (1.05)
	9.2	1.21cx (0.06)	1.27by (0.09)	1.01bx (0.09)	1.00bx (0.10)	0.068cx (0.01)	0.062by (0.02)	3.13bx (0.21)	2.80by (0.34)	17.66bx (1.86)	17.39cx (2.01)	.	.

* Means within a column followed by similar letters (a, b, c) for a given dependent variable are not significantly different (P>0.05) for that independent variable. Means within a row followed by similar letters (x or y) for a given dependent variable are not significantly different (P>0.05) due to starch source. Values in parentheses are standard deviation. ER is expansion ratio, UD is unit density, WAI is water absorption index, WSI is water solubility index, PDI is pellet durability index, L/D is length-to-diameter ratio of the die.

Table 5. Interaction effects due to starch source, moisture content, die temperature, screw speed and die L/D on extrudate physical properties (p values).*

Independent variables and interactions	ER (-)	UD (kg/m³)	SV (m/s)	WAI (-)	WSI (%)	PDI (%)
Starch	<.0001	0.0003	<.0001	<.0001	0.7628	<.0001
MC	<.0001	<.0001	<.0001	<.0001	<.0001	<.0001
Tdie	<.0001	<.0001	<.0001	<.0001	<.0001	<.0001
Speed	<.0001	0.1904	<.0001	0.0037	<.0001	<.0001
L/D die	<.0001	<.0001	<.0001	<.0001	<.0001	0.1173
Starch*MC	0.0011	0.0709	<.0001	<.0001	<.0001	<.0001
Starch*Tdie	<.0001	<.0001	<.0001	0.0041	<.0001	<.0001
Starch*Speed	0.1294	<.0001	<.0001	<.0001	<.0001	<.0001
Starch*L/D die	<.0001	0.0051	<.0001	<.0001	0.0015	0.0002
MC*Tdie	<.0001	<.0001	<.0001	0.0025	<.0001	<.0001
MC*Speed	<.0001	0.0015	<.0001	0.0041	0.0003	<.0001
MC*L/D die	0.0061	<.0001	<.0001	<.0001	0.0002	<.0001
Tdie*Speed	0.9351	0.5716	<.0001	0.0001	<.0001	<.0001
Tdie*L/D die	<.0001	<.0001	0.3124	<.0001	<.0001	0.4163
Speed*L/D die	0.5090	0.3242	<.0001	0.0004	<.0001	0.0007
Starch*MC*Tdie	<.0001	<.0001	<.0001	<.0001	0.0194	<.0001
Starch*MC*Speed	0.0006	0.2487	<.0001	<.0001	<.0001	<.0001
Starch*MC*L/D die	<.0001	0.0011	<.0001	0.0012	<.0001	.
Starch*Tdie*Speed	0.3710	0.0013	<.0001	0.0661	<.0001	<.0001
Starch*Tdie*L/D die	0.0008	0.0346	<.0001	0.5730	<.0001	.
Starch*Speed*L/D die	0.1755	<.0001	<.0001	0.0158	<.0001	.
MC*Tdie*Speed	0.1539	0.0302	<.0001	<.0001	<.0001	<.0001
MC*Tdie*L/D die	0.0553	<.0001	<.0001	<.0001	<.0001	<.0001
MC*Speed*L/D die	0.0003	0.2030	<.0001	0.0263	<.0001	0.8807
Tdie*Speed*L/D die	0.2703	0.0472	<.0001	0.0009	<.0001	0.0013
Starch*MC*Tdie*Speed	0.1026	0.0401	<.0001	0.0038	<.0001	<.0001
Starch*MC*Tdie*L/D die	0.0156	<.0001	<.0001	<.0001	<.0001	.
Starch*MC*Speed*L/D die	0.0446	0.0003	<.0001	<.0001	<.0001	.
Starch*Tdie*Speed*L/D die	0.0413	0.0205	<.0001	0.0012	<.0001	.
MC*Tdie*Speed*L/D die	0.1373	0.0002	<.0001	<.0001	<.0001	0.4805
Starch*MC*Tdie*Speed*L/D die	0.0036	<.0001	<.0001	<.0001	<.0001	.

* MC is moisture content, Tdie is temperature of the die, ER is expansion ratio, UD is unit density, WAI is water absorption index, WSI is water solubility index, PDI is pellet durability index, L/D is length-to-diameter ratio of the die.

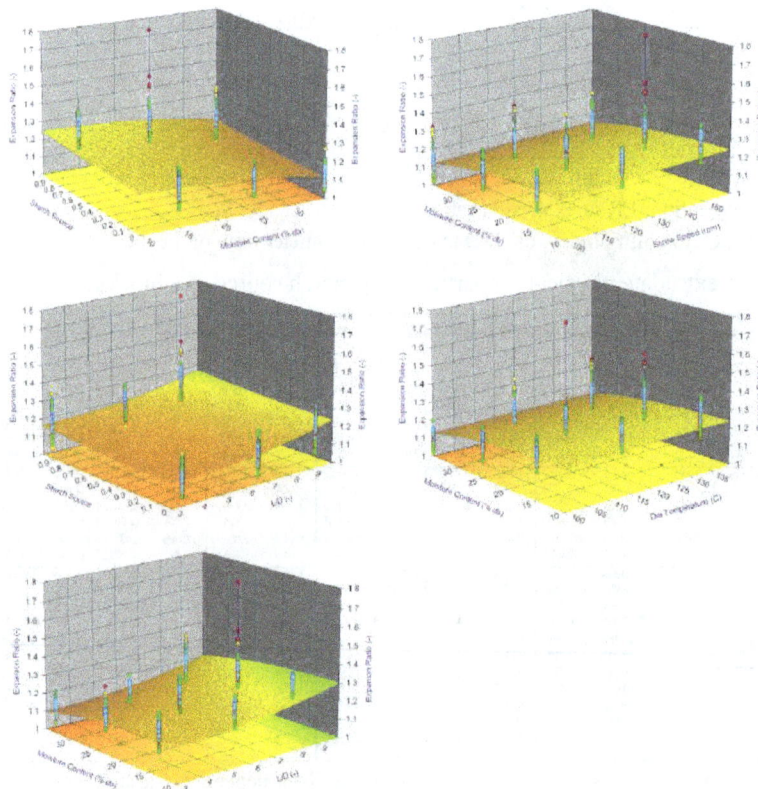

Figure 1. Treatment combination effects on expansion ratio. (Starch source 0 = 0% amylopectin, 1 = 100% amylopectin).

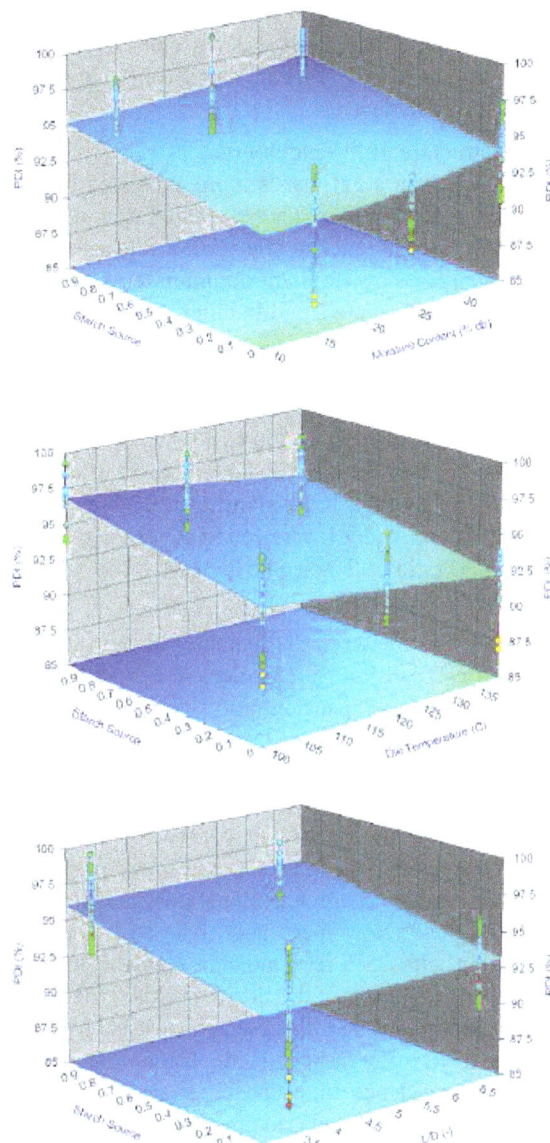

Figure 2. Treatment combination effects on pellet durability index (PDI). (Starch source 0 = 0% amylopectin, 1 = 100% amylopectin).

3.1 Extrudate Properties

Expansion ratio (ER)

Expansion ratio is a decisive factor for aqua feeds. The degree of expansion upon exiting the die impacts the unit density, and thus the floatability of the feed. Fragility and hardness of extrudates are also affected by ER (Kannadhason et al., 2009a, Rosentrater et al., 2009a).

The proportion of amylopectin in starch and the feed moisture content mainly affect expansion. Studies on the effects of different starch sources on DDGS-based extrudates showed increasing expansion for blends with higher amylopectin proportions of the starch (Kannadhason et al., 2009a). Amylopectin is considered to have swelling properties whereas amylose acts as a diluent (Tester and Morrison, 1990). Regarding the main treatment effects (Table 3), the impact of higher amylopectin portion was detected by a 2.5% increase of ER when using the blend with the 100% amylopectin proportion compared to the one with 30%. Investigating different starch sources, the highest expansion ratio was observed at 1.65 when using cassava starch (83% amylopectin) in combination with 20% DDGS, whereas the lowest ER at 1.09 had been recorded for the 72% amylopectin corn starch in combination with 40% DDGS (Kannadhason et al., 2009a).

Feed moisture is a key factor in extrusion processing which highly impacts expansion and density. Due to changes in the molecular structure of amylopectin, moisture reduces the melt elasticity and expansion decreases whereas density increases (Ding et al., 2005). This could be observed in the main treatment effects (Table 3) where an increase in moisture content from 15 to 35% yielded a 6.6% decrease in ER from 1.22 to 1.14. The blends with the Hylon VII as the starch source reached significantly higher ER than the Waxy I (100% amylopectin) blends at the same MC level (Table 4).

During gelatinization, the crystalline structure of the starch granules is destroyed and the amylopectin turns into an amorphous state. The starch granules form a gel that is more expanded than their initial state. At the same time the amylose (if present) leaches out of the amylopectin and forms a continuous gel phase outside the granules (Hermansson and Svegmark, 1996; Yu and Christie, 2005). Shear forces and temperature affect the gel formation. However, different starch sources show different optimum temperatures for expansion (Chinnaswamy and Hanna, 1988). For the main treatment effects, a small but significant increase in ER from 1.17 to 1.20 was observed when increasing the screw speed from 100 to 160 rpm (Table 3). Increasing the temperature from 100 to 120°C yielded also a small but significant increase in ER from 1.17 to 1.19. The blends with Hylon VII resulted in significantly lower ER with increasing temperature whereas the Waxy I (100% amylopectin) blends increased in ER with increasing temperature only from 100 to 120°C. The Waxy I blends reached significantly higher ER than the Hylon VII blends. For both blends ER increased significantly with increasing screw speed. Raising the screw speed from 130 to 160 resulted in no significant difference in ER for the Hylon VII blends (Table 4). The highest ER for Hylon VII and Waxy I were found at 1.31 and 1.43 for the treatment combination at 15% MC, 120°C, 160 rpm and 9.2 L/D ratio of the die and 25% MC, 140°C, 160 rpm, and 6.6 L/D ratio of the die, respectively.

Radial expansion was calculated as the ratio between the diameter of the extrudate and the die, after extrusion and drying. Longitudinal and volumetric expansion were neglected. For all blends, ER did not change when the die L/D ratio increased from 3.4 to 6.6, whereas expansion ratio significantly increased by 6.9% with an increase in die L/D ratio (decrease in die diameter) from 6.6 to 9.2 (Table 3). Again, ER for the Waxy 1 blends was significantly higher than the Hylon VII blends (Table 4). Several interactions existed between independent variables (Table 4). Regarding the treatment combination effects, ER decreased with higher moisture content in combination with lower amylopectin content, lower screw speed, lower temperature setting, and lower L/D ratio, respectively. Figure 1 shows these behaviours, and indicates non-linear relationships between these variables. The highest ER was found at 1.43 with the 100% amylopectin starch, containing 35% feed moisture content, at highest temperature setting (140°C), highest screw speed and highest L/D ratio of the die. The lowest ER was detected at 1.05 for the same blend but with 25% feed moisture content, lowest temperature setting (100°C), lowest screw speed (100 rpm) and lowest L/D ratio of the die (3.4). Also, ER significantly increased with higher screw speed. This is in line with the assumption that increased shear leads to higher expansion due to increased gelatinization unless a certain shear degree is reached. Both, die nozzle diameter and length determine an increase or decrease in back pressure, residence time and shear as the dough passes the die section. This produces a sudden drop from high pressure to atmospheric pressure impacting the strength of water flashing-off at the die and applying enormous tensile forces while the melt expands. Forces and temperature drop lead to changes in the material states. For fully expansion an optimum amount of pressure is required which is achieved by an optimum die L/D ratio (Ganjyal and Hanna, 2007).

Unit density (UD)

Unit density determines the density for a single extrudate. Generally, it is inversely related to the expansion ratio. Unit density can predict the floatability of extrudates which is mandatory for top feeding fish, as Nile tilapia, to ensure food supply and reduce feed loss and water pollution.

All main treatment effects significantly affected the unit density, except for screw speed and die L/D ratio where at least two treatments did not show significant differences. Changes due to the starch source were small but significant. The Hylon VII blends had mostly lower UD values than the Waxy I (100% amylopectin) blends. Initially, UD declined from 1.02-0.94 kg/m³ when increasing MC from 15-25%, but then increased significantly when enhancing the MC from 25-35% (Table 3 and Table 4). These differences do not conform with the changes in ER what may be related to the higher standard deviation of UD. Increasing the temperature from 100-140°C yielded a significant reduction in UD by 11.5% (Table 3). This agrees with Ding et al. (2005) findings that feed moisture and temperature significantly affect extrudate density.

The highest UD was recorded at 1.18 kg/m³ for the treatment combination with Waxy I, at highest MC (35%), lowest temperature setting (100°C), highest die L/D ratio of the die (9.2) and medium screw speed (130 rpm). The lowest UD was also observed for the high amylopectin blends at 0.76 kg/m³ at medium MC (25%), highest temperature (140°C), highest screw speed (160 rpm) and highest L/D ratio of the die (9.2) setting. For the same treatment combination as the lowest UD was determined, the highest expansion ratio was observed. Blends with high amylopectin content starch sources in combination with appropriate moisture content, temperature, and screw speed, resulted in increased gel formation and thus higher expansion. As anticipated, the significance of higher temperature resulting in lower unit density was evident in this study. These findings are in complete agreement with the results of Kannadhason et al. (2009b) and Rosentrater et al. (2009a). An inverse relationship between UD and ER was not clearly detected which could be ascribed to the relatively high standard deviation of the UD data.

Sinking velocity (SV)

The sinking velocity relates to the stability and floatability of extrudates. It can conclude about the porosity of an extrudate and how fast it absorbs water and sinks. Therefore, sinking velocity is related to the unit density, expansion ratio and the biochemical and biomechanical changes that occur inside the extruder barrel. All main treatment effects showed significant differences in SV (Table 3). Regarding the main effects, the blend containing Waxy I as a starch source had 10.5% significantly lower SV than the Hylon VII blends. No clear pattern could be observed on changes in SV due to moisture content, screw speed, and die L/D ratio. Increasing the processing temperature from 100 to 140°C yielded a significant decrease in SV by 25.6%, which is in line with the findings of Kannadhason et al. (2009b) and Rosentrater et al. (2009a, 2009b). Changes in SV are only related to changes in UD for moisture content and temperature. Increasing the screw speed from 130 to 160 rpm resulted in a 9.5% significant increase in SV, whereas the increase in screw speed from 100 to 130 rpm yielded a decrease of 1.4%. The slowest SV value was observed at 0.002 m/s for the Waxy I diet containing at 25% moisture content, highest temperature (140°C), highest screw speed setting (160 rpm) and highest L/D ratio of the die (9.2). This setting complies with the treatment combination, which had the highest expansion and evidently leading to higher floatability. The same observation was made by Kannadhason et al. (2009a) and Rosentrater et al. (2009a). Unexpectedly, the fastest SV at 0.102 m/s could also be observed at the settings as for the fastest SV but at medium screw speed (130 rpm), and lowest die L/D ratio (3.4). Interaction effects among all independent variables were observed, except for temperature and L/D. This showed the big impact of screw speed on SV.

Water absorption and water solubility index (WAI and WSI)

Materials based on starch absorb water due to the abundance of hydroxyl group forming hydrogen bonds with water. Gelatinization involves the destruction of the crystalline-like structure of starch (Pizzoferrato et al., 1999) and expands the water absorption and water solubility properties (Colonna and Mercier, 1983).

The water absorption index reflects the weight of gel that is retained per unit weight of dry sample, whereas the water solubility index describes the percent of dry sample of the supernatant (Anderson et al., 1969). Both properties depend on the amylose-to-amylopectin ratio, moisture content, screw speed, and die geometry (Colonna and Mercier, 1983), which was observed in this study. However, the main treatment effects showed a significant increase of 6.9% in WAI when raising the amylopectin ratio from 30 to 100%, whereas the WSI yielded no significant difference with changes in starch source. Similar results were reported by Mani and Bhattacharya for WAI (1998). The increase in WAI with higher amylopectin ratio can be ascribed to the escalating gelatinization and debranching of the amylopectin structure yielding an expanded matrix capable of holding more water. Other than the starch source, feed moisture, and temperature had significant effects on WAI, which increased with higher moisture content and temperature. This can be explained by the gel-forming capacity of macromolecules and the water-binding capacity of hydrophilic groups for gelatinized corn starch (Gomez and Aguilera, 1983). Regarding the main treatment effects due to the starch source, Hylon VII blends reached significantly higher values than Waxy I (100% amylopectin) blends (Table 4).

The WAI of the blends decreased with lower temperatures indicating a more dense structure of the extrudate with a lower water holding capacity. The lower WAI values with reducing moisture content can be ascribed to the starch dextrinization that occurs at low-moisture contents below 20% during high-shear extrusion cooking, whereas gelatinization predominates at moisture contents higher than 20% (Gomez and Aguilera, 1984). The highest WAI was detected at 3.6 for the treatment combination of Waxy I blends, at 35% moisture content, 140°C, 130 rpm and L/D ratio of the die of 3.4. The lowest WAI was observed at 2.3 for the treatment combination of the Hylon VII blends at 35% moisture content, 120°C, 130 rpm and L/D ratio of the die of 3.4.

WSI is a measure of starch dextrinization (Bhatnagar and Hanna, 1994) and depends on available solubles that increase with starch degradation (Jin et al., 1995). In contrast to WAI, the starch source showed no significant effect on WSI and with higher screw speeds from 100 to 160 rpm WSI significantly decreased by 5.1% (Table 3). These unexpected results are also contradictory to other findings where WSI increased with higher screw speed starting at 150 and 180 rpm, respectively (Gomez and Aguilera, 1983; Jin et al., 1995; Iwe, 1998) and can be ascribed to relatively high standard deviations in this study. They same may apply for changes in L/D of the die from 3.4 to 9.2 where a significant decrease of WSI by 10.4% occurred. The findings for highest and lowest WSI confirm the assumption that relatively high standard deviations may have impacted the results. The treatment combination with the 100% amylopectin starch at 25% moisture content, 120°C, 130 rpm and L/D of 3.4 showed the highest WSI at 24.2%, whereas the lowest WSI was found at 10.75% for 30% amylose starch at 35% moisture content, 120°C, 130 rpm and L/D ratio of 3.4.

Pellet durability index (PDI)

Resistance against destructive and abrasive forces during transportation, handling, and storage is a highly desired property of extrudates in order to maintain their value and quality (Rosentrater et al., 2005). Fines, such as dust, will not be consumed by fish and have to be minimized to avoid economic losses and minimize water pollution (Sørensen et al., 2010).

Durability is tested by simulating the mechanical handling of extrudates during tumbling test and is defined by possible fines produced. It is not only dependent on the gelatinization of starch but also on the heat treatment and moisture content of the blends. This is reflected in the main treatment effects of each independent factor on pellet durability index (Table 3). A significant increase in PDI from 92.85 to 95.98% was observed between Hylon VII and Waxy I (100% amylopectin) blends. Decreasing the temperature setting from 140 to 100°C, resulted in an increase of PDI from 92.98 to 95.13% for all blends. There were no significant differences detected when changing the L/D of the die from 3.4 to 6.4. The treatment combination effects illustrated that PDI steadily increased with higher moisture content and higher amylopectin portion. As shown in Figure 2, the graphs show an increase in PDI with lower temperature and using Waxy I starch (100% amylopectin). Figure 2 shows these behaviours, and indicates non-linear relationships between these variables. Changes in PDI in combination with starch source and L/D ratio, demonstrated an increase only when the amylopectin content was raised, but not with changes in L/D ratio only. This was also reflected in the main treatment effects. The highest PDI was found at 99.3% when Waxy I blends were used containing 25% moisture content, using a L/D ratio of 3.4 at 100°C and 130 rpm screw speed. The lowest PDI was detected with the Hylon VII starch (30% amylopectin), with 15% MC, at 140°C, 100 rpm, and L/D ratio of 3.4. Overall, all extrudates yielded good pellet durability indices that were significantly affected by the starch source, moisture content, and temperature settings. These observations are in accordance with findings done by Kannadhason et al. (2009a, 2009b) and confirm better stabilizing properties with higher amylopectin portions of starch. Heat treatment and available water also highly affected the gelatinization of starch and thus the cohesion and stability of the extrudates.

4. Conclusions

The incorporation of either of two starch sources with varying amylopectin portions in combination with distillers dried grains with solubles, soy, and other ingredients was investigated to determine their effects on resulting binding capacity. Feed ingredients (moisture content and amylopectin-amylose ratio) and processing conditions (die temperature, L/D, and screw speed) were modified to examine their effects on the properties of the resulting extrudates. Altering the amylopectin portions and the starch sources, respectively, had significant effects on all measured parameters. Moisture content, temperature setting, and screw speed, significantly affected most of the extrudate properties, whereas L/D ratio of the die showed only some effects on the resulting extrudates. Unexpectedly, expansion ratio was low. As anticipated, ER increased with higher amylopectin portion and higher screw speed, whereas it declined with higher moisture content. Highest PDI values were achieved by using the starch source with the highest amylopectin portion, highest moisture content, and lowest temperature setting. The starch source with higher amylopectin in combination with adequate moisture content, temperature, and screw speed demonstrated to be the best choice for better quality extrudates in terms of durability and binding capacity.

Acknowledgements

The authors thank the North Central Agricultural Research Laboratory, USDA-ARS, Brookings, South Dakota, for funding, facilities, equipment, and supplies.

References

AACC. (2000). Method 44-19, Moisture-air oven method, drying at 135°C. AACC Approved Methods. 10th Ed. American Association of Cereal Chemists, St. Paul, MN.

Anderson, R. A., Conway, H. F., Pfeifer, V. F., & Griffin, L. E. J. (1969). Gelatinization of corn grits by roll-and extrusion cooking. *Cereal Science Today, 14*, 4-7, 11-12.

ASAE. (2004). Engineering Standards, Practices and Data. American Society of Agricultural and Biological Engineers, St. Joseph, MI. ISBN 1-892769-38-7.

Ayadi, F., Rosentrater, K. A., & Muthukumarappan, K. (2009). A review of alternative protein sources in aquaculture feeds. ASABE Paper No 1008496. St. Joseph, MI, American Society of Agricultural and Biological Engineers.

Bals, B., Dale, B., & Balan, V. (2006). Enzymatic hydrolysis of distiller's dry grain and solubles (DDGS) using ammonia fiber expansion pretreatment. *Energy and Fuels, 20*(6), 2732-2736. http://dx.doi.org/10.1021/ef060299s

Bhatnagar, S., & Hanna, M. A. (1994). Amylose-lipid complex formation during single-screw extrusion of various corn starches. *Cereal Chemistry, 71*(6), 582-587.

Cheng, Z. J., & Hardy, R. W. (2004). Nutritional value of diets containing distiller's dried grain with solubles for rainbow trout, Oncorhynchus mykiss. *Journal of Applied Aquaculture, 15*(3/4), 101-113. http://dx.doi.org/10.1300/J028v15n03_08

Chevanan, N., Muthukumarappan, K., Rosentrater, K. A., & Julson, J. L. (2007). Effect of die dimensions on extrusion processing parameters and properties of DDGS-based aquaculture feeds. *Cereal Chemistry, 84*(4), 389-398. http://dx.doi.org/10.1094/CCHEM-84-4-0389

Chevanan, N., Rosentrater, K. A., & Muthukumarappan, K. (2008). Effect of DDGS, moisture content, and screw speed on physical properties of extrudates in single-screw extrusion. *Cereal Chemistry, 85*(2), 132-139. http://dx.doi.org/10.1094/CCHEM-85-2-0132

Chevanan, N., Muthukumarappan, K., & Rosentrater, K. A. (2009). Extrusion studies of aquaculture feed using distillers dried grains with solubles and whey. *Food and Bioprocess Technology, 2*(2), 177-185. http://dx.doi.org/10.1007/s11947-007-0036-8

Chevanan, N., Rosentrater, K. A., & Muthukumarappan, K. (2010). Effects of processing conditions on single screw extrusion of feed ingredients containing DDGS. *Food and Bioprocess Technology, 3*(1), 111-120. http://dx.doi.org/10.1007/s11947-008-0065-y

Chinnaswamy, R., & Hanna, M. A. (1988). Relationship between amylose content and extrusion-expansion properties of corn starches. *Cereal Chemistry, 65*(2), 138-143.

Colonna, P., & Mercier, C. (1983). Macromolecular modifications of manioc starch components by extrusion-cooking with and without lipids. *Carbohydrate Polymers, 3*(2), 87-108. http://dx.doi.org/10.1016/0144-8617(83)90001-2

Coyle, S. D., Mengel, G. J., Tidwell, J. H., & Webster, C. D. (2004). Evaluation of growth, feed utilization, and economics of hybrid tilapia, Oreochromis niloticus x Oreochromis aureus, fed diets containing different protein sources in combination with distillers dried grains with solubles. *Aquaculture Research, 35*(4), 365-370. http://dx.doi.org/10.1111/j.1365-2109.2004.01023.x

Ding, Q. B., Ainsworth, P., Tucker, G., & Marson, H. (2005). The effect of extrusion conditions on the physicochemical properties and sensory characteristics of rice-based expanded snacks. *Journal of Food Engineering, 66*(3), 283-289. http://dx.doi.org/10.1016/j.jfoodeng.2004.03.019

Ganjyal, G. M., & Hanna, M. A. (2004). Effects of extruder die nozzle dimensions on expansion and micrographic characterization during extrusion of acetylated starch. *Starch/Stärke, 56*(3-4), 108-117. http://dx.doi.org/10.1002/star.200300200

Gomez, M. H., & Aguilera, J. M. (1983). Changes in the starch fraction during extrusion-cooking of corn. *Journal of Food Science, 48*(2), 378-381. http://dx.doi.org/10.1111/j.1365-2621.1983.tb10747.x

Gomez, M. H., & Aguilera, J. M. (1984). A physicochemical model for extrusion of corn starch. *Journal of Food Science, 49*(1), 40-43. http://dx.doi.org/10.1111/j.1365-2621.1984.tb13664.x

Hermansson, A. M., & Svegmark, K. (1996). Developments in the understanding of starch functionality. *Trends in Food Science and Technology, 7*(11), 345-353. http://dx.doi.org/10.1016/S0924-2244(96)10036-4

Himadri, K. D., Tapani, M. H., Myllymaki, O. M., & Malkikki, Y. (1993). Effects of formulation and processing variables on dry fish feed pellets containing fish waste. *Journal of the Science of Food and Agriculture, 61*(2), 181-187. http://dx.doi.org/10.1002/jsfa.2740610208

Iwe, M. O. (1998). Effects of extrusion cooking on functional properties of mixtures of full-fat soy and sweet potato. *Plant Foods for Human Nutrition, 53*(1), 37-46. http://dx.doi.org/10.1023/A:1008095703026

Jacques, K. A., Lyons, T. P., & Kelsall, D. R. (2003). *The Alcohol Textbook* 4th ed. p.379. Nottingham, United Kingdom, Nottingham University Press.

Jin, Z., Hsieh, F., & Huff, H. E. (1995). Effects of soy fiber, salt, sugar and screw speed on physical properties and microstructure of corn meal extrudate. *Journal of Cereal* Science, *22*(2), 185-194. http://dx.doi.org/10.1016/0733-5210(95)90049-7

Kannadhason, S., Muthukumarappan, K., & Rosentrater, K. A. (2009a). Effect of starch sources and protein content on extruded aquaculture feed containing DDGS. *Food and Bioprocess Technology* (in press). DOI 10.1007/s11947-008-0177-4. http://dx.doi.org/10.1007/s11947-008-0177-4

Kannadhason, S., Muthukumarappan, K., & Rosentrater, K. A. (2009b). Effects of ingredients and extrusion parameters on aquafeeds containing DDGS and tapioca starch. *Journal of Aquaculture Feed Science and Nutrition, 1*(1), 6-21.

Kannadhason, S., Rosentrater, K. A., & Muthukumarappan, K. (2010). Twin screw extrusion of DDGS-based aquaculture feeds. *Journal of the World Aquaculture Society, 41*(51), 1-15. http://dx.doi.org/10.1111/j.1749-7345.2009.00328.x

Keetels, C. J. A. M., Oostergetel, G. T., & van Vliet, T. (1996a). Recrystallization of amylopectin in concentrated starch gels. *Carbohydrate Polymers, 30*(1), 61-64. http://dx.doi.org/10.1016/S0144-8617(96)00057-4

Keetels, C. J. A. M., van Vliet., T. & Walstra, P. (1996b). Gelation and retrogradation of concentrated starch systems, 1, Gelation. *Food Hydrocolloids, 10*(3), 343-353. http://dx.doi.org/10.1016/S0268-005X(96)80011-7

Klopfenstein, T. J., Erickson, G. E., & Bremer, V. R. (2008). Board-invited review, Use of distillers by-products in the beef cattle feeding industry. *Journal of Animal Science, 86*(5), 1223-1231. http://dx.doi.org/10.2527/jas.2007-0550

Lai, L. S., & Kokini, J. L. (1991). Physicochemical changes and rheological properties of starch during extrusion (A review). *Biotechnology Progress, 7*(3), 251-266. http://dx.doi.org/10.1021/bp00009a009

Lim, C., Garcia, J. C. Yildirim-Aksoy, M., Klesius, P. H., Shoemaker, C. A., & Evans, J. J. (2007). Growth response and resistance to *Streptococcus iniae* of Nile tilapia, *Oreochromis niloticus*, fed diets containing distiller's dried grains with solubles. *Journal of the World Aquaculture Society, 38*(2), 231-237. http://dx.doi.org/10.1111/j.1749-7345.2007.00093.x

Lim, C., Yildirim-Aksoy, M., & Klesius, P. H. (2009). Growth response and resistance to *Edwardsiella ictaluri* of channel catfish, *Ictalurus punctatus*, fed diets containing distiller's dried grains with solubles. *Journal of the World Aquaculture Society, 40*(2), 182-193. http://dx.doi.org/10.1111/j.1749-7345.2009.00241.x

Liu, X., Yu, L., Liu, H., Chen, L., & Li, L. (2009). Thermal decomposition of corn starch with different amylose/amylopectin ratios in open and sealed systems. *Cereal Chemistry, 86*(4), 383-385. http://dx.doi.org/10.1094/CCHEM-86-4-0383

Lumpkins, B. S., Batal, A. B., & Dale, N. M. (2004). Evaluation of distillers dried grains with solubles as a feed ingredient for broilers. *Poultry Science, 83*(11), 1891-1896. http://dx.doi.org/10.1093/ps/83.11.1891

Mani, R., & Bhattacharya, M. (1998). Properties of injection moulded starch/synthetic polymer blends-III. Effect of amylopectin to amylose ratio in starch. *European Polymer Journal, 34*(10), 1467-1475. http://dx.doi.org/10.1016/S0014-3057(97)00273-5

Manners, D. J. (1989). Recent developments in our understanding of amylopectin structure. *Carbohydrate Polymers, 11*(2), 87-112. http://dx.doi.org/10.1016/0144-8617(89)90018-0

NSFI. (2008). Hylon VII. Technical service bulletin. Bridgewater, NJ, National Starch and Chemical Company. Available at http://eu.foodinnovation.com/docs/HYLONVII.pdf. Accessed 6 May 2010.

Pizzoferrato, L., Rotilio, G., & Paci, M. (1999). Modification of structure and digestibility of chestnut starch upon cooking, A solid state 13C CP MAS NMR and enzymatic degradation study. *Journal of Agricultural and Food Chemistry, 47*(10), 4060-4063. http://dx.doi.org/10.1021/jf9813182

Robinson, E. H., & Li., M. H. (2008). Replacement of soybean meal in channel catfish, *Ictalurus punctatus*, diets with cottonseed meal and distiller's dried grains with solubles. *Journal of the World Aquaculture Society, 39*(4), 521-527. http://dx.doi.org/10.1111/j.1749-7345.2008.00190.x

Rosentrater, K. A., Richard, T. L., Bern, C. J., & Flores, R. A. (2005). Small-scale extrusion of corn masa by-products. *Cereal Chemistry, 82*(4), 436-446. http://dx.doi.org/10.1094/CC-82-0436

Rosentrater, K. A., & Muthukumarappan, K. (2006). Corn ethanol coproducts, Generation, properties, and future prospects. *International Sugar Journal, 108*(1295), 648-657.

Rosentrater, K. A., Muthukumarappan, K., & Kannadhason, S. (2009a). Effects of ingredients and extrusion parameters on aquafeeds containing DDGS and potato starch. *Journal of Aquaculture Feed Science and Nutrition, 1*(1), 22-38.

Rosentrater, K. A., Muthukumarappan, K., & Kannadhason, S. (2009b). Effects of ingredients and extrusion parameters on properties of aquafeeds containing DDGS and corn starch. *Journal of Aquaculture Feed Science and Nutrition, 1*(2), 44-60.

Schingoethe, D. J., Kalscheur, K. F., Hippen, A. R., & Garcia, A. D. (2009). Invited review, The use of distillers products in dairy cattle diets. *Journal of Dairy Science, 92*(12), 5802-5813. http://dx.doi.org/10.3168/jds.2009-2549

Sørensen, M., Morken, T., Kosanovic, M., & Øverland, M. (2010). Pea and wheat starch possess different processing characteristics and affect physical quality and viscosity of extruded feed for Atlantic salmon. *Aquaculture Nutrition* (in press). http://dx.doi.org/10.1111/j.1365-2095.2010.00767.x

Stein, H. H., & Shurson, G. C. (2009). Board-invited review, The use and application of distillers dried grains with solubles in swine diets. *Journal of Animal Science, 87*(4), 1292-1303. http://dx.doi.org/10.2527/jas.2008-1290

Tester, R. F., & Morrison, W. R. (1990). Swelling and gelatinization of cereal starches. I. Effects of amylopectin, amylose, and lipids. *Cereal Chemistry, 67*(6), 551-557.

Thiex, N. (2009). Evaluation of analytical methods for the determination of moisture, crude protein, crude fat, and crude fiber in distillers dried grains with solubles. *Journal of AOAC International, 92*(1), 61-73.

Thompson, K. R., Rawles, S. D., Metts, L. S., Smith, R., Wimsatt, A., Gannam, A. L., Twibell, R. G., Johnson, R. B., Brady, Y. J., & Webster, C. D. (2008). Digestibility of dry matter, protein, lipid, and organic matter of two fishmeals, two poultry by-product meals, soybean meal, and distiller's dried grains with solubles in partial diets for sunshine bass, *Morone chrysops x M. saxatilis. Journal of the World Aquaculture Society, 39*(3), 352-363. http://dx.doi.org/10.1111/j.1749-7345.2008.00174.x

Webster, C. D., Tidwell, J. H., & Goodgame, L. S. (1993). Growth, body composition, and organoleptic evaluation of channel catfish fed diets containing different percentages of distillers' grains with solubles. *The Progressive Fish-Culturist, 55*(2), 95-100. http://dx.doi.org/10.1577/1548-8640(1993)055<0095:GBCAOE>2.3.CO;2

Webster, C. D., Tidwell, J. H., Goodgame-Tiu, L. S., & Yancey, D. H. (1995). Evaluation of distillers grains with solubles as an alternative plant protein in aquaculture diets. In *Nutrition and Utilization Technology in Aquaculture*, p.192. C.E. Lim and D.J. Sessa, eds. Champaign, IL: AOCS Press.

Wu, V. W., Rosati, R. R., & Brown, P. B. (1996). Effect of diets containing various levels of protein and ethanol coproducts from corn on growth of tilapia fry. *Journal of Agricultural and Food Chemistry, 44*(6), 1491-1493. http://dx.doi.org/10.1021/jf950733g

Yu, L., & Christie, G. (2005). Microstructure and mechanical properties of orientated thermoplastic starches. *Journal of Materials Science, 40*(1), 111-116. http://dx.doi.org/10.1007/s10853-005-5694-1

5

Formulation of Senescent Plantain Dish with Various Local Cereal and Leguminous Flours for Feeding Rats: Growth Performance Evaluation

Kouadio N. Joseph[1], Akoa E. Edwige[1], Kra K. A. Séverin[1] & Niamké L. Sébastien[1]

[1]Laboratoire de Biotechnologies, UFR Biosciences, Département de Biochimie, Université Félix Houphouët-Boigny, 22 BP 582 Abidjan 22, Côte d'Ivoire

Correspondence: Kra K. A. Séverin, Laboratoire de Biotechnologies, UFR Biosciences, Département de Biochimie, Université Félix Houphouët-Boigny, 22 BP 582 Abidjan 22, Côte d'Ivoire. E-mail: kra_severin@yahoo.fr

Abstract

The aim of this study was to valorize senescent plantain. Therefore, a traditional dish named Dockounou was prepared with a mixture of senescent plantain and various millet, soybean, sorghum, cassava, maize or rice flours. The growth performance of several Wistar rats feed by Dockounou was followed. Thus, batches of rats were fed for 15 days with three formulations (F1, F2, F3) in proportion of 90:10, 80:20 and 75:25 (senescent plantain dough/flours) obtained after two cooking modes (dry cooking: baked ; wet cooking: boiled). The effects of these formulations were compared to control diet (C. diet). Beyond the control diet, rats fed with the soybean baked Dockounou presented, the best following growth parameters: weight gain (2.82 to 4.19 g/d), food intake (8.92 to 9.72 g/d), feed efficiency (0.10 to 0.42), proteins intake (8.28 to 19.67), proteins efficiency (0.13 to 3.15). The physicochemical and nutritive characteristics of soybean baked Dockounou were as follow: ash (2.93 ±0.15 %), proteins (10.62±0.59 %), carbohydrates (15.46±1.53 %), calcium (232.04 – 558.20 mg/100g), potassium (313.97 – 385.11 mg/100g), magnesium (42.40 – 72.22 mg/100g), sodium (211.24 – 303.85 mg/100g) and phosphorus (330.70 – 433.71 mg/100g). Also, the study showed that, two formulations, 80:20 and 75:25, have really impact on rats growth. These results suggest that soybean baked Dockounou with important proportions, 80 % and 75 %, of senescent plantain dough can be effectively used in the diet of laboratory Wistar rats regarding the good zoological performances there are obtained.

Keywords: senescent plantain, baked and boiled Dockounou, rats feed, growth performance

1. Introduction

Plantain provides more than 25 % of the carbohydrate requirements for over 70 million people in Africa (IITA, 1998). In Ivory Coast, plantain is the third crop (1 624 354 tons/year) after yam (5 731 719 tons/year) and cassava (2 436 495 tons/year) (FAOSTAT, 2013). In spite of this importance, plantain is a highly perishable foodstuff due to ethylene production associated with rudimentary conditions of harvesting, transport and storage (Rodríguez et al., 1999; Emaga, Wathelet, & Paquot, 2008). This perishability leads to post-harvest losses estimated at 35 to 60 % of the annual production (Atanda et al., 2011). Indeed, the production of plantain depends on season marked by a period of high abundance which goes from September to April (Kuperminc, 1998; Sery, 1988). During this period, the difficulties of plantains conservation make the fruits, while ripening at the ambient temperature undergo depreciation, and also qualitative and quantitative degradations throughout the distribution chain (Chia & Huggins, 2003). In order to reduce the post-harvest losses, the ripe fruits of plantain are often processed by Ivorian farmer women into traditional dish called Dockounou (N'guessan, Yao, & Kehe, 1993; Lassois, Jean-Pierre, & Haïssam, 2009; Dzomeku, Dankyi, & Darkey, 2011; Honfo, Tenkouano, & Coulibaly, 2011). Indeed, the Dockounou is a baked or boiled mixture of over ripe banana paste and cereals flours (Akoa et al., 2012 & 2013). Several studies have been performed on this food primarily designed for human consumption. In this way, Kra et al. (2013) and Kouadio et al. (2014) have showed that Dockounou is an energetic dish for human. Moreover, the reports of Akoa et al. (2014) have revealed the optimized parameters for this food preparation with best hygienic and nutritive qualities. As far as animal feeding is concerned, preliminary studies of Kouadio et al. (2015) highlighted the Wistar rats interest for this dish. Indeed, rats are experimental mammals, mostly used in biomedical, behavioral, toxicological and nutritional research projects

(Baker, Lindsey, & Weisbroth, 1979). Therefore, using rats in experimental design could give an accurate response to some issues related to nutritional aspects of Dockounou dish. Furthermore, the use of senescent plantain in laboratory rats diet can contribute to reduce plantain post-harvest losses. Thus, several formulations using feed ingredients such as cereal or leguminous flours in mixture with various proportions of senescent plantain dough were investigated in order to evaluate their effect on the Wistar rats growth performance and contribute therefore to valorize the fruits at the senescent stage.

2. Materials and Methods

2.1 Sampling

Material used for the Dockounou production composed of senescent plantain fingers (*Musa paradisiaca*), grains of maize (*Zea mays*), rice (*Oriza sativa*), millet (*Pennisetum americanum*), sorghum (*Sorghum bicolor*), soybean (*Glycine max*) and roots of cassava (*Manihot esculenta Crantz*). The whole previous material were bought on the market of Adjame (in Abidjan - Côte d'Ivoire).

As for the animal material, it consisted in a total of one hundred and eighty five (185) young Wistar rats (60 to 75 g) with 50±3 days age. They were kindly provided by the animals' barn of the UFR Biosciences of Felix Houphouët-Boigny University (Abidjan, Côte d'Ivoire).

2.2 Production of Grains and Roots Flours

2.2.1 Sorghum, Millet, Maize and Soybean Flours

Sorghum, millet, maize and soybean flours were prepared using the method described by Lombor, Umoh, & Olakumi (2009). The grains were cleaned and soaked in clean tap water and covered container. The soaked grains were allowed to ferment at room temperature (27 °C) for 24 h. After fermentation, the water was removed and the grains rinsed with 500 mL of water and oven dried at 80 °C for three (3) hours. The soybean grains were sorted, cleaned and blanched at 100 °C for 10 min. The blanched grains were, dehulled and rinsed with 500 mL of water in order to remove the seed coat. The rinsed seed were then dried in oven at 80 °C for five (5) h. Each dried sample was separately milled and sieved with a 100 μm particle size sieve.

2.2.2 Cassava Flour

Method described by Younoussa et al. (2013) was used. The cassava roots were peeled, cut in small pieces and washed before being sun-dried for three to seven (3-7) days. After drying, the pieces were then ground to obtain the flour which was sifted with 100 μm size sieve.

2.2.3 Rice Flour

Rice flour was obtained using the method described by Akoa et al. (2014). The rice grains were soaked in water for two (2) hours. After drying, the grains were crushed in a home clean wooden mortar and the flour was sieved (100 μm).

2.3 Preparation of Dockounou

Methods described by Kra et al. (2013) and Akoa et al. (2014) were used in the processing step. Fruits of senescent plantain were washed, peeled and crushed in a traditional mortar to obtain homogenous dough. Three (3) types of formulations (plantain dough: flour) were obtained. Thus, the first formulation F1 (90:10) was made with 90 % of senescent plantain dough and 10 % of flour. The second formulation F2 (80:20) was achieved with 80 % of senescent plantain dough and 20 % of flour and the third formulation F3 (75:25) was achieved with 75 % of senescent plantain dough and 25 % of flour. Each sample obtained was fermented during four (4) hours before being wrapped (in portions of 100 g) in leaves of *Thaumatococcus danieilii*. One part of samples has been boiled (1 h) and the second part was baked in oven at 150 °C for one (1) h.

2.4 Animals Feeding

The experiments were conducted with young Wistar rats from the animals' house of the UFR Biosciences of the Felix Houphouët-Boigny University (Abidjan, Côte d'Ivoire). The average temperature of the room was 26 °C, and the humidity was 70 %, with 12 hours of daylight. Thirty seven (37) groups of five (5) Wistar rats each were used. They were divided as follow: one (1) group of five (5) young rats was fed with control diet (C), and for each formulation (F1, F2, F3), twelve (12) groups of five (5) young rats were fed with baked and boiled Dockounou (six groups for each cooking mode). The control diet (see Appendix A) used in this study was a granular provided by SFACI Company. Rats were acclimatized for three (3) days during which, they were fed with control diet and thereafter, fed with different experimental diets. Rats were disposed in individual screened bottomed cages designed separately to feed "*ad libitum*" for fifteen (15) days.

2.5 Experimental Design for Data Statement

Method described by Adrian, Rabache, & Fragne (1991) was used to determine the growth parameters. During the experimental period, the feed intake was measured daily while the body weight, length of body, tail and shin were measured each 3 days between 7h30 and 8h30 am. Each food was weighed before being given to rats and the following day, the leftover food was also weighed, in order to determine the food intake. The dry matter of each feed was determined according to AOAC (1990). The table in Appendix B shows the formulas used to determinate growth parameters.

During the experimental period, faeces of each rat were collected and the nitrogen (N) balance was evaluated by using the Kjeldahl method (AOAC, 1990).

2.6 Physicochemical and Biochemical Analysis of Dockounou

Physicochemical and biochemical characteristics of Dockounou involving best zoological performances were determined. Moisture, dry matter and ash content were determined by using the AOAC (1990) methods while the pH was directly measured with a pH-meter (Roucaire, Metr Ohm 632, Germany). Total titrable acidity was evaluated on the puree of Dockounou obtained from 40 g of paste. Raw proteins contents were measured by the Kjeldahl method according to BIPEA (1976) using 6.25 as conversion factor. Lipid content was determined after extraction with hexane in a Soxtherm system during 6 h (BIPEA, 1976). Total sugars contents were evaluated with the method of Dubois et al. (1956). As for carbohydrate and starch contents, they were calculated by the expression described by Coulibaly (2008):

Carbohydrate content = 100 - (Moisture (%) + Ash (%) + Fat (%) + Proteins (%))

Starch content = 0.9 (Carbohydrate (%) - total sugars (%)).

Energy value was also calculated using the relation described by Atwater and Rosa (1899).

2.7 Mineral Analysis of Dockounou

Minerals contents were determined by the AOAC (1990) method. Five (5) grams of samples were digested with a mixture of concentrated nitric acid, sulfuric acid and perchloric acid (10:0.5:2, v/v) and analyzed using an atomic absorption spectrophotometer (GBC 904AA; Germany). Total phosphorus was determined as orthophosphate by the colorimetric ascorbic acid method (APHA, 2001). After acid digestion and neutralization using phenolphthalein as indicator reagent, the absorbance was read at 880 nm in a spectrometer (Spectronic 21 D, Miltonroy, New York, USA). Blank analysis was performed before, in the same conditions using distilled water.

2.8 Statistical Analysis

Statistical analysis of data was done by the one way Analysis of Variance (ANOVA) using software IBM SPSS Statistics version 17.0. Differences between means were tested using the Duncan Multiple Range Test with 5 % level of significant difference and figures were drawn on EXCELL 2013 Software.

3. Results

3.1 Weight Gain

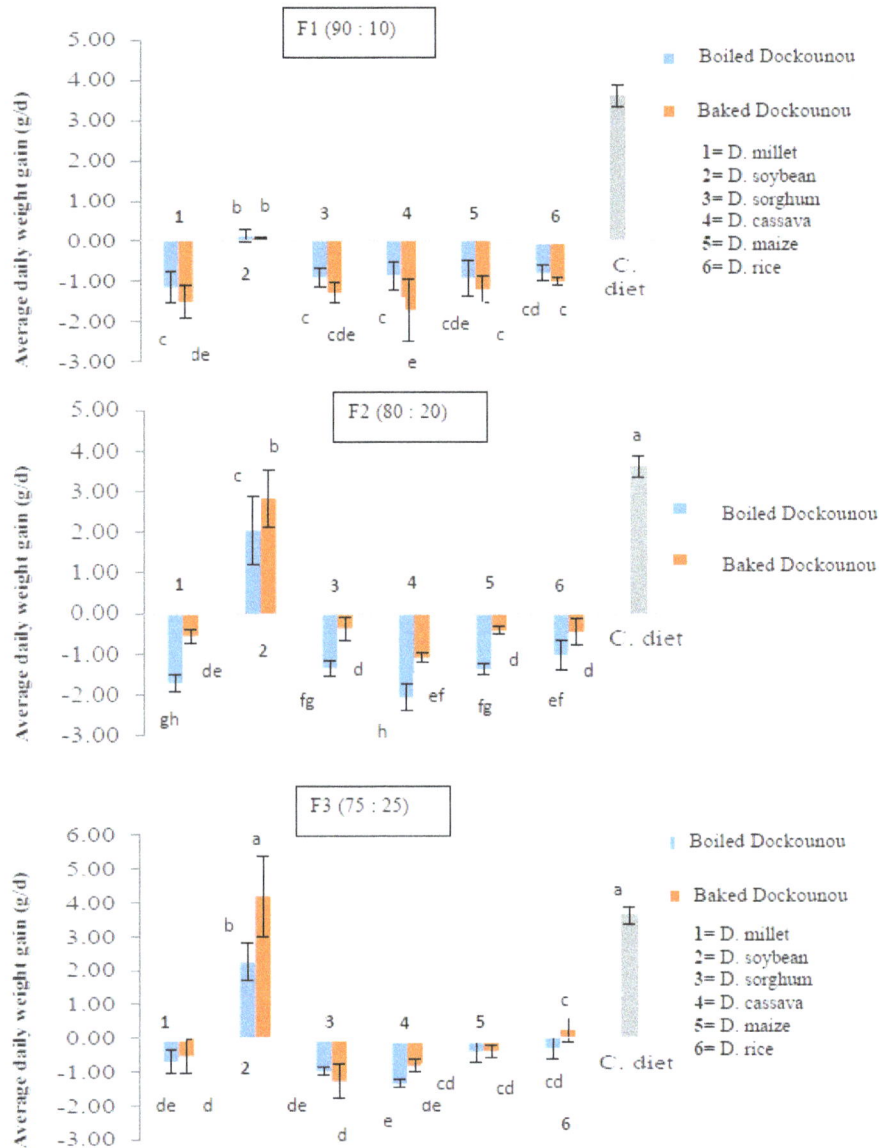

Figure 1. Weight gain per day of rats fed with various types of boiled and baked Dockounou for F1, F2 and F3 formulations and control diet

Data are represented as Means ± SD (n=5). Means with no common letter (a, b, c, d) are significantly different (p<0.05)

The weight gain of Wistar rats fed with F1, F2 and F3 formulations of boiled and baked Dockounou are presented in Figure 1. The results showed that the weight of rats fed with the control diet, range was higher (3,34±0,27 g/d) than those fed with formulated diets F1, F2 and F3. Considering these formulated diets, only the D soybean diet allowed slight gain of weight in rats, in all the formulations with the following order: F1 < F2 < F3. The values obtained for F1 were 0.09±0.02 g/d for baked D. soybean and 0.13±0.17 g/d for boiled D. soybean. For the formulation F2 the values were closed to 2.82±0.71 g/d (baked D. soybean) and 2.05±0.84 g/d for boiled D. soybean. The reported values for the F3 formulation were 4.19±1.2 g/d for baked D. soybean and 2.25±0.55 g/d for boiled D. soybean. Contrary to tweight gain after feeding D. soybean diet, the feed of D. millet, D. sorghum, D. cassava, D. maize and D. rice diets induced a decrease of rat's weight.

3.2 Dry Matter Intake

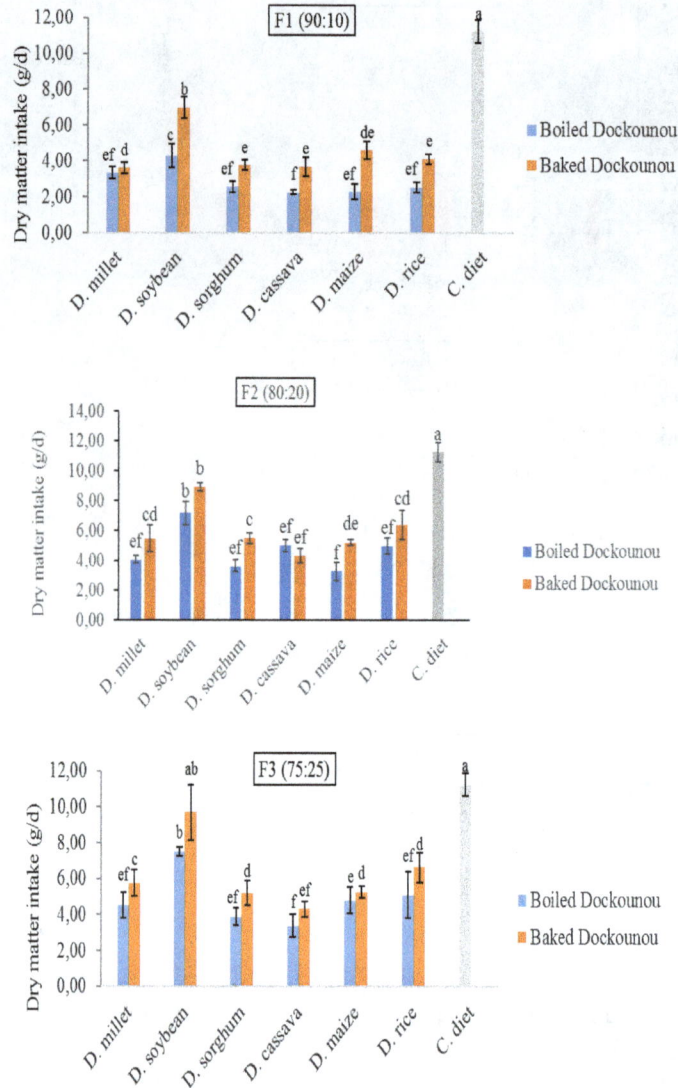

Figure 2. Dry matter intake by rats with F1, F2 and F3 formulations of boiled and baked dockounou and control diet. DMI= dry matter intake

Data are represented as Means ± SD (n=3). Means with no common letter (a, b, c, d) are significantly different (p<0.05)

Dry matter intake of rats is given in Figure 2. For all formulations, feed intake is higher for baked Dockounou, than boiled Dockounou. Except for the C diet where the highest intake noted for F2 and F3 formulations was recorded by the baked D soybean. The registered values were 8.92±0.28 g/d and 9.72±1.54 g/d for F2 and F3 formulations, respectively. However, C diet registered the most important intake, followed by D. soybean.

3.3 Food Efficiency Ratio

Table 1. Food efficiency ratios of rats fed with baked and boiled Dockounou diets

Type of formulation	Diets	Boiled Dockounou	Baked Dockounou
Reference food	Control	0.25±0.02a	0.30±0.04a
	D. Millet	-0.33±0.10a	-0.42±0.14a
	D. Soybean	0.02±0.03a	0.10±0.07a
F1	D. Sorghum	-0.34±0.06a	-0.34±0.08a
[90:10]	D. cassava	-0.40±0.15a	-0.50±0.30a
	D. Maize	-0.38±0.12a	-0.26±0.08a
	D. Rice	-0.31±0.06a	-0.24±0.03a
	D. Millet	-0.42±0.07b	-0.10±0.04a
	D. Soybean	0.25±0.10a	0.31±0.07a
F2	D. Sorghum	-0.36±0.05b	-0.06±0.05a
[80:20]	D. cassava	-0.63±0.13b	-0.25±0.02a
	D. Maize	-0.34±0.03b	-0.08±0.02a
	D. Rice	-0.20±0.06b	-0.07±0.05a
	D. Millet	-0.15±0.06a	-0.10±0.09a
	D. Soybean	0.29±0.06b	0.42±0.06a
F3	D. Sorghum	-0.26±0.04b	-0.09±0.08a
[75:25]	D. cassava	-0.40±0.07b	-0.19±0.60a
	D. Maize	-0.88±0.06b	-0.07±0.03a
	D. Rice	-0.31±0.06a	0.03±0.05a

Values are mean ± standard deviation of five determinations. Values with different letters in the same line of each parameter indicate statistical difference ($p < 0.05$). D: Dockounou, F: Formulation

Food efficiency ratios of all diets are reported in Table 1. The statistical analysis showed that there is no significant difference at 5 % level between food efficiency ratios of boiled and baked Dockounou for F1 formulation. Nevertheless, a significant difference ($p<0.05$) was observed between food efficiency ratios of boiled and baked Dockounou for F2 and F3 formulations. At the whole, food efficiency ratios of baked D soybean (0.31±0.07 to 0.42±0.06) are higher than those of boiled D soybean (0.25±0.10 to 0.29 ± 0.06) for F2 and F3 formulations. However, the food efficiency ratios of D soybean for both formulations were higher than those of the others Dockounou.

3.4 Physical Parameters of Growth

Table 2. Length of body, shin and tail of young rats fed with control diet and various types of baked and boiled Dockounou in the three formulations F1, F2, F3

Type of formulations	Diets	Length of body (cm)		Length of shin (cm)		Length of tail (cm)	
		Boiled	Baked	Boiled	Baked	Boiled	Baked
Reference food	Control	2.28±0.07a	2.28±0.07a	0.57±0.15a	0.57±0.15a	1.47±0.29a	1.47±0.29a
	D. Rice	-		0.12±0.02a	0.12±0.02a	-	
	D. Soybean	0.26±0.18b	0.82±0.69a	0.04±0.08b	0.54±0.11a	0.30±0.18b	0.88±0.71a
F1	D. Sorghum			-		-	
[90 :10]	Cassava, Maize,Millet	-					
	D. Rice	-	0.02±0.04a	0.12±0.03a	0.08±0.01a	-	
	D. Soybean	0.92±0.56b	2.02±0.47a	0.22±0.10b	0.44±0.11a	1.18±0.60a	1.32±0.66a
F2	D. Sorghum	-		-		-	
[80 :20]	Cassava, Maize,Millet	-					
	D. Rice	-	0.54±0.18a	0.06±0.01a	-	-	0.2±0.01a
	D. Soybean	2.70±0.48a	2.76±0.30a	0.44±0.21a	0.40±0.20a	0.9± 0.30a	1.3±0.50a
F3	D. Sorghum	-		-		-	
[75 :25]	Cassava, Maize,Millet						

Values are mean ± standard deviation of five determinations. Values with different letters in the same line of each parameter indicate statistical difference ($p < 0.05$). D: Dockounou, F: Formulation, - : any variation of the parameter measured during experimental period.

Results showed (Table 2) that only the D soybean diets resulted in a visible variation of length of body, shin and tail of young rats followed by the rice Dockounou diet which had just given a relatively weak variation of these parameters. There are significantly different ($p<0.05$) between both cooking method and type of formulations, mainly as F1 and F2 are concerned. The higher length of body was obtained for rats fed with the C diet (2.28±0.07 cm), followed by those of the D soybean (0.26±0.18 cm < 0.92±0.56 cm < 2.76±0.30 cm in F1, F2 and F3 formulations, respectively). The rats body length evolution was similar with both boiled and baked D soybean in the F3 formulation (2.70±0.48 cm and 2.76±0.30 cm for boiled and baked D soybean, respectively). Rats fed with the baked D rice, in 75:25 proportions (F3) sleazy, obtained an increase in length of body estimate to 0.54±0.18 cm. About the other diets (D. millet, D. sorghum, D. cassava and D. maize),, there was no increase in length of body, tail and shin for rats whatever the cooking mode was.

3.5 Proteins Intake

Table 3. Proteins intake of rats fed with baked and boiled Dockounou diets

Type of formulations	Diets	Boiled Dockounou (g)	Baked Dockounou (g)
Reference food	Control	25.25±1.45a	25.25±1.45a
	D. Millet	0.31±0.05b	1.82±0.15a
	D. Soybean	1.15±0.38b	8.28±0.72a
F1	D. Sorghum	0.25±0.05b	1.33±0.10a
[90:10]	D. cassava	0.16±0.02b	1.34±0.19a
	D. Maize	0.20±0.09b	1.51±0.16a
	D. Rice	0.22±0.04b	1.53±0.11a
	D. Millet	0.50±0.04b	2.76±0.44a
	D. Soybean	5.65±0.64b	14.22±0.44a
F2	D. Sorghum	0.80±0.03b	1.95±0.12a
[80:20]	D. cassava	0.22±0.02b	1.27±0.14a
	D. Maize	0.58±0.09b	1.71±0.06a
	D. Rice	0.63±0.06b	2.36±0.35a
	D. Millet	0.52±0.10b	2.98±0.39a
	D. Soybean	5.95±0.49b	19.67±3.18a
F3	D. Sorghum	0.36±0.04b	3.32±0.44a
[75:25]	D. cassava	0.22±0.03b	1.30±0.13a
	D. Maize	0.72±0.09b	2.25±0.14a
	D. Rice	0.58±0.09b	3.58±0.44a

Values are mean ± standard deviation of five determinations. Values with different letters in the same line of each parameter indicate statistical difference (p < 0.05). D: Dockounou, F: Formulation

Table 3 presents the amount of total proteins intake in rats. At the whole, it was observed that the contents of proteins ingested by rats fed with baked Dockounou were higher than those of rats fed with boiled Dockounou. The proteins intakes for the baked preparations were 14.22 ±0.44 g and 19.67 ±3.18 g for D soybean in F3 and F2 formulations, respectively. Whatever, the amount of proteins intake of rats increased with incorporation of flours. The amount of proteins intake by rats fed with C diet (25.25±1.45 g) was higher than those of baked D soybean (8.28±0.72 g < 14.22±0.44 g < 19.67±3.18 g for F1, F2 and F3 formulations, respectively).

3.6 Proteins Efficiency Ratios

Table 4. Proteins efficiency ratios of rats fed with the control, baked and boiled Dockounou diets

Type of formulations	Diets	Boiled Dockounou	Baked Dockounou
Reference food	Control	2.15±0.37a	2.15±0.37a
	D. Millet	-53.34±11.71b	-12.57±4.25a
	D. Soybean	1.46±1.67a	0.13±0.93a
F1	D. Sorghum	-54.63±4.28b	-14.60±3.60a
[90:10]	D. cassava	-78.90±24.73b	-20.29±12.35a
	D. Maize	-66.46±12.13b	-12.12±3.68a
	D. Rice	-54.41±9.65b	-10.02±1.31a
	D. Millet	-51.67±5.92b	-3.15±1.27a
	D. Soybean	4.73±1.94a	2.96±0.72a
F2	D. Sorghum	-33.72±5.02b	-2.88±2.33a
[80:20]	D. cassava	-158.40±38.71b	-12.95±1.45a
	D. Maize	-35.86±3.17b	-3.74±0.82a
	D. Rice	-24.39±8.24b	-2.98±0.72a
	D. Millet	-20.67±12.13b	-2.97±2.73a
	D. Soybean	5.71±1.38a	3.15± 0.44b
F3	D. Sorghum	-42.46±6.99b	-2.26±2.00a
[75:25]	D. cassava	-93.34±11.05b	-9.71±3.14a
	D. Maize	-8.74±4.40b	-2.61±1.37a
	D. Rice	-8.21±4.17b	0.88±1.40a

Values are mean ± standard deviation of five determinations. Values with different letters in the same line indicate statistical difference (p < 0.05). D: Dockounou, F: Formulation

Proteins efficiency ratios of different formulations for baked and boiled Dockounou are shown in Table 4. The whole values in the table were negative, except for those obtained with C. diet and D. soybean which were positive. The statistical analysis revealed that there was no significant difference (p>0.05) between values of proteins efficiency ratios of baked and boiled D soybean for F1 and F2 formulations. These values in the both formulations were higher than those of the C diet. However, for the F3 formulation, proteins efficiency ratios of boiled Dockounou were higher than those of baked Dockounou (5.71±1.38 > 3.15±0.44).

3.7 Proteins Content of Formulated Dockounou

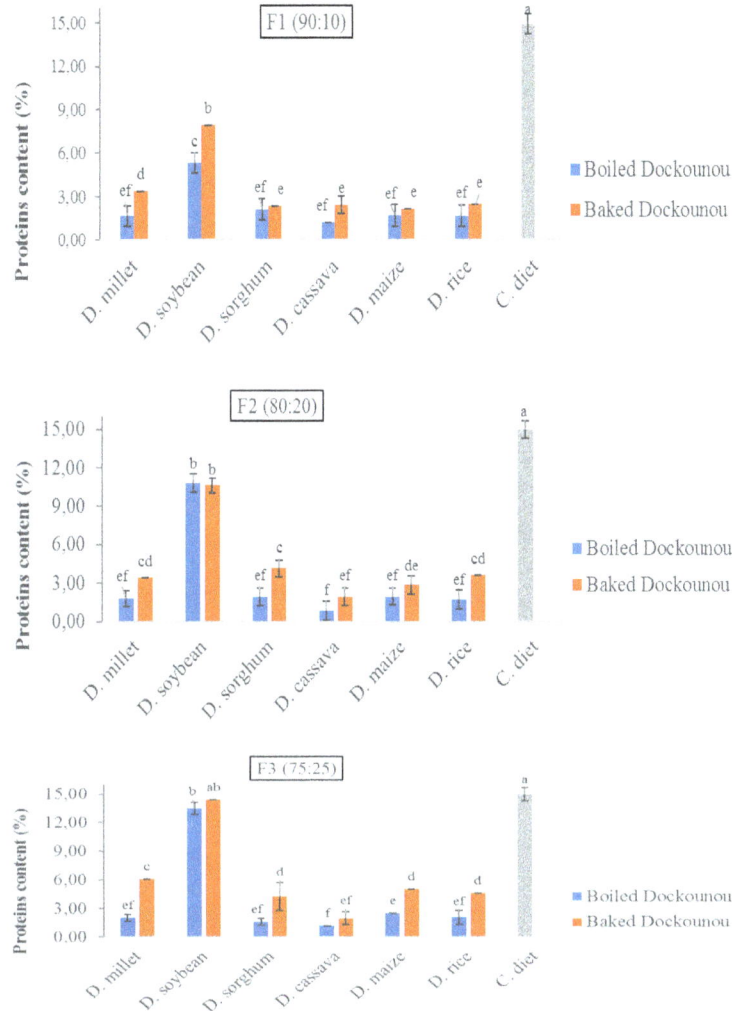

Figure 3. Proteins content of various Dockounou for F1, F2 and F3 formulations

Data are represented as Means ± SD (n=3). Means with no common letter (a, b, c, d) are significantly different (p<0.05)

Proteins content of Dockounou for F1, F2 and F3 formulations are presented in Figure 3. Values were significantly different (p<0.05) between cooking methods and between formulations. Proteins contents of types of Dockounou ranged from 1.23±0.00 % to 14.44±0.00 % for D cassava and D soybean in F1 and F3 formulations, respectively. Control diet recorded the highest protein content, followed by those of baked and boiled D soybean in F2 and F3 formulations. All the other contents were at far slighter and statistically similar.

3.8 Nutritive and Physicochemical Parameters of Soybean Dockounou

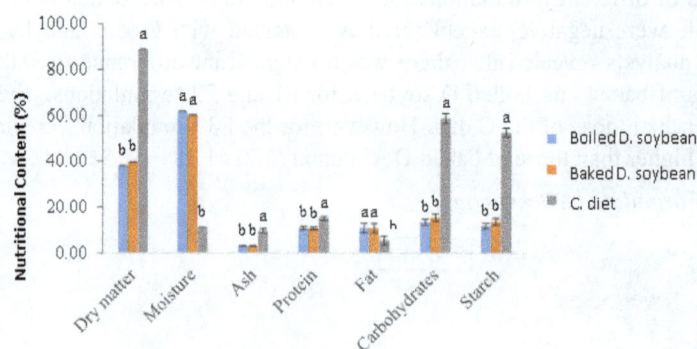

Figure 4. Nutritive properties of baked and boiled D soybean and C diet

Data are represented as Means ± SD (n=3). Means with no common letter (a, b, c, d) are significantly different (p<0.05) for each parameter.

Table 5. Physicochemical and biochemical properties of baked and boiled D soybean and C diet

Parameters	Boiled soybean Dockounou	Baked soybean Dockounou	C diet
Acidity (meq/100 g)	116.67±2.89a	124.33±12.10a	113.33±11.55a
pH	5.70±0.10b	5.60±0.10b	6.20±0.10a
Energy (Kcal)	165.67±10.60b	171.35±9.98b	293.11±7.46a
Total sugars (g)	0.23±0.02b	0.29±0.03a	0.21±0.03b

Values are mean ± standard deviation of five determinations. Values with different letters in the same line of each parameter indicate statistical difference (p < 0.05). D: Dockounou, F: Formulation

The nutritive parameters of soybean Dockounou are presented in Figure 4. Results revealed significant difference (p < 0.05) between the parameters of the whole diets (C diet and the boiled and baked D soybean). Generally, the C diet recorded higher contents of dry matter, ash, proteins, total carbohydrate and starch than boiled and baked D soybean. However, both Dockounou registered the same amount of ash, proteins, fat, total carbohydrate and starch. Data in table 5 also revealed significant differences between the three types of Dockounou, as far as the pH, total sugar content and energy are concerned. If Control diet registered the most important pH (6.20±0.10) and energy (293.11±7.46), it also topped the slightest amount of total sugars (0.21±0.03). All the diets presented statistically, similar acidity, but the pH (5.70±0.10 and 5.60±0.10) of both boiled and baked D. soybean were lower than that of the Control diet. Acidity. Baked D. soybean is richer in total sugars (0.29±0.03 g) than boiled D. soybean (0.23±0.02 g) which is less energetic (165.67±10.60 Kcal) than the first (171.35±9.98 Kcal).

3.9 Minerals Composition of the Diets

Table 6. Minerals composition of soybean baked and boiled Dockounou and C diet

Parameters	Soybean boiled Dockounou	Soybean baked Dockounou	C diet
Ca	558.20±21.73b	232.04±12.08c	1944,05±182,13a
Cu	0.64±0.02b	0.71±0.03b	2.5±0.23a
Co	8.75±0.34b	4.85±0.25c	13.41±1.26a
Fe	1.86±0.07b	1.64±0.08b	9.11±0.85a
K	313.97±12.21b	385.11±20.05b	1025.40±96.06a
Mg	42.40±1.65b	72.22±3.72c	191.05±17.89a
Mn	trace	trace	3.43±0.32a
Na	211.24±8.22c	303.85±15.78b	389.85±36.52a
Pb	0.14±0.00b	0.15±0.01b	0.24±0.02a
P	330.70±11.80c	433.71±46.10b	642.14±74.95a
Zn	trace	trace	4.29 ± 0.39a

Values are mean ± standard deviation of triplicate determinations. Values with different letters in the same line indicate statistical difference (p < 0.05).

The whole diets contained many various minerals such as calcium, sodium, phosphorus, magnesium, manganese, zinc, cobalt, iron, copper, but also lead (Pb). The mineral content of soybean Dockounou and C diet, showed no significant difference (p < 0.05) between copper (Cu), iron (Fe), potassium (K) and lead (Pb) contents of soybean baked Dockounou and those boiled D soybean as reported in the Table 6. At the whole, the C diet had higher mineral contents than soybean Dockounou, but Ca, Co and Mg contents of boiled D soybean are higher than those of baked D soybean.

4. Discussion

The length and weight are parameters that assess the growth and well-being of individual for a particular time (Morgane et al., 1993; Turlejski, 1996), but more especially appreciate of the nutritional state (WHO, 1995). Rats fed with Dockounou which were being incorporated the soybean flour had given best results zoological parameters at the end of the experiment. According to Ijarotimi & Oluwalana (2013), the weight gain is influenced by the quality of food composition and the quantity of food intakes by rats. Soybean is rich in protein and fats (Bender & Bender, 1999) and this could explained the fact that Dockounou without soybean could not allow Wistar rats growth. This fact highlights the importance to add nutrient-rich ingredients in formulations for better growth.

The consumption increased with proportion of soybean flour added and the cooking method. Thereby, an increase in weight of rats was also a link to these conditions. Udensi et al. (2012) reported the same observation in their study. In fact, soybean is a good source of protein and is generally used to enrich the foods poor in nutrients. The high food intake for baked D soybean might suggest that rats found this food more acceptable than the others as reported by Onweluzo & Nwabugwu (2009). The food efficiency ratio means the efficiency of food digestion, in other words, the weight gain obtained per unit of food consumed. Results of this parameter confirmed easly the lost or the gain of weight of rats according the diet.

The same applies to proteins intake and proteins efficiency ratio for which the statistical analysis revealed that there was no significant difference between baked and boiled Dockounou but proteins efficiency ratio of boiled Dockounou were higher than that baked. According to Ahenkora et al. (1997), the nature of the processing medium influenced nutritional qualities in the products. Indeed, the cooking mode influences proteins content. With the dry cooking, the food becomes dry and proteins which it contains are made difficult to digest by young rats. In contrast, with the wet cooking, the presence of water in the food makes easier the digestion of proteins by young rats. These proteins in a wet food are more accessible. Swaminathan & Gangwar (1961) estimated that the losses of nutrients are due to leaching into the cooking water. Generally, nutritional qualities of plantain fruits varied with the cooking processing method employed (Baiyeri et al., 2011). Thus, proteins content of the boiled D soybean were more ingested by young rats than proteins of baked D soybean. For the other types of formulated Dockounou, weak proteins intake could be explained by the fact that these cereal foods and roots of cassava are less rich in proteins and more important source of energy (Gomez, 1985; Tonukari, 2004). The high proteins intake, and protein efficiency ratio of baked D soybean could be due to his high food intake by the animals in the one hand and the other hand, by the content of soybean. Proteins content of the baked D soybean determined in our study highlight the wealth of soybean in protein. Indeed, this content of proteins is high than the others types of Dockounou. Moreover, the studies showed that a high food intake resulted in a sharp increase in proteins intake because proteins is required for proper growth, healthy living, maintenance and production tissues and cells of the young body (Messina, 1999). Also, the addition of legumes such as soybean to foods, increases the nutritional benefits by increasing of proteins content of food (Gomez, 1985; Nnam, 2001; Perez et al., 2008).

The quantity of proteins recommended in the diet of rats adults is least 12 % (Rogers, 1979). Consequently, the proteins content of soybean Dockounou in 80:20 formulation could be used for feeding adult rats. Concerning minerals, the recommended dietary allowance (RDA) (FAO, 2004) are: calcium (1000 mg/day) and phosphorus (800 mg/day). Thus, the soybean boiled Dockounou could cover 55 % of calcium RDA and 41 % of phosphorus RDA. While, the soybean baked Dockounou could cover 23 % of calcium RDA and 54 % of phosphorus RDA. Calcium and Phosphorus are associated for growth and maintenance of bones, teeth and muscles (Turan et al., 2003).

5. Conclusion

The present study showed that the cooking method could affect nutrients intake of Dockounou. The baking processing provides also,a better nutrients intake than the boiling one. However, soybean baked Dockounou formulated in 80:20 and 75:25 proportions used to feed the rats and which provide them a high physical and biological performance are the best diets. The proportions of senescent plantain (80 and 75 %) in this diet reveled that feeding rats with Dockounou could help to valorize the very ripe banana in order to reduce its post-harvest losses.

Acknowledgement

This work was conducted at Félix Houphouet-Boigny University (Biotechnology Laboratory, Biosciences Department). It was also supported by a PhD financial support from the Higher Education and Scientific Research Ministry of Côte d'Ivoire.

References

Adrian, J., Rabache, M., & Fragne, R. (1991). Technique d'analyse nutritionnelle, *in*: Lavoisier Tec et Doc (Eds). Principes de techniques d'analyse, Paris, 451-478.

Ahenkora, K., Kye, A., Marfo, K., & Banful, B. (1997). Nutritional composition of false horn *pantu pa* plantain during ripening and processing. *Afr. Crop Sci. J., 5*, 243-248.

Akoa, E. E. F., Kra, K. A. S., Mégnanou, R-M., Akpa, E. E., & Ahonzo, N. L. S. (2012). Sensorial characteristics of senescent plantain empiric dish (Dockounou) produced in Côte d'Ivoire. *J. Food Res., 1*(4), 150-159. https://doi.org/10.5539/jfr.v1n4p150

Akoa, E. E. F., Kra, K. A. S., Mégnanou, R-M., Kouadio, N. J., & Niamké, L. S. (2014). Optimization of Dockounou Manufacturing Process Parameters. *Sust. Agri. Res., 3*(1), 67-75. https://doi.org/10.5539/sar.v3n1p67

Akoa, E. E. F., Megnanou, R-M., Kra, K. A. S., & Ahonzo-Niamke, L. S. (2013). Technical variation in the processing of dockounou, a traditional plantain derivate dish of Côte d'Ivoire. *Ame. J. Res. Com., 1*(5), 81-97.

AOAC (Association of Official Analytical Chemists). (1990). Official methods of analysis Ed. Washington DC, p. 684.

AOAC (Association of Official Analytical Chemists)..(1980). Official method of analysis, 13th ed.,Washington DC, USA, 376-384.

APHA (American Public Health Association), (2001). Phosphorus Determination using the Colorimetric Ascorbic Acid Technique. CEE 453: Laboratory Research in Environmental Engineering. 64-69.

Atanda, S. A., Pessu, P. O., Agoda, S., Isong, I. U., & Ikotun, I. (2011). The concepts and problems of post-harvest food losses in perishable crops. *Afr. J. Food Sci., 5*(11), 603-613.

Atwater, W., & Rosa, E. (1899). A new respiratory calorimeter and the conservation of energy in human body. *II-physical, 9*, 214-251.

Baiyeri, K. P., Aba, S. C., Otitoju, G. T., & Mbah, O. B. (2011). The effects of ripening and cooking method on mineral and proximate composition of plantain (*Musa* sp. AAB cv. 'Agbagba') fruit pulp. *Afr. J. Biotech., 10*(36), 6979-6984.

Baker, H. J., Lindsey, J. R., & Weisbroth, S. H. (1979). The laboratory rat, vol. I, Biology and diseases. A Subsidiary of Harcourt Brace Jovanovich Publishers, Academy Press, New York, London, Toronto, Sidney, San-francisco.

Bender, D. A., & Bender, A. E. (1999). Bender's dictionary of nutrition and food technology. Woodland pub., New York, 302-303. https://doi.org/10.1201/9781439834381

BIPEA (Bureau Interprofessionnel d'Etudes Analytiques), (1976). Dans : Recueil de méthodes d'analyse des communautés européennes. BIPEA, Gennevilliers, 51-52.

Chia, C. L., & Huggins, C. A. (2003). Bananas. Community Fact sheet BA-3(A) Fruit. Hawaii Cooperative Extension Service, CTAHR, University of Hawaii.

Coulibaly, N. (2008). Caractérisation physicochimique, rhéologique et analyse sensorielle des fruits de quelques cultivars de bananiers (Musa AAB, AAAA, AAAB). Thèse de Doctorat, UFR des Sciences et Technologies des Aliments, Université d'Abobo-Adjamé, Côte d'Ivoire, p.180.

Dubois, M., Gilles, K., Hamilton, J., Rebers, P., & Simith, F. (1956). Colorimetric method for determinations of sugars and related substances. *Anal. Chim., 280*, 350-356. https://doi.org/10.1021/ac60111a017

Dzomeku, B. M., Dankyi, A. A., & Darkey, S. K. (2011). Socioeconomic importance of plantain cultivation in Ghana. *The J. Ani. Plant Sci., 21*(2), 269-273.

Emaga, H. T., Wathelet, B., & Paquot, M. (2008). Changements texturaux et biochimiques des fruits du bananier au cours de la maturation, Leur influence sur la préservation de la qualité du fruit et la maîtrise de la maturation. *Biotech. Agro. Soci. Environ., 12*(1), 89-98.

FAO (Food and Agriculture Organization of the United Nations), (2004). Human vitamin and mineral requirements. FAO Ed., p. 361.

FAOSTAT (Food and Agriculture Organization of the United Nations Statistics), (2013). Food and Agricultural Organization, Statistics division, Crops, Rome, Italy.

Gomez, M. H. (1985). Development of a food of intermediate moistness from extracts of corn and soybean. *Archivos Latinoamericanos de Nutrición, 35*(2), 306-314.

Honfo, F. G., Tenkouano, A., & Coulibaly, O. (2011). Banana plantain-based foods consumption by children and mothers in Cameroon and Southern Nigeria: A comparative study. *Afr. J. Food Sci., 5*(5), 287-291.

IITA (International Institute of Tropical Agriculture), (1987). Root, Tuber and plantain improvement program: Annual Report, Onne, Nigeria, p. 39.

IITA (International Institute of Tropical Agriculture), (1998). Plantain and Banana Improvement Program: Annual Report for 1997, Onne, Nigeria, p.19.

Ijarotimi, O. S., & Oluwalana, (2013). Chemical Compositions and Nutritional Properties of Popcorn-Based Complementary Foods Supplemented With *Moringa oleifera* Leaves Flour. *J. Food Res., 2*(6), 117-132. https://doi.org/10.5539/jfr.v2n6p117

Kouadio, N. J., Rose-Monde, M., Eric, A., Severin, K., & Sebastien, N. (2014). In Vitro Digestibility of Dockounou a Traditional Plantain Derivate Dish of Côte d'Ivoire. *Am. J. Bio. Sci., 2*(6), 211-216. https://doi.org/10.11648/j.ajbio.20140206.14

Kouadio, N. J., Rose-Monde, M., Eric, A., Edwige, E. A., Séverin, K. K., & Sébastien, L. S. (2015). Impact of the nutritional supply of Dockounou with millet, soybean, cassava, sorghum flours in Wistar rat growth. *Inter. J. Inn. App. Stud., 10*(2), 576-583.

Kra, K. A. S., Akoa, E., Megnanou, R-M., Yeboue, K., Akpa, E. E., & Niamke, L. S. (2013). Physicochemical and nutritional characteristics assessment of two different traditional foods prepared with senescent plantain. *Afr. J. Food Sci., 7*(3), 51-55. https://doi.org/10.5897/AJFS12.087

Kuperminc, O. (1988). Saisonnalité et commercialisation de la banana plantain en Côte d'Ivoire. *Fruits, 43*(6), 359-368.

Lassois, L., Jean-Pierre, B., & Haïssam, J. (2009). La banane: de son origine à sa commercialisation. Univ. Liège - Gembloux Agro-Bio Tech. Plant Pathology Unit, Passage des Déportés, 2. B-5030 Gembloux (Belgium). *Biotech. Agro. Soci. Environ., 13*(4), 575-586.

Lombor, T. T., Umoh, E. J., & Olakumi, E. (2009). Proximate composition and organoleptic properties of complementary food formulated from millet (*Pennisetum psychostchynum*), soybeans (*Glycine max*) and crayfish (Euastacusspp). *Pak. J. Nut., 8*(10), 1676-1679. https://doi.org/10.3923/pjn.2009.1676.1679

Messina, M.J. (1999). Legumes and soybeans: overview of their nutritional profile and health effects. *Am. J. Clin. Nut., 70*, 439-450.

Morgane, P. J., Austin-Lafrance, R. J., Bronzino, J., Tonkiss, J., Diaz-Cintra, S., Cintra, L., Kemper, T., & Galler, J. R. (1993). Prenatal malnutrition and development of the brain. *Neu. Biobe. Rev.,* 17, 91-128. https://doi.org/10.1016/S0149-7634(05)80234-9

N'guessan, A., Yao, N., & Kehe, M. (1993). La culture du bananier plantain en Côte d'ivoire. Spécial bananes II: systèmes de production du bananier plantain. *Fruits, 48*(2), 133-143.

Nnam, N. M. (2001). Comparison of the protein nutritional value of food blends based onsorghum, bambara groundnut and sweet potatoes. *Int. J. Food Sci. Nut., 52*, 25-29. https://doi.org/10.1080/09637480020027246

Onweluzo, J. C., & Nwabugwu, C. C. (2009). Development and Evaluation of Weaning Foods from Pigeon Pea and Millet. *Pak. J. Nut., 8*(6), 725-730. https://doi.org/10.3923/pjn.2009.725.730

Pérez, A., González, R. J., Drago, S. R., Carrara, C., De Greef, D. M., & Torres, R. L. (2008). Extrusion cooking of a maize-soybean mixture: Factors affecting expanded product characteristics and flour dispersion viscosity. *J. Food Eng., 87*(3), 333-340. https://doi.org/10.1016/j.jfoodeng.2007.12.008

Rodríguez, M. J. L., Rodríguez, S. A., & Belalcázar, C. S. (1998). Importancia Socioeconómica del Cultivo del Plátano en la Zona Central Cafetera (Segunda Versión) Oficina Regional de Planeación - Corpoica, Regional Nueve, Manizales, marzo.

Rogers, H. E. (1979). Nutrition, *in*: The laboratory rat, vol I, biology and diseases, Baker H. J., Lindsey J.R. and Weisbroth S.H. eds. Academy Press, New York, NY, 123-152. https://doi.org/10.1016/b978-0-12-074901-0.50013-7

Sery, D. G. (1988). Rôle de la banane plantain dans l'économie ivoirienne. *Fruits, 43*(2), 73-78.

Swaminathan, K., & Gangwar, B. M. L. (1961). Cooking losses of vitamin C in Indian potato varieties. *Indian Potato J., 3*, 86-91.

Tonukari, N. J. (2004). Cassava and the future of starch. *Electronic J. Biotech., 7*(1), 5-8. https://doi.org/10.2225/vol7-issue1-fulltext-9

Turan, M., Kordali, S., Zengin, H., Dursun, A., & Sezen, Y. (2003). Macro and micro-mineral content of some wild edible leaves consumed in Eastern Anatolia. *Plant Soil Sci., 53*, 129-137. https://doi.org/10.1080/090647103100095

Turlejski, K. (1996). Evolutionary ancient roles of serotonin: long lasting regulation of activity and development. *Acta Neur. Exp., 56*(6), 19-36.

Udensi, A. A., Odom, T. C., Nwaorgu, O. J., Emecheta, R. O., & Ihemanma, C. A. (2012). Production and evaluation of the nutritional quality of weaning food formulation from roasted millet and Mucuna cochinchinesis. *Sky J. Food Sci., 1*(1), 1-5.

WHO (World Health Organization), (1995). Physical Status: The Use and Interpretation of Anthropometry. Technical Report Series, 854, Geneva, OMS, p. 452.

Younoussa, D., Momar, T. G., Mama, S., Praxède, G., Amadou, D., Jean-Paul, B., & Georges, L. (2013). Importance nutritionnelle du manioc et perspectives pour l'alimentation de base au Sénégal (synthèse bibliographique). *Biotech. Agro. Soci. Environ., 17*(4), 634-643.

Appendix A. Composition of control diet (SFACI company).

Components	Proportion
Protein	15 %
Fat	3,5 %
Cellulose	12 %
Carbohydrate	58 %
Minerals	11,2 %
Vitamin A	15000 UI/Kg
Vitamin D	3000 UI/Kg
Vitamin E	10 mg/Kg

Appendix B. Formulas for growth parameters determination.

Parameters	Mathematical formulas
Food intake (FI)	Given food – Left-over food
Dry mater (DM)	100 - [((wet weight - dry weight) / wet weight) × 100]
Dry mater intake (DMI)	FI × DM
Average weight gain (AWG)	Final weight - Initial weight
Average daily weight gain (ADWG)	AWG / days
Feed efficiency (FE)	AWG / DMI
Total protein intake (TPI)	DMI × % protein of diet
Protein efficiency (PE)	AWG / TPI

6

Consumer Acceptability and Descriptive Characterization of Fresh and Dried King Oyster (*Pleurotus eryngii*) and Hedgehog (*Hydnum repandum*) Mushrooms

Elisa A. S. F. Boin[1,2], Cláudia M. A. M. Azevedo[1], João M. S. A. Nunes[2] & Manuela M. Guerra[1]

[1] Estoril Higher Institute for Tourism and Hotel Studies, Av. Condes de Barcelona, 801, Estoril, 2769-510, Portugal

[2] BLC3 Association –Technology and Innovation Campus, Rua Nossa Senhora da Conceição, 2, Lagares, Oliveira do Hospital, 3405-155, Portugal

Correspondence: Elisa A. S. F. Boin, BLC3 Association –Technology and Innovation Campus– Dept. of Agriculture and Food Technologies, Rua Nossa Senhora da Conceição, 2, Lagares, Oliveira do Hospital, 3405-155, Portugal. E-mail: elisa.boin@blc3.pt

Abstract

King oyster *(Pleurotus eryngii)* and hedgehog (*Hydnum repandum)* mushrooms have great commercial interest due to their nutraceutical and nutritional properties, besides being new products on the market. The aim of this study was to evaluate the acceptability and characterize descriptively fresh and dried *Pleurotus eryngii* and *Hydnum repandum* mushrooms. Raw mushrooms were analyzed by descriptive tests and cooked mushrooms were analyzed by hedonic, discriminative and descriptive tests. Descriptive analysis was performed by QDA[TM] method with a semi-trained panel. Acceptability as a guide to consumer trends was assessed as hedonic tests with 20 untrained judges to evaluate appearance, aroma, texture, flavor, and purchase decision. To evaluate the influence of the drying process in sensory characteristics, mushroom risottos were compared by discriminative analysis. Raw fresh hedgehog mushroom was mainly characterized by the presence of teeth, cap waviness and intensity of aroma. Raw dried *H. repandum* was mainly depicted by the presence of teeth and wrinkles, crunchiness and hardness. Well-defined gills and velvet touch characterized raw fresh *P. eryngii*. Dried *P. eryngii* mushroom was crunchy and had different colors of cap and stem. All cooked mushrooms presented average hardness and were slightly umami, watery, chewy and had some umami aftertaste. Cooked *Hydnum repandum* presented high intensity of aroma and bitter aftertaste. Fresh and dried *Pleurotus eryngii* were well accepted, as well as fresh *H. repandum*. Dried *H. repandum* had low acceptability scores, being thus not recommended to be consumed sautéed, but in sauces or risottos.

Keywords: Acceptability, descriptive analysis, *Hydnum repandum*, mushrooms, *Pleurotus eryngii*, sensory analysis

1. Introduction

Edible mushrooms have become very popular due to their high nutritional value, low calorie, fat and sodium values, and for being cholesterol-free (Valverde, Hernández-Pérez, & Paredes-López, 2015). They are constituted approximately of 90% of water and their dry matter is composed by 35-70% of carbohydrates, 15-35% proteins and less than 5% of lipids. Moreover, mushrooms also contain a significant amount of vitamins (B1, B2, B3, B9, B12, C, D, E and ß-carotene) and minerals (Ca, K, Mg, P, Cu, Fe, Mn, Se and Zn) (Pereira, Barros, Martins, & Ferreira, 2012).

Sensory analysis of *Hydnum repandum* mushroom is still unexplored, since the literature on this mushroom focus mainly on metal bioaccumulation (Severoglu, Sumer, Yalcin, Leblebici, & Aksoy, 2013), radioactivity (García, Alonso, & Melgar, 2015; Kalač, 2001) and antioxidants (Fernandes et al., 2013; Heleno, Barros, Sousa, Martins, & Ferreira, 2010). Antimicrobial and antitumoral effects of this mushroom were also reported regarding this mushroom (Ozen, Darcan, Aktop, & Turkekul, 2011).

Pleurotus eryngii mushrooms contain antifungal properties, antioxidant components (Mishra et al., 2013) and a high amount of dietary fibers (approximately 35%, of which 88% are insoluble) (Manzi, Marconi, Aguzzi, &

Pizzoferrato, 2004). This species is also considered a unique and delicious mushroom (Li et al., 2014, 2015) due to its non-volatile taste components (Beluhan & Ranogajec, 2011). Its nutritional composition and taste compounds have also been investigated (Li et al., 2014; Mishra et al., 2013). Sensory analysis of *Pleurotus eryngii* mushrooms has been used as a complement of quality and shelf life studies (Li et al., 2013), and with products developed with *P. eryngii* (Jeong & Shim, 2004; Sung, Kim, & Kang, 2008).

Dehydration is a process that increases shelf life of food, stabilizes microbiological activity, reduces package size and avoids the need of refrigeration, making commercialization much easier and therefore being one of the most used methods for preservation of fruits and vegetables (Li et al., 2015). Thus, it is important to investigate if mushroom drying process lowers their acceptability and in which ways it changes their sensory characteristics.

Characterizing a product through consumer's perception is central for marketing, product development and consumer acceptance (Lee & O'Mahony, 2005). This study aimed at evaluating the acceptability and describing the sensory characteristics of fresh and dried *Pleurotus eryngii* and *Hydnum repandum* mushrooms.

In the present study, the definitions of mushroom forms used were:

Fresh – fresh mushroom, not processed or preserved;

Dried –mushroom preserved by hot air drying;

Raw – uncooked mushroom, may be in dried or fresh form;

Cooked – mushroom prepared by cooking, may be fresh (before cooked), or dried (in this case, rehydrated prior to cooking).

2. Materials and Methods

In the present study, raw mushrooms were characterized by descriptive analysis and cooked mushrooms were evaluated by three sensory tests: hedonic, to perceive acceptability as a guide for consumer trends; discriminative, to evaluate if the drying process significantly changed sensory characteristics when the mushrooms were included in a recipe; and descriptive, for descriptive characterization. All tests were performed for both mushroom species *(P. eryngii* and *H repandum*) in fresh and dried forms. Sensory analyses of cooked mushrooms were performed with 10 g of samples at 30 °C served in three-digit coded white plastic cups and with white plastic spoons. Water and low-salt toasts were available for mouth cleansing.

2.1 Samples

Hydnum repandum mushrooms were collected in the Inner Center region of Portugal, in natural habitats of pines (*Pinus pinaster* and *Pinus pinea*), oaks (*Quercus* sp.) and hazelnut trees (*Corylus avellana*), from November/2014 to January/2015. *Pleurotus eryngii* mushrooms were produced by Voz da Natureza, Lda., and BLC3 Association (Oliveira do Hospital, Portugal). All collected and produced mushrooms were cleaned using paper towels and brushes before use. Mushrooms were sliced (approximately 0.8 mm) and hot air dried in convective oven (ON-01E, Lab Companion JeioTech, South Korea) at 40 °C for 24 hours and then stored in sealed plastic bags. Fresh mushrooms were kept under refrigeration at 3 °C and tests were performed up to three days after mushroom picking.

2.2 Panel Selection and Training

In order to form internal panels for acceptability and characterization studies, an online questionnaire was performed on 26 employees from BLC3 Association, which were divided in two different panels according to their answers: a non-trained panel for the hedonic tests, and a semi-trained panel for descriptive and discriminative analyses. Questions included socio-demographic data and requisites for participating in the panels. Requisites for the non-trained panel were: not being sick or having a chronic disease that could affect sensory perception, not being pregnant, interest in performing the tests, availability and liking the analyzed product. For the semi-trained panel, the requisites also included interest in being trained, not being colorblind and having a good capacity of oral communication (Stone & Sidel, 2009).

A discriminative triangular test was performed according to ISO 4120:2004 (ISO, 2004) for the final selection of the semi-trained panel to evaluate the aptitude of the judges to discriminate concentrations of the basic tastes. To accomplish this evaluation, two different concentrations of each of the following solutions were used: sucrose (12 and 24 g/L), citric acid (0.6 and 1.2 g/L), monosodium glutamate (1 and 2 g/L), caffeine (0.27 and 0.54 g/L) and sodium chloride (2 and 4 g/L) for sweet, sour, umami, bitter and salt tastes, respectively.

The panel was semi-trained over seven sessions of two hours each, in a total of 14 hours. The training included explanations about sensory tests, the most common mistakes in sensory tests, and taste and odor physiology.

Out of the 26 interviewed individuals, 20 fulfilled the requisites to participate in hedonic tests. Therefore, ten women and ten men, aged between 21-45 years old, composed the untrained panel. Twelve individuals fulfilled the requisites to be trained, but only eight passed the selection test and composed the semi-trained panel. Semi-trained judges were aged between 18-35 years old.

2.3 Descriptive Characterization of Raw Mushrooms

Descriptive tests were carried out with 8 semi-trained judges, using QDA^{TM} method (Stone, Sidel, Oliver, Woolsey, & Singleton, 2004) for mushroom characterization. Descriptor generation was carried out individually by species (*P. eryngii* and *H. repandum*) and treatment (dried and fresh), due to mushroom morphology. Descriptors of raw mushrooms were generated for appearance, aroma (3-5 cm from the nose) and non-oral texture. Descriptor selection criterion was to be cited by more than half of the panel. The descriptors used for raw mushrooms are defined in Table 1.

Table 1. Descriptors and definition used in sensory analysis for raw mushrooms

Treatment	Mushroom	Descriptor	Definition
Fresh	*Hydnum repandum*	Presence of teeth	Presence of teeth under the cap
		Fragility of teeth	Ease of teeth to detach from the cap
		Moldy	Aroma of mold
		Hardness	Resistance to deform with touch
		Cap waviness	Waviness of the cap
		Intensity of aroma	Intensity of aroma in general
		Difference of hardness	Hardness difference between cap and stem
Dried	*Hydnum repandum*	Presence of teeth	Presence of teeth under the cap
		Fragility of teeth	Ease of teeth to detach from the cap
		Moldy	Aroma of mold
		Hardness	Resistance to deform with touch
		Crunchiness	Crunchy texture
		Wrinkles	Wrinkled surface
Fresh	*Pleurotus eryngii*	Difference of color	Color difference between cap and stem
		Gill definition	Ease to visualize the definitions of the gills
		Cap definition	Ease to visualize the definitions of the cap
		Velvet touch	Velvet touch
		Stem hardness	Resistance of stem to deform with touch
		Intensity of aroma	Intensity of aroma in general
Dried	*Pleurotus eryngii*	Difference of color	Color difference between cap and stem
		Crunchiness	Crunchy texture

For descriptive tests, the selected descriptors were presented in a sheet of paper with 10 cm unstructured line scales. The judges placed a mark in the scale indicating the perceived intensity. Later, that mark was converted to a number representing the distance of the mark from the zero end of the scale (values between 0.0-10.0). Therefore, descriptors that presented higher values were more characteristic. The descriptors attributed to both fresh and dried treatments were also compared to verify if dehydration culminated into sensory changes.

2.4 Acceptability and Descriptive Characterization of Cooked Mushrooms

2.4.1 Acceptability

Internal hedonic tests were carried out for cooked *Hydnum repandum* and *Pleurotus eryngii* mushrooms with 20 untrained judges in order to evaluate acceptability as a guide to consumer trends (Kemp et al., 2009). Fresh samples were sliced and sautéed with olive oil and salt (1:10:100 salt:oil:mushroom, m/m/m). Dried samples were rehydrated for 30 min in hot water (80 °C) and also sautéed with olive oil and salt (1:10:100 salt:oil:mushroom, m/m/m). Appearance, aroma, texture, flavor and global acceptance were analyzed using a Likert scale (1-9): "Dislike extremely", "Dislike very much", "Dislike moderately", "Dislike slightly", "Neither like nor dislike", "Like slightly", "Like moderately", "Like very much" and "Like extremely". Purchase decision was also evaluated with a Likert scale (1-5): "Very unlikely", "Unlikely", "Neither likely nor unlikely", "Likely" and "Very likely".

2.4.2 Discriminative Tests

Eight semi-trained individuals assessed mushroom risottos to evaluate if the drying process significantly changed sensory characteristics when the mushroom was included in a dish. Tests were performed on *H. repandum* and *P. eryngii* species. For fresh mushrooms, the risotto was prepared by mixing rice and mushrooms simultaneously and cooking them until ready. Dried mushrooms were rehydrated for 30 min in hot water (80 °C) before being added to rice to be cooked. Triangle tests were performed according to ISO 4120:2004 (ISO, 2004).

2.4.3 Descriptive Characterization

Descriptor generation for cooked mushrooms was carried out for mushrooms in general, as fresh and dried cooked *P. eryngii* and *H. repandum* were presented altogether. Descriptors of cooked mushrooms were generated for appearance and aroma (3-5 cm from the nose) before being tasted and for texture and flavor during and after tasting samples. Descriptive selection criterion and tests were carried out likewise raw mushrooms. The selected descriptors for cooked mushrooms are defined in Table 2.

Table 2. Descriptors and definition used in sensory analysis for cooked mushrooms

Descriptor	Definition
Intensity of aroma	Intensity of aroma in general
Hardness	Resistance to deform with touch
Umami	Umami basic taste – Reference: monosodium glutamate
Wateriness	Water released from mushroom while chewing
Chewiness	Hard texture at first and then elastic – Reference: bubble gum
Toughness	Resistance to mastication
Bitter aftertaste	Bitter basic taste – Reference: caffeine
Umami aftertaste	Umami basic taste – Reference: monosodium glutamate

2.5 Statistical Analysis

Values of hedonic tests are presented as medians, first and third percentiles (P25 and P75, respectively), as Likert scales were ordinal. Descriptive results are presented as means and standard deviations, since line scales result in ratio variables. However, the number of judges for descriptive tests was small and therefore data was considered non-parametric. Thus, data from hedonic and descriptive tests was analyzed by Friedman test and sum of ranks for multiple comparisons with a significance level of 5% using R-package *agricolae* (Mendiburu, 2014) and R software 3.2.1 (R Core Team, 2015). Binomial test with 1/3 chance of success was performed for discriminative tests.

3. Results and Discussion

3.1 Descriptive Characterization of Raw Mushrooms

In the present study, raw mushrooms were described according to species (*H. repandum* and *P. eryngii*) and treatment (dried and fresh), with the aim of characterizing these mushrooms in raw and cooked forms. Means and standard deviation of similar and different descriptors for fresh and dried raw mushrooms are shown in Table 3 and Table 4 for *H. repandum* and *P. eryngii*, respectively. Judges considered that four descriptors of fresh *H. repandum* mushroom were similar to the dried form: presence of teeth, fragility of teeth, moldy aroma and hardness. The other descriptors were representative of the treatments: cap waviness, intensity of aroma and difference of hardness between cap and stem for fresh *H. repandum* mushroom; and crunchiness and wrinkles for dried *H. repandum* mushroom.

According to judges, fresh hedgehog mushroom was characterized mainly by the presence of teeth, cap waviness and intensity of aroma, as may be seen by the higher values in Table 3. Fragility of teeth, moldy aroma, difference of hardness between cap and stem, and hardness were less characteristic. Dried *H. repandum* was mostly characterized by the presence of wrinkles and teeth, crunchiness and hardness. These aspects are recurrent in dried mushrooms and, for this reason, dehydration and rehydration of mushrooms have been studied (Dinani, Hamdami, Shahedi, & Havet, 2015; Reyes et al., 2014). Aroma loss in dried mushroom was depicted by a lower value of the moldy aroma in comparison to the fresh mushroom, besides the inexistence of the descriptor intensity of aroma in the dried mushroom. Presence and fragility of teeth had lower scores in dried *H. repandum* mushroom in comparison to the fresh form. Hardness scores were higher in dried *H. repandum*, as Kotwaliwale et al. (2007) observed in oyster mushrooms.

Table 3. Descriptive results of raw *Hydnum repandum* (mean ± standard deviation, n=8 judges)

	Fresh *Hydnum repandum*		Dried *Hydnum repandum*	
Similar descriptors	Presence of teeth	$9.3 \pm 0.5^{a,A}$	Presence of teeth	$8.1 \pm 1.4^{ab,B}$
	Fragility of teeth	$6.5 \pm 2.9^{c,A}$	Fragility of teeth	$5.2 \pm 3.0^{bc,B}$
	Moldy	$5.4 \pm 3.4^{c,A}$	Moldy	$3.8 \pm 3.5^{c,B}$
	Hardness	$4.9 \pm 2.2^{c,B}$	Hardness	$7.9 \pm 1.8^{abc,A}$
Different descriptors	Cap waviness	8.2 ± 1.6^{b}	Crunchiness	8.0 ± 2.1^{a}
	Intensity of aroma	7.7 ± 2.3^{b}	Wrinkles	8.1 ± 1.6^{ab}
	Difference of hardness	5.0 ± 2.6^{c}		

Values within the same column with different lower case letters are significantly different (p<0.05). For similar descriptors, values within the same row with different upper case letters are significantly different (p<0.05).

Judges considered that only difference of color between cap and stem was a descriptor of both fresh and dried *P. eryngii*. The other descriptors were representative of the treatments: gill definition, cap definition, velvet touch, stem hardness and intensity of aroma for fresh *P. eryngii* mushroom; and crunchiness for dried *P. eryngii* mushroom.

All descriptors, except intensity of aroma, presented high values for fresh *Pleurotus eryngii* and thus were characteristic (Table 4). Gill definition and velvet touch obtained the highest scores, between 8.1-9.0. Difference of color between cap and stem, cap definition and stem hardness were scored between 7.2-8.1. Although intensity of aroma was low scored, it may also be used as an attribute, as the presence of other aroma would be considered off-odors. For dried *P. eryngii* mushroom, crunchiness and difference of color between cap and stem were scored between 6.2-7.9. Comparing dried and fresh *P. eryngii* mushrooms, sensory analysis indicated that the drying process did not alter the difference of color between cap and stem.

Table 4. Descriptive results of raw *Pleurotus eryngii* (mean ± standard deviation, n=8 judges)

	Fresh *Pleurotus eryngii*		Dried *Pleurotus eryngii*	
Similar descriptors	Difference of color	$8.1 \pm 1.0^{b,A}$	Difference of color	$6.2 \pm 2.7^{a,A}$
Different descriptors	Gill definition	9.0 ± 0.7^{a}	Crunchiness	7.9 ± 2.3^{a}
	Cap definition	8.1 ± 1.6^{b}		
	Velvet touch	8.1 ± 1.4^{ab}		
	Stem hardness	7.2 ± 1.5^{b}		
	Intensity of aroma	3.5 ± 2.8^{c}		

Values within the same column with different letters are significantly different (p<0.05). For similar descriptors, values within the same row with different upper case letters are significantly different (p<0.05).

3.2 Acceptability and Descriptive Characterization of Cooked Mushrooms

Values of the hedonic tests for *Hydnum repandum* and *Pleurotus eryngii* mushrooms are shown in Table 5 as medians and percentiles.   Although the internal panel was smaller than usual consumer panels, it produced results as guides to consumer trends, which may indicate the viability of commercialization of the products assessed. Nonetheless, further consumer hedonic tests should be assessed in order to obtain a more significant evaluation.

Neither dehydration nor mushroom species influenced on appearance or aroma acceptability (Table 5), parameters in which both mushrooms had good scores with median values of "Like slightly" and "Like moderately". Aroma may be lost after dehydration (Rivera, Blanco, Salvador, & Venturini, 2010) or due to cooking process (MacLeod & Panchasara, 1983). The judges may have not perceived differences between dried and fresh mushrooms due to aroma loss during cooking and therefore their acceptability was not altered. Other hypothesis is that judges perceived those differences, yet it did not change their acceptability.

Table 5. Acceptability of cooked *H. repandum* and *P. eryngii* (n=20 judges). Values in Median (P25;P75)

Acceptability attribute	Hydnum repandum		Pleurotus eryngii	
	Fresh	Dried	Fresh	Dried
Appearance (1.0-9.0)	7.0 (6.0; 7.0)[ab]	6.0 (5.0; 7.0)[b]	7.0 (7.0; 8.0)[a]	7.0 (6.75; 8.0)[a]
Aroma (1.0-9.0)	6.5 (6.0; 7.0)[a]	6.0 (4.75; 7.0)[a]	7.0 (6.0; 7.0)[a]	6.0 (5.0; 7.25)[a]
Texture (1.0-9.0)	7.0 (6.0; 8.0)[a]	6.0 (3.0; 6.0)[c]	7.0 (6.0; 8.0)[ab]	6.5 (4.0; 8.0)[b]
Flavor (1.0-9.0)	6.0 (5.5; 7.0)[b]	4.0 (2.0; 5.25)[c]	7.0 (6.0; 8.0)[a]	7.0 (6.0; 7.25)[ab]
Global acceptance (1.0-9.0)	6.5 (5.0; 7.0)[b]	4.5 (3.0; 6.0)[c]	7.0 (6.0; 8.0)[a]	7.0 (6.0; 7.25)[ab]
Purchase decision (1.0-5.0)	4.0 (2.75; 4.0)[a]	2.0 (2.0; 3.0)[b]	4.0 (3.0-5.0)[a]	4.0 (2.75-5.0)[a]

Values within the same row with different letters are significantly different (p<0.05).

Even with wrinkling and darkening that may be caused by dehydration (Kotwaliwale, Bakane, & Verma, 2007; Reyes, Mahn, Cubillos, & Huenulaf, 2013), median values for appearance did not significantly differ between dried and fresh mushrooms (Table 5), probably due to the darkening of all mushrooms and water loss during cooking process. Li et al. (2013) obtained color acceptability values of 2.6 out of 5.0 and aroma acceptability 3.1 out of 5.0 on fresh *P. eryngii*, slightly lower values than those obtained in the present study. Dehydration process significantly reduced acceptability of *H. repandum* mushroom on texture, flavor and global, in contrast to *P. eryngii*, that did not show significant differences between fresh and dried samples. Fresh king oyster had "Like moderately" evaluation for flavor and global acceptance, as fresh hedgehog had "Like slightly". On the other hand, dried mushrooms presented greater difference of scores for global acceptance, since king oyster was scored as "Like moderately" and hedgehog as "Dislike slightly".

Hot air drying may cause cellular damage and structural shrinkage, resulting in loss of rehydration capacity (García-Segovia, Andrés-Bello, & Martínez-Monzó, 2011), and Krokida and Marinos-Kouris (2003) observed that there were pores from *Agaricus bisporus* and vegetables that were not filled during rehydration after hot air drying. Although increasing of hardness and chewiness has been observed in *Pleurotus* sp. during drying process (Kotwaliwale et al., 2007), in this study, dried *P. eryngii* had no significant difference from fresh in texture acceptability (Table 5). These results support the descriptive analysis performed, which provided no differences between dried and fresh samples regarding chewiness and hardness (Table 6).

Flavor acceptability of dried *H. repandum* was significantly lower than fresh mushroom (Table 5), which may have been caused by a combination of toughness, intensity of aroma and bitter aftertaste, as seen in the results of descriptive analysis of cooked mushrooms (Table 6). Dried *H. repandum* also had a significant lower median value for global acceptance, and purchase decision was "Unlikely" (Table 5). This shows that consumer trends for dried *H. repandum* are not favorable when the mushroom is sautéed. However, mushrooms may be consumed as an ingredient of a recipe (sauces, with meat, etc.). In this way, the parameters that presented lower acceptability when the mushrooms were assessed only sautéed may be not perceived in a sauce or risotto, as occurred in the discriminative analysis, presenting a possibility of market for dried mushrooms. Another possible application of this mushroom is through the development of new products where texture and flavor intensity can be studied and controlled, using sensory characteristics perceived by consumers to drive a formulation (Raz et al., 2008).

Table 6. Descriptive characterization of cooked *P. eryngii* and *H. repandum* mushrooms (mean ± standard deviation)

Descriptor	Fresh *H. repandum*	Dried *H. repandum*	Fresh *P. eryngii*	Dried *P. eryngii*
Intensity of aroma	4.2 ± 2.3^a	4.7 ± 2.4^a	2.0 ± 1.5^b	2.0 ± 1.6^b
Hardness	5.1 ± 2.3^a	5.7 ± 2.8^a	5.7 ± 3.0^a	6.6 ± 2.3^a
Umami	6.0 ± 2.9^a	5.3 ± 3.7^a	6.0 ± 2.3^a	3.7 ± 2.4^a
Wateriness	4.7 ± 3.6^a	3.4 ± 3.0^a	4.8 ± 3.0^a	4.2 ± 2.9^a
Chewiness	4.8 ± 2.9^a	4.1 ± 2.5^a	3.7 ± 2.3^a	4.6 ± 3.0^a
Toughness	3.2 ± 2.8^b	5.9 ± 2.4^a	3.9 ± 2.6^b	6.4 ± 2.6^a
Bitter aftertaste	6.8 ± 2.9^a	6.3 ± 2.9^a	1.9 ± 1.7^b	2.1 ± 2.4^b
Umami aftertaste	4.2 ± 2.4^a	3.7 ± 2.6^a	3.2 ± 2.6^a	3.0 ± 2.1^a

Values within the same row with different letters are significantly different (p<0.05).

In the present study, the judges did not discriminate dried from fresh cooked mushrooms in a risotto (p=0.273) in discriminative analysis, neither *H. repandum* nor *P. eryngii*. Reyes, Mahn and Vasquéz (2014) noted that dried and fresh button mushrooms (*Agaricus bisporus*) were not significantly different regarding color, shrinkage, texture and thermal efficiency. Also, discriminative tests were carried out to compare fresh and frozen truffle aroma (*Tuber melanosporum*), and in all cases they were discriminated, showing that freezing influenced organoleptic properties in truffles (Culleré, Ferreira, Venturini, Marco, & Blanco, 2013) and pointing that dehydration may be a better alternative to mushroom conservation.

Cooked mushrooms did not present significant difference in what concerns mushroom species and drying process for hardness, umami taste, wateriness, chewiness, and umami aftertaste (Table 6). For all mushrooms tested, hardness indicated that there was some resistance to touch, corroborating previous published studies (Kim et al., 2009; Pogoń, Jaworska, Duda-Chodak, & Maciejaszek, 2013). Wateriness and chewiness indicated, respectively, the release of a certain amount of water during mastication and some plastic consistence (Table 6), as observed in Kotwaliwale et al. (2007) with *Pleurotus ostreatus*. The values of umami aftertaste reveal that these descriptive attributes are present, however not too evidently. Intensity of aroma and bitter aftertaste distinguished cooked mushroom species by being almost inexistent in *P. eryngii* and more prominent in *H. repandum* (Table 6), and did not vary with dehydration, hence being intrinsic to the *Hydnum* mushrooms.

Natural taste of *P. eryngii* is due to sweet and umami (aspartic and glutamic acids, and 5'- nucleotides) components, and bitter is camouflaged by sweet components (Li et al., 2014). As shown in Table 6, only toughness, out of the eight descriptors, was significantly different in relation to treatment (dried and fresh). These values were higher in dried mushrooms due to low moisture content in rehydrated samples (García-Segovia et al., 2011). Drying conditions can be optimized to avoid flavor loss and reduction of rehydration, such as temperature, cutting size and other possible factors that may influence mushroom texture.

The global evaluation of *Pleurotus eryngii* indicates its gastronomic quality, as cited in previous studies (Li et al., 2014, 2015; Mishra et al., 2013). Fresh *Pleurotus eryngii* and *Hydnum repandum*, as well as dried *P. eryngii*, showed to be marketable and had positive global acceptance, being good products to the Portuguese market, as well as other countries, as they are little explored so far. Dried *H. repandum*, however, is not recommended to be consumed sautéed, but in dishes as sauces or risottos. Furthermore, all their nutritional and nutraceutical properties (Li et al., 2014; Mishra et al., 2013) fulfill what the consumer has been searching recently: products with gastronomical value and nutraceutical benefits.

4. Conclusion

The present study aimed at evaluating the acceptability and characterizing descriptively fresh and dried *Pleurotus eryngii* and *Hydnum repandum* mushrooms. Raw fresh hedgehog mushroom was mainly characterized by the presence of teeth, cap waviness and intensity of aroma. Raw dried *H. repandum* was mainly characterized by the presence of teeth and wrinkles, crunchiness and hardness. Moldy aroma, presence and fragility of teeth had lower scores in comparison to the fresh form. Hardness scores were higher for raw dried *H. repandum*. Well-defined gills and velvet touch were the strongest attributes of raw fresh *P. eryngii*. Dried *P. eryngii* mushroom was characterized as crunchy and possessed cap and stem with different colors.

Regarding acceptability, fresh and dried *Pleurotus eryngii* presented high scores in all acceptability attributes and strong purchase decisions. Fresh *Hydnum repandum* presented acceptability scores similar to *P. eryngii*, as well as "Likely" purchase decision. Dried *H. repandum* is not recommended to be consumed sautéed due to its low acceptability scores. Further studies using larger panels on hedonic tests should be applied in order to obtain a more significant evaluation.

All cooked mushrooms presented average hardness and were slightly umami, watery, chewy and had certain umami aftertaste. *H. repandum* presented higher intensity of aroma and bitter aftertaste than *P. eryngii*. The descriptors that were more characteristic may be used as attributes for marketing and values may be also used for standardization on production batches. The descriptive characterization and acceptability study implement the knowledge on sensory analysis of mushrooms and create new possibilities for investigation of fungi and its use on culinary.

Acknowledgments

The authors acknowledge the financial support of COMPETE/QREN/EU (Project "Value MicotecTruf" QREN/COMPETE/24845/2012 and Project "Fruit ECO-Drying Line" QREN/COMPETE/23266/2012) and Estoril Higher Institute for Tourism and Hotel Studies.

References

Beluhan, S., & Ranogajec, A. (2011). Chemical composition and non-volatile components of Croatian wild edible mushrooms. *Food Chemistry*, *124*(3), 1076–1082. http://dx.doi.org/10.1016/j.foodchem.2010.07.081

Culleré, L., Ferreira, V., Venturini, M. E., Marco, P., & Blanco, D. (2013). Chemical and sensory effects of the freezing process on the aroma profile of black truffles (Tuber melanosporum). *Food Chemistry*, *136*(2), 518–525. http://dx.doi.org/10.1016/j.foodchem.2012.08.030

Dinani, S. T., Hamdami, N., Shahedi, M., & Havet, M. (2015). Quality assessment of mushroom slices dried by hot air combined with an electrohydrodynamic (EHD) drying system. *Food and Bioproducts Processing*, *94*, 572–580. http://dx.doi.org/10.1016/j.fbp.2014.08.004

Fernandes, Â., Barros, L., Barreira, J. C. M., Antonio, A. L., Oliveira, M. B. P. P., Martins, A., & Ferreira, I. C. F. R. (2013). Effects of different processing technologies on chemical and antioxidant parameters of Macrolepiota procera wild mushroom. *LWT - Food Science and Technology*, *54*(2), 493–499. http://dx.doi.org/10.1016/j.lwt.2013.06.027

García-Segovia, P., Andrés-Bello, A., & Martínez-Monzó, J. (2011). Rehydration of air-dried Shiitake mushroom (Lentinus edodes) caps: Comparison of conventional and vacuum water immersion processes. *LWT - Food Science and Technology*, *44*(2), 480–488. http://dx.doi.org/10.1016/j.lwt.2010.08.010

García, M. A., Alonso, J., & Melgar, M. J. (2015). Radiocaesium activity concentrations in macrofungi from Galicia (NW Spain): Influence of environmental and genetic factors. *Ecotoxicology and Environmental Safety*, *115*, 152–158. http://dx.doi.org/10.1016/j.ecoenv.2015.02.005

Heleno, S. A., Barros, L., Sousa, M. J., Martins, A., & Ferreira, I. C. F. R. (2010). Tocopherols composition of Portuguese wild mushrooms with antioxidant capacity. *Food Chemistry*, *119*(4), 1443–1450. http://dx.doi.org/10.1016/j.foodchem.2009.09.025

ISO. (2004). ISO 4120:2004 - Sensory Analysis - Methodology - Triangle test.

Jeong, C. H., & Shim, K. H. (2004). Quality Characteristics of Sponge Cakes with Addition of Pleurotus eryngii Mushroom Powders. *Journal of the Korean Society of Food Science and Nutrition*, *33*(4), 716–722.

Kalač, P. (2001). A review of edible mushroom radioactivity. *Food Chemistry*, *75*(1), 29–35. http://dx.doi.org/10.1016/S0308-8146(01)00171-6

Kim, M.-Y., Chung, L.-M., Lee, S.-J., Ahn, J.-K., Kim, E.-H., Kim, M.-J., … Song, H.-K. (2009). Comparison of free amino acid, carbohydrates concentrations in Korean edible and medicinal mushrooms. *Food Chemistry*, *113*(2), 386–393. http://dx.doi.org/10.1016/j.foodchem.2008.07.045

Kotwaliwale, N., Bakane, P., & Verma, A. (2007). Changes in textural and optical properties of oyster mushroom during hot air drying. *Journal of Food Engineering*, *78*(4), 1207–1211. http://dx.doi.org/10.1016/j.jfoodeng.2005.12.033

Krokida, M. ., & Marinos-Kouris, D. (2003). Rehydration kinetics of dehydrated products. *Journal of Food Engineering, 57*(1), 1–7. http://dx.doi.org/10.1016/S0260-8774(02)00214-5

Lee, H.-S., & O'Mahony, M. (2005). Sensory evaluation and marketing: measurement of a consumer concept. *Food Quality and Preference, 16*(3), 227–235. http://dx.doi.org/10.1016/j.foodqual.2004.04.013

Li, P., Zhang, X., Hu, H., Sun, Y., Wang, Y., & Zhao, Y. (2013). High carbon dioxide and low oxygen storage effects on reactive oxygen species metabolism in Pleurotus eryngii. *Postharvest Biology and Technology, 85,* 141–146. http://dx.doi.org/10.1016/j.postharvbio.2013.05.006

Li, W., Gu, Z., Yang, Y., Zhou, S., Liu, Y., & Zhang, J. (2014). Non-volatile taste components of several cultivated mushrooms. *Food Chemistry, 143,* 427–431. http://dx.doi.org/10.1016/j.foodchem.2013.08.006

Li, X., Feng, T., Zhou, F., Zhou, S., Liu, Y., Li, W., ... Yang, Y. (2015). Effects of drying methods on the tasty compounds of Pleurotus eryngii. *Food Chemistry, 166,* 358–364. http://dx.doi.org/10.1016/j.foodchem.2014.06.049

MacLeod, A. J., & Panchasara, S. D. (1983). Volatile aroma components, particularly glucosinolate products, of cooked edible mushroom (Agaricus bisporus) and cooked dried mushroom. *Phytochemistry, 22*(3), 705–709. http://dx.doi.org/10.1016/S0031-9422(00)86966-6

Manzi, P., Marconi, S., Aguzzi, A., & Pizzoferrato, L. (2004). Commercial mushrooms: nutritional quality and effect of cooking. *Food Chemistry, 84*(2), 201–206. http://dx.doi.org/10.1016/S0308-8146(03)00202-4

Mendiburu, F. (2014). agricolae: Statistical Procedures for Agricultural Research. Retrieved from www.cran-r-project.org/package=agricolae

Mishra, K. K., Pal, R. S., ArunKumar, R., Chandrashekara, C., Jain, S. K., & Bhatt, J. C. (2013). Antioxidant properties of different edible mushroom species and increased bioconversion efficiency of Pleurotus eryngii using locally available casing materials. *Food Chemistry, 138*(2-3), 1557–1563. http://dx.doi.org/10.1016/j.foodchem.2012.12.001

Ozen, T., Darcan, C., Aktop, O., & Turkekul, I. (2011). Screening of Antioxidant, Antimicrobial Activities and Chemical Contents of Edible Mushrooms Wildly Grown in the Black Sea Region of Turkey. *Combinatorial Chemistry & High Throughput Screening, 14*(2), 72–84. http://dx.doi.org/10.2174/138620711794474079

Pereira, E., Barros, L., Martins, A., & Ferreira, I. C. F. R. (2012). Towards chemical and nutritional inventory of Portuguese wild edible mushrooms in different habitats. *Food Chemistry, 130*(2), 394–403. http://dx.doi.org/10.1016/j.foodchem.2011.07.057

Pogoń, K., Jaworska, G., Duda-Chodak, A., & Maciejaszek, I. (2013). Influence of the Culinary Treatment on the Quality of Lactarius deliciosus. *Foods, 2*(2), 238–253. http://dx.doi.org/10.3390/foods2020238

R Core Team. (2015). R: A Language and Environment for Statistical Computing. Vienna: R Foundation for Statistical Computing.

Raz, C., Piper, D., Haller, R., Nicod, H., Dusart, N., & Giboreau, A. (2008). From sensory marketing to sensory design: How to drive formulation using consumers' input? *Food Quality and Preference, 19*(8), 719–726. http://dx.doi.org/10.1016/j.foodqual.2008.04.003

Reyes, A., Mahn, A., Cubillos, F., & Huenulaf, P. (2013). Mushroom dehydration in a hybrid-solar dryer. *Energy Conversion and Management, 70,* 31–39. http://dx.doi.org/10.1016/j.enconman.2013.01.032

Reyes, A., Mahn, A., & Vásquez, F. (2014). Mushrooms dehydration in a hybrid-solar dryer, using a phase change material. *Energy Conversion and Management, 83,* 241–248. http://dx.doi.org/10.1016/j.enconman.2014.03.077

Rivera, C. S., Blanco, D., Salvador, M. L. M. L., & Venturini, M. E. M. E. (2010). Shelf-life extension of fresh Tuber aestivum and Tuber melanosporum truffles by modified atmosphere packaging with microperforated films. *Journal of Food Science, 75*(4), 225–233. http://dx.doi.org/10.1111/j.1750-3841.2010.01602.x

Severoglu, Z., Sumer, S., Yalcin, B., Leblebici, Z., & Aksoy, A. (2013). Trace metal levels in edible wild fungi. *International Journal of Environmental Science and Technology, 10*(2), 295–304. http://dx.doi.org/10.1007/s13762-012-0139-2

Stone, H., & Sidel, J. L. (2009). Food Science and Technology International Series. In *Wine Tasting* (Vol. 28 Suppl 1, pp. 485–487). Elsevier. http://dx.doi.org/10.1016/B978-0-12-374181-3.00017-X

Stone, H., Sidel, J., Oliver, S., Woolsey, A., & Singleton, R. C. (2004). Sensory Evaluation by Quantitative Descriptive Analysis. In M. C. Gacula (Ed.), *Descriptive Sensory Analysis in Practice* (pp. 23–34). Trumbull, Connecticut, USA: Food & Nutrition Press, Inc. http://dx.doi.org/10.1002/9780470385036.ch1c

Sung, S.-Y., Kim, M.-H., & Kang, M.-Y. (2008). Quality Characteristics of Noodles Containing Pleurotus eryngii. *Korean Journal of Food and Cookery Science, 24*(4), 405–411.

Valverde, M. E., Hernández-Pérez, T., & Paredes-López, O. (2015). Edible Mushrooms: Improving Human Health and Promoting Quality Life. *International Journal of Microbiology, 2015*, 1–14. http://dx.doi.org/10.1155/2015/376387

7

Bacteriology and Meat Quality of Moisture Enhanced Pork from Retail Markets in Canada

Karola R. Wendler[1,2], Francis M. Nattress[1,3], Jordan C. Roberts[1], Ivy L. Larsen[1] & Jennifer L. Aalhus[1]

[1]Agriculture & Agri-Food Canada, Lacombe Research Centre, Canada

[2]Present address: Delacon Biotechnik GmbH, Weissenwollfstr. Steyregg, Austria

[3]Retired

Correspondence: Jennifer Aalhus, Lacombe Research and Development Centre, Agriculture & Agri-Food Canada, 6000 C & E Trail, Lacombe, AB, T4L 1W1, Canada. E-mail: jennifer.aalhus@arg.gc.ca

Abstract

Packages of moisture-enhanced and conventional pork chops were collected from six Canadian retail stores on five sampling days. The composition of injection brines differed between retailers, but all contained polyphosphates and salt as main ingredients. Meat quality characteristics and bacteriology were analyzed from collected meat samples. Moisture enhanced chops had a higher pH and a higher water holding capacity than conventional. Juiciness and overall tenderness were improved in moisture enhanced chops. The surfaces of moisture enhanced chops were discoloured; the chops were darker and displayed less colour saturation. Total numbers of aerobes, psychrotrophs and lactic acid bacteria were not affected by moisture enhancement but numbers of *Enterobacteriaceae*, pseudomonads and *Brochothrix thermosphacta*, bacteria frequently associated with microbial spoilage, were approximately 1 log CFU·g^{-1} higher in moisture enhanced samples. This work shows moisture enhancement with injection brines containing salt and phosphates can result in a more palatable product.

Keywords: pork, moisture enhancement, retail survey, meat quality

1. Introduction

Moisture enhancement involves the injection of brine solutions containing polyphosphates and salt into meat; these procedures have been widely used in the meat industry to improve water holding capacity (Bendall, 1954; Detienne & Wicker, 1999) and eating quality of meat products (Cannon, McKeith, Martin, Novakofski, & Carr, 1993; Jones, Carr, & McKeith, 1987; Prestat et al., 2002b; Vote et al., 2000). A major portion of pork produced in the United States and some in Canada undergoes moisture enhancement procedures (Mandell & McEwen, 2011). Moisture enhancement of fresh pork is thought to improve eating quality by increasing juiciness and tenderness (Brewer, Gusse, & McKeith, 1999; Sheard, Nute, Richardson, Perry & Taylor, 1999; Smith, Simmons, McKeith, Bechtel, & Brady, 1984; Wynveen et al., 2001). Additionally, injection of brine may improve tolerance to some cooking abuse (Baublits, Meullenet, Sawyer, Mehaffey & Saha, 2006).

Although the needle injection applied during the moisture enhancement process appears to improve meat palatability, this process disrupts the integrity of the meat surface. Such disruptions carry a risk of introducing bacteria from the meat surfaces into the interior of the muscle. Similar contamination has been shown for the blade tenderization of beef inoculated with *E. coli* O157:H7 (Phebus, Mardsen, Thippareddi, Sporing, & Ortega, 2002), for needle-injected pork inoculated with *Salmonella* spp. (Kastner et al., 2001), and for needle-injected chicken breast inoculated with *Clostridium perfringens* (Mead & Adams, 1979). However, there is only limited information on the bacteriology of moisture-enhanced meat cuts at retail, and how bacterial counts compare to conventional meat (Banks, Wang, & Brewer, 1998; Bohaychuk & Greer, 2003; Wen, Li, & Dickson, 2014) and further research regarding the food safety and bacterial spoilage of moisture-enhanced products is necessary.

The aim of the present study, therefore, was to determine meat quality of conventional and moisture-enhanced pork products of different suppliers available at Canadian retail outlets and to enumerate bacteria in raw meat.

2. Method

2.1 Sampling of Pork Products

Fifty packages of moisture-enhanced and 38 packages conventional pork chops were purchased from six Canadian stores (3 retailers) on five sampling days.   Stores 1-4 were from Western Canada and stores 5-6 were from Central Canada. The pork chops were boneless thick, grill-style chops and all products were in-store packaged. One retailer, however, did not offer boneless conventional chops; therefore only moisture enhanced chops from that store were included in analyses. Moisture-enhanced chops differed in ingredients and volume of injection brine, according to the specifications of the manufacturers, but all contained polyphosphates and salt as main ingredients. Purchased meat was transported to the Agriculture and Agri-Food Research and Development Centre, Lacombe, AB, Canada and stored overnight at 2°C. Samples were subsequently analysed for physico-chemical, sensory, and bacteriological meat quality characteristics.

2.2 Physico-chemical Meat Quality

Meat colour was measured using a Minolta Chroma Meter CR 300 (Minolta Canada Inc., Mississauga, ON, Canada), recording CIE L* (brightness), a* (red-green axis), and b* (yellow-blue axis) values (CIE, 1976). Results were transformed to hue angle (H_{ab} = arctan [b*/a*]) and chroma (C_{ab} = [a*^{2} + b*^{2}]$^{0.5}$). Chops were subjectively evaluated for muscle colour (5-point scale: 1 = extremely pale, 5 = extremely dark), surface discolouration (7-point scale: 1 = no surface discolouration, 7 = complete surface discolouration), and retail appearance (7-point scale: 1 = extremely undesirable, 7 = extremely desirable) by three experienced raters; these evaluation scales have been previously used by Nattress and Baker (2003).   Temperature adjusted pH was measured using an Accumet ATC temperature probe (Fisher Scientific, Edmonton, AB, Canada) and an Accumet 1002 pH meter equipped with an Orion Ingold electrode (Urdorf, Switzerland).   Meat was homogenized using a Robot Coupe Blixer BX3 (Robot Coupe USA Inc., Jackson, MS, USA) and homogenates were analysed for expressible juice (a measure of water holding capacity) by centrifugation at 16,000 RPM for 15 min at 2°C. Moisture content of homogenized meat was determined as the weight loss after heating samples at 102°C for 24 h (Juárez et al., 2009). Crude protein content was first measured from dried samples using a Leco Nitrogen/Protein Determinator CN2000 (Leco Corp. St. Joseph, MI, USA), protein content from fresh meat was then calculated (Method 992.15) (Association of Official Analytical Chemists [AOAC], 1995b). Crude fat was extracted from dried samples at 105°C using a Soxtec Extraction Unit/Service Unit HT-1043 (FOSS Tecator, Höganäs, Sweden) and petroleum ether as the solvent (Method 960.39) (AOAC, 1995a).

Two chops out of each package were grilled to an internal temperature of 71°C on an electric grill (surface temperature 210°C) for determination of maximum shear force and sensory evaluation following procedures described by Juárez et al. (2009). Briefly, cylindrical cores (1.9 cm in diameter) were removed from one cooked chop parallel to the longitudinal axis of the muscle fibre and cores were sheared perpendicular to the muscle fibre with a Warner-Bratzler Shear head attached to an Instron Material Testing System (Model 4301, Instron; Burlington, ON, Canada).

The second chop was trimmed of all fat and cut into 1.9 cm cubes. Trained panellists rated cooked samples for initial and overall tenderness, initial and sustained juiciness, pork and other flavour intensities using a nine-point descriptive scale (9 = extremely tender, extremely juicy, extremely intense for pork flavour, extremely intense for other flavours; 1 = extremely tough, extremely dry, extremely bland for pork flavour, extremely bland for other flavour). Flavour desirability and overall palatability were rated using nine-point hedonic scale (9 = extremely desirable; 1 = extremely undesirable); these factors have been previously used to characterize pork sensory characteristics by Prieto et al. (2015).

2.3 Bacteriological Analysis

Samples (10 g) of surface and subsurface muscle from chops were homogenized using a Stomacher Lab-Blender Model "400" (Seward Laboratory, London, England) following Nattress and Jeremiah (2000). Dilutions used for plating, selective media, and incubation procedures described by Nattress and Jeremiah (2000) were used to enumerate: pseudomonads, *Brochothrix thermosphacta*, lactic acid bacteria (LAB), and *Enterobacteriaceae*. Total aerobic bacteria and psychotrophs were enumerated on Plate Count Agar and plates were incubated for 2 d at 25°C and 10 at 4°C respectively (Difco Laboratories, Becton Dickinson Microbiology Systems, Sparks, MD, USA). The lower limit for enumeration of *Enterobacteriaceae* was log 1 cfu·g^{-1} and log 2 cfu·g^{-1} for other bacteria.

2.4 Statistical Analysis

Statistical analyses were carried out with SAS (SAS, 2009) applying a mixed model procedure with meat type (conventional or moisture enhanced) as a fixed factor; sampling day and store were included as a random factors. LS-Means were tested for significant differences by Scheffe's multiple comparison. Where appropriate, Pearson's correlation values were calculated amongst factors attempting to establish relationships of quality measures. Additionally, summary statistics (minimum, maximum, and mean) were calculated for bacteria counts from each store.

3. Results

3.1 Physico-chemical Meat Quality

Overall, the majority of the physico-chemical meat quality traits were significantly different (P < 0.05) between moisture-enhanced and conventional samples of retail pork (Table 1). Moisture-enhanced pork had a higher pH (P<0.001), a lower expressible juice (P = 0.002), and lower shear force (P = 0.001) (0.5 kg lower) than conventional samples.

Objective measures of colour showed enhanced chops were darker (L*: P < 0.001), meat colour was less saturated (Chroma: P < 0.001), and less red (Hue: P = 0.049) than in conventional chops. Moisture-enhanced pork was also rated higher in subjective scoring for surface discoloration during retail display vs. conventional pork (P=0.005). Additionally, subjective retail appearance of moisture enhanced pork was less desirable (P=0.016).

Moisture-enhanced pork contained more water (P < 0.001) and less protein (P < 0.001) than conventional pork. Fat content was not significantly different between moisture enhanced and conventional chops, however, tended to be numerically lower in moisture-enhanced chops (P = 0.116). From the chops cooked for the taste panel, cooking losses were significantly lower for moisture enhanced pork vs. conventional pork (P = 0.035).

There was significant variation in many of the physico-chemical traits amongst stores reflecting a range in pork quality which included variations in pH, colour attributes and moisture content of the muscle.

Table 1. Physico-chemical meat quality characteristics

Physico-chemical characteristic	Control		Moisture Enhanced		
	mean	sem[c]	mean	sem	P
Shear (kg)	4.85[a]	0.12	4.31[b]	0.11	0.001
pH	5.84[b]	0.04	6.07[a]	0.03	<0.001
Expressible Juice mg·g^{-1}	93.2[a]	6.1	67.3[b]	5.3	0.002
Cook loss mg·g^{-1}	270.4[a]	7.3	249.7[b]	6.3	0.035
Objective Colour					
L*	53.70[a]	0.62	49.90[b]	0.54	<0.001
Chroma	11.23[a]	0.31	9.11[b]	0.27	<0.001
Hue	40.13[a]	1.00	37.47[b]	0.88	0.049
Subjective Retail Scores					
Surface colour	2.94[b]	0.06	3.35[a]	0.06	<.0001
Surface discoloration	1.63[b]	0.14	2.16[a]	0.12	0.005
Retail appearance	5.20[a]	0.18	4.60[b]	0.16	0.016
Crude Nutrient Composition					
Moisture, mg.g^{-1}	724.5[b]	1.5	744.3[a]	1.3	<0.001
Crude fat, mg.g^{-1}	21.7	1.7	18.1	1.5	0.116
Crude protein, mg.g^{-1}	251.0[a]	1.4	228.2[b]	1.2	<0.001

[a,b]Least squares means (LSMean) within a row without a common superscript were significantly different (*P* < 0.05)

[c]Standard error of means

3.2 Sensory Meat Quality

Overall palatability of the pork chops was improved by moisture-enhancement (P<0.001) (Table 2). Moisture-enhanced pork was also rated as more tender and juicier than conventional pork. Pork flavour intensity however, was significantly lower (P<0.001) and other flavour intensity was significantly higher in enhanced vs. conventional chops (P<0.001) (Table 2).

Table 2. Panel ratings for moisture enhanced and conventional pork

	Control		Moisture Enhanced		
Sensory characteristic	mean	sem[c]	mean	sem	P
Initial tenderness	4.82[b]	0.17	6.48[a]	0.15	<0.001
Overall tenderness	5.02[b]	0.15	6.44[a]	0.13	<0.001
Initial juiciness	4.48[b]	0.15	5.72[a]	0.13	<0.001
Sustained juiciness	4.28[b]	0.14	5.61[a]	0.12	<0.001
Pork flavour intensity	5.49[a]	0.17	3.54[b]	0.15	<0.001
Other flavour intensity	1.66[b]	0.26	4.69[a]	0.23	<0.001
Overall palatability	4.86[b]	0.11	5.58[a]	0.10	<0.001

[a,b]Least squares means (LSMean) within a row without a common superscript were significantly different ($P < 0.05$)

[c]Standard error of means

3.3 Bacteriology

The numbers of aerobes, psychrotrophs and lactic acid bacteria were not significantly different between conventional and moisture enhanced pork (Table 3), with values ranging from 3.0 log $CFU \cdot g^{-1}$ to 7.8 log $CFU \cdot g^{-1}$. *Enterobacteriaceae* and pseudomonads, were significantly higher in moisture enhanced vs. conventional pork (*Enterobacteriaceae* P<0.001, pseudomonads P=0.002), however the differences were less than 1 log $CFU \cdot g^{-1}$ (Table 3). The numbers of *B. thermosphacta* were 1 log $CFU \cdot g^{-1}$ higher in moisture enhanced than in conventional chops (*B. thermosphacta* P = 0.002). In addition to differences between moisture enhanced and conventional chops, mean bacterial counts were different between stores (Table 3). Store 5 had lower numbers of all groups of bacteria, while stores 2 and 4 tended to have higher numbers of bacteria.

Table 3. Bacteria counts log (cfu·g^{-1}) for moisture enhanced and conventional pork

Set type		N	aerobes	Pseudomonads	*B. thermosphacta*	lactic acid bacteria	psychrotrophs	*Enterobacteriaceae*
Pork	C[c]	38	5.18	2.96[b]	3.18[b]	5.09	5.18	2.10[b]
	M[d]	50	5.31	3.82[a]	4.18[a]	5.20	5.22	2.94[a]
Store	1	14	5.47	3.56	3.93	5.44	5.42	3.05
	2	16	5.83	4.29	4.78	5.79	5.78	3.40
	3	16	4.89	3.11	3.54	4.72	5.01	2.43
	4	14	5.96	3.26	3.69	5.90	5.85	3.19
	5	18	4.48	3.07	2.82	4.36	4.38	1.35
	6	10	5.00	3.44	3.92	4.80	4.84	2.17

[a,b]Least squares means (LSMean) between C and M without a common superscript were significantly different ($P < 0.05$)

[c]conventional pork

[d]moisture enhanced

4. Discussion

4.1 Physico-chemical Characteristics and Eating Quality

Meat pH was significantly higher in enhanced chops than in conventional; this is similar to effects reported in pork from previous studies involving polyphosphate injections (Detienne & Wicker, 1999; Smith et al., 1984). Higher pH, further from the isoelectric point of the myofibril proteins, results in a greater charge allowing higher levels of water binding to the protein matrix (Huff-Lonergan & Lonergan, 2005).

However, there was some variability in pH of enhanced products among the different retailers (ranging from 5.99-6.37); this may have been caused by differences in final concentration of polyphosphates in meat. These differences could result from either lower concentration of polyphosphate in injection brines or a lower injection volume. Injection volume can vary according to the recipes of manufacturers, but usually, meat is injected between 7% to 10 % of fresh weight (Miller, 1998). At higher injection volumes, retail appearance can be impaired due to higher package purge; this has been shown for loins pumped to 12% or 18% of fresh weight (Brashear, Brewer, Meisinger, & McKeith, 2002).

Moisture enhanced pork also had higher moisture and lower protein content in the present study. It has been shown the composition of brine injections is important when considering changes in water and protein content; for example, injection of pure water, does not increase water content in meat as a large proportion of injected water is lost as package purge (Sheard et al., 1999). Higher moisture content in enhanced chops, therefore, is likely the result of less package purge and higher water holding capacity of myofibril proteins due to the increase in meat pH and ionic strength through the addition of polyphosphates and salt. As well, Offer and Trinick (1983) reported a swelling of myofibril proteins following exposure to polyphosphates due to phosphate binding to myosin, preventing cross-bridge formation.

Moisture enhanced pork in the present study had a lower expressible juice. Similarly, Cannon et al. (1993) reported significantly less free water and lower purge loss after marinating pork with sodium tripolyphosphate. In addition to the effects of a higher pH and phosphate binding to myosin, salt also leads to swelling of muscle fibres by increasing ionic strength (Offer & Trinick, 1983). An interaction of salt and polyphosphate in the injection brines for chicken meat has been reported to improve water holding capacity (Shults & Wierbicki, 1973). Additionally, packaging purge was lower for pork injected with combinations of sodium tripolyphosphates and salt, rather than a single ingredient alone, confirming the additive effect on water holding capacity (Detienne & Wicker, 1999). Thus, salt included in injection brines likely would have contributed to the low proportion of expressible juice in moisture-enhanced pork of the present study.

Moisture enhanced pork had significantly lower cooking losses than conventional pork. In agreement with the present results, pork loins injected with combinations of salt and tripolyphosphate (Detienne & Wicker, 1999; Apple, Dikeman, Simms, & Kuhl, 1991) or marinated with tripolyphsophate (Cannon et al., 1993) have been shown to have lower cooking losses than conventional meat. The reported effects of polyphosphates on cook loss in pork, however, are not entirely consistent; cooking loss has also been shown to slightly decrease (Sheard et al., 1999) or display no change after polyphosphate injection (Sutton, Brewer, & McKeith, 1997).

Moisture enhancement also had an effect on sensory quality traits of pork chops. The improved moisture content of the enhanced chops was reflected in higher taste panel scores for juiciness. In accordance, palatability was closely correlated with initial juiciness (r = 0.76), sustained juiciness (r = 0.79), initial tenderness (r = 0.73), and overall tenderness (r = 0.77). Improved perception of juiciness for moisture-enhanced pork is in agreement with Cannon et al. (1993), where pork was marinated with polyphosphate. Additionally, cook loss was moderately correlated with panel scores for initial (r = -0.64), and sustained juiciness (r = -0.63), while the correlations between moisture content in chops and subjective juiciness attributes were weaker (initial r = 0.42, sustained juiciness r = 0.51). Thus, higher moisture content in meat does not necessarily improve juiciness if water does not remain in the meat during the cooking process.

Improved tenderness in pork treated with dilutions of polyphosphates, with or without salt, has also been reported in previous studies (Cannon et al., 1993; Sheard et al., 1999; Smith et al., 1984). The higher taste-panel scores for tenderness in moisture enhanced pork for the present study corresponded with the lower shear force values of enhanced chops. Further, subjective scores for overall tenderness, were moderately correlated with shear force (r = -0.51). The hypothesized mechanisms for improvement in tenderness by polyphosphates are similar to those affecting moisture content (i.e. changes to pH and ionic strength). The swelling of the muscle fibres caused by increased pH and ionic strength (Bendall, 1954) lead to disruption of myofibril protein structure and thus increased tenderness (Offer & Trinick, 1983; Voyle, Jolley, & Offer, 1986).

Tripolyphosphate use in injection brines has been observed to decrease pork flavour intensity and increase abnormal flavour intensity in pork (Sheard et al., 1999). A similar effect was noted in the present study, where enhanced chops were rated lower for pork flavour but higher for other flavour intensity. Other authors, however, found no effect of polyphosphate injection on off-flavour and pork flavour intensity (Brewer et al., 1999; Smith et al., 1984). Thus, it cannot be excluded that differences in flavour obtained in the present study were caused by ingredients of the injection brines other than polyphosphates.

Objective meat colour characteristics were affected by moisture enhancement. Enhanced pork chops were lower in L*-value and chroma, additionally, surface colour was subjectively rated as darker. These results are in agreement with some previous studies (Banks et al., 1998; Glaeser, Nattress, Greer, Gibson, & Aalhus, 2005; Prestat, Jensen, McKeith, & Brewer, 2002a). In contrast, Sutton et al. (1997) found polyphosphate injection had no effect on meat colour of pork. The higher water holding capacity of enhanced meat is thought to result in less light scattering on free water molecules and thus darker meat (Fernández-López, Sayas-Barberá, Pérez-Alvarez, & Aranda-Catalá, 2004). Additionally, the lower chroma of enhanced chops has been associated with changes to the oxidation state of myoglobin in higher salt concentrations (Fernández-López et al., 2004).

Corresponding to the objective measures of lower saturation and darker surface colour, enhanced pork was given higher subjective scores for surface discolouration than conventional pork. Surface discolouration in injected chops was mainly the result of grey or dark surface areas along the path of the injection needles (Figure 1). So-called two-toned meat is one of the major quality defects identified in needle-injected fresh meat products (Miller, 1998) and is thought to be the result of disruption of myofibril structure by high salt concentration around the injection sites (Voyle et al., 1986). Meat colour has been identified as highly important for consumer choice (Dransfield et al., 2005). As such, higher discoloration in enhanced chops may negatively affect the purchase intent of consumers and will likely contribute to their perception of quality. Corresponding to this, enhanced chops in the present study were rated lower for retail appearance. It has been observed certain additives containing antioxidants (e.g.: hydrolyzed plasma protein or oregano oil) when included in injections brines, improve measures of colour stability relative to loins only injected with salt and polyphosphate (Scramlin et al., 2010; H. Seo, J. Seo, & Yang, 2016). However a detailed study on methods to alleviate striping, including changes to the brine, enhancement level, or tumbling following injection, was unsuccessful (Gooding et al., 2009).

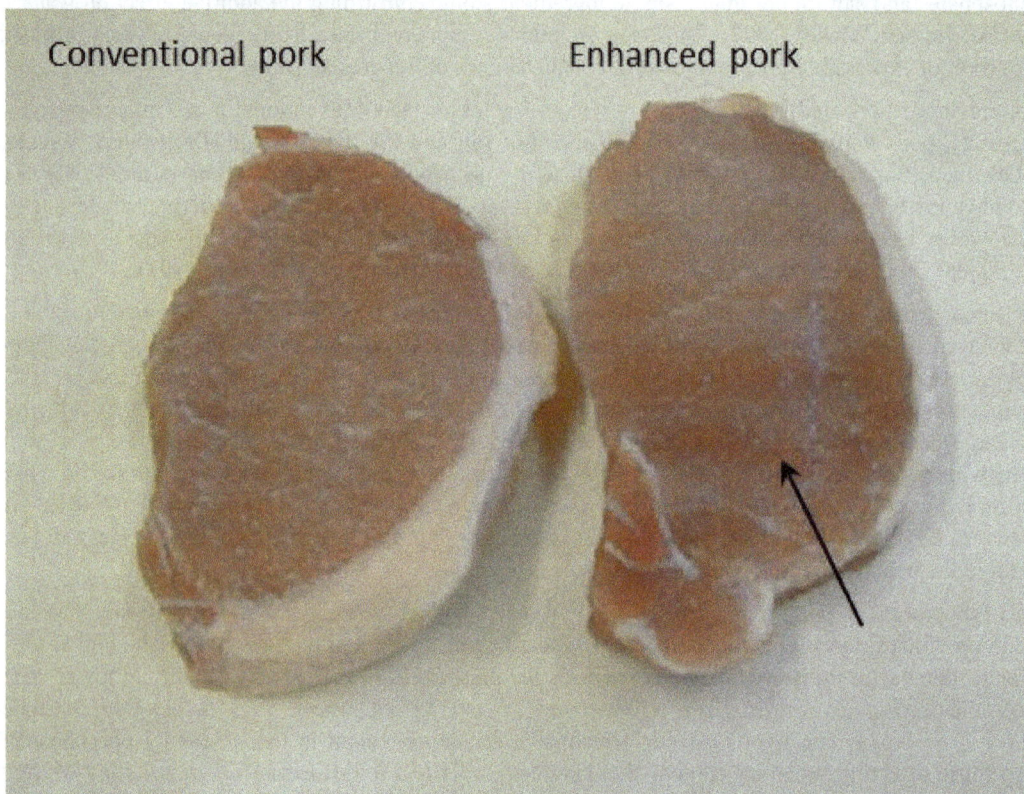

Figure 1. Surface discoloration in moisture-enhanced pork chops. Arrow pointed at region of discolouration

4.2 Bacteriology

Bacterial load in injection brines have been shown to increase with processing time by up to 1.5 log $CFU \cdot ml^{-1}$ through the duration of the moisture-enhancement process (Greer, Nattress, Dilts, & Baker, 2004). Additionally, Bohaychuk and Greer (2003) in their analysis of commercial pork loins observed 2 log $CFU \cdot g^{-1}$ higher numbers of psychrotrophic bacteria in moisture-enhanced pork when compared to non-injected product; numbers of lactic acid bacteria, *Enterobacteriaceae* and *B. thermosphacta* were not different. However, in cooked chops the moisture enhancement process did not result in increased risk for survival of the pathogenic bacteria *Salmonella typhimurium* or *Campylobacter jejuni* (Wen et al., 2014). In the current study, there were no significant differences in numbers of aerobes, psychrotrophs and lactic acid bacteria but increases in numbers (0.84-1 log $CFU \cdot g^{-1}$) of *Enterobacteriaceae*, *B. thermosphacta* and pseudomonads were observed. Increases in these bacteria resulting from moisture enhancement could be associated with bacterial spoilage (Borch, Kant-Muermans, & Blixt, 1996). However, it is currently unknown whether the bacterial increases observed here would result in increases to spoilage of commercial importance. Storage trials would be required to determine whether these differences in bacterial levels would impact storage life of the meat. However, Bohaychuk and Greer (2003) concluded there was no evidence of a compromised storage life for moisture-enhanced pork loins when compared to non-injected pork.

There was some variability in bacterial counts between stores (Table 3), with the mean counts of *Enterobacteriaceae* and *B. thermosphacta* being most variable. The range in mean counts of all bacteria measured here between stores were greater than 1 log cfu·g^{-1}. This may indicate the hazard analysis critical control point (HACCP) procedures, among the retailers and suppliers could contribute to overall differences in meat bacterial levels.

5. Conclusions

Moisture enhanced pork had higher levels of some bacteria associated with spoilage. It is currently unknown how these increases to bacteria affect shelf-life of moisture enhanced pork. Future research should explore how spoilage rates are affected by moisture enhancement under different retail conditions.

Many factors associated with eating quality of pork were improved by moisture enhancement in the present study. The improvements to eating quality of moisture enhanced chops included subjective panel scores for juiciness, palatability, and tenderness. Additionally, moisture enhanced chops had a lower shear force. Despite the improvements to the textural factors associated with eating quality and higher overall palatability, enhanced chops were rated lower for pork flavour intensity and had higher ratings of other flavour intensity.

Moisture enhanced pork displayed a higher surface discolouration and lower colour saturation. Since colour is an important consideration when the consumer chooses a product, further means to reduce striping should be considered in producing moisture enhanced pork for retail markets.

Acknowledgments

Funding for this project was provided by the Canadian Meat Research Institute, Ontario Pork, Genex Swine Group and the AAFC Matching Investment Initiative. The authors gratefully acknowledge the valuable technical assistance of Lynda Baker, Bryan Dilts, Glynnis Croken, Fran Costello, Lorna Gibson, and Rhona Thacker. The authors also thank Mr. Bill Collier (Ontario Pork) for assistance with the meat sampling.

References

Apple, J. K., Dikeman, M. E., Simms, D. D., & Kuhl, G. (1991). Effects of synthetic hormone implants, singularly or in combinations, on performance, carcass traits, and longissimus muscle palatability of Holstein steers. *Journal of Animal Science, 69*(11), 4437-4448. http://dx.doi.org//1991.69114437x

Association of Official Analytical Chemists (1995a). Method 960.39. Fat (crude) or ether extract in meat final action. AOAC, Gaithersburg, MD, 39, 2.

Association of Official Analytical Chemists (1995b). Method 992.15. Crude protein in meat and meat products including pets. Combustion method. Final Action 1992. or Ether extract in meat final action. AOAC, Gaithersburg, MD, 39, 6-7.

Banks, W. T., Wang, C., & Brewer, M. S. (1998). Sodium lactate/sodium tripolyphosphate combination effects on aerobic plate counts, ph and color of fresh pork longissimus muscle. *Meat Science, 50*(4), 499-504. http://dx.doi.org/10.1016/S0309-1740(98)00064-3

Baublits, R. T., Meullenet, J. F., Sawyer, J. T., Mehaffey, J. M., & Saha, A. (2006). Pump rate and cooked temperature effects on pork loin instrumental, sensory descriptive and consumer-rated characteristics. *Meat Science, 72*(4), 741-750. http://dx.doi.org/10.1016/j.meatsci.2005.10.006

Bendall, J. R. (1954). The swelling effect of polyphosphates on lean meat. Journal of the Science of Food and Agriculture, 5(10), 468-475. http://dx.doi.org/10.1002/jsfa.2740051005

Bohaychuk, V. M., & Greer, G. G. (2003). Bacteriology and storage life of moisture-enhanced pork. *Journal of Food Protection, 66*, 293-299.

Borch, E., Kant-Muermans, M. L., & Blixt, Y. (1996). Specific spoilage organismsbacterial spoilage of meat and cured meat products. *International Journal of Food Microbiology, 33*(1), 103-120. http://dx.doi.org/10.1016/0168-1605(96)01135-X

Brashear, G., Brewer, M. S., Meisinger, D., & McKeith, F. K. (2002). Raw material pH, pump level and pump composition on quality characteristics of pork. *Journal of Muscle Foods, 13*(3), 189-204. http://dx.doi.org/10.1111/j.1745-4573.2002.tb00330.x

Brewer, M. S., Gusse, M., & McKeith, F. K. (1999). Effects of injection of a dilute phosphate-salt solution on pork characteristics from PSE, normal and DFD carcasses. *Journal of Food Quality, 22*(4), 375-385. http://dx.doi.org/10.1111/j.1745-4557.1999.tb00171.x

Cannon, J. E., McKeith, F. K., Martin, S. E., Novakofski, J., & Carr, T. R. (1993). Acceptability and shelf-life of marinated fresh and precooked pork. *Journal of Food Science, 58*(6), 1249-1253. http://dx.doi.org/10.1111/j.1365-2621.1993.tb06158.x

Cie. (1976). International commission on illumination, colorimetry. CIE Publication 15 colourimetry., Bureau Central de la CIE, Paris, France.

Detienne, N. A., & Wicker, L. (1999). Sodium chloride and tripolyphosphate effects on physical and quality characteristics of injected pork loins. *Journal of Food Science, 64*(6), 1042-1047. http://dx.doi.org/10.1111/j.1365-2621.1999.tb12278.x

Dransfield, E., Ngapo, T. M., Nielsen, N. A., Bredahl, L., Sjödén, P. O., Magnusson, M., ... Nute, G. R. (2005). Consumer choice and suggested price for pork as influenced by its appearance, taste and information concerning country of origin and organic pig production. *Meat Science, 69*(1), 61-70. http://dx.doi.org/10.1016/j.meatsci.2004.06.006

Fernández-López, J., Sayas-Barberá, E., Pérez-Alvarez, J. A., & Aranda-Catalá, V. (2004). Effect of sodium chloride, sodium tripolyphosphate and pH on color properties of pork meat. *Color Research and Application, 29*(1), 67-74. http://dx.doi.org/10.1002/col.10215

Glaeser, K. R., Nattress, F. M., Greer, G. G., Gibson, L. L., & Aalhus, J. L. (2005). Quality and bacteriology of moisture-enhanced and conventional pork products available at retail markets in Western Canada. Canadian Meat Science Association News, 03/2005, 7-12.

Gooding, J. P., Holmer, S. F., Carr, S. N., Rincker, P. J., Carr, T. R., Brewer, M. S., ... Kilefer, J. (2009). Characterization of striping in fresh, enhanced pork loins. Meat Science, 81(2), 364-371.

Greer, G. G., Nattress, F., Dilts, B., & Baker, L. (2004). Bacterial contamination of recirculating brine used in the commercial production of moisture-enhanced pork. *Journal of Food Protection, 67*(1), 185-188.

Huff-Lonergan, E., & Lonergan, S. M. (2005). Mechanisms of water-holding capacity of meat: The role of postmortem biochemical and structural changes. *Meat Science, 71*(1), 194-204. http://dx.doi.org/10.1016/j.meatsci.2005.04.022

Jones, S. L., Carr, T. R., & McKeith, F. K. (1987). Palatability and storage characteristics of precooked pork roasts. *Journal of Food Science, 52*(2), 279-281. http://dx.doi.org/10.1111/j.1365-2621.1987.tb06592.x

Juárez, M., Caine, W. R., Larsen, I. L., Robertson, W. M., Dugan, M. E. R., & Aalhus, J. L. (2009). Enhancing pork loin quality attributes through genotype, chilling method and ageing time. *Meat Science, 83*(3), 447-453. http://dx.doi.org/10.1016/j.meatsci.2009.06.016

Kastner, C. L., Phebus, R. K., Thippareddi, H., Mardsen, J. L., Karr Getty, K. J., Danler, B., ... Schwenke, J. R. (2001). Meat technology and processing - Current issues and trends. 47th ICoMST 2001, 68-71.

Mandell, I., & McEwen, P. (2011). A literature review on the effects of moisture enhancement on pork quality. Ontario Pork.

Mead, G., & Adams, B. (1979). Microbiological aspects of polyphosphate injection in the processing and chill storage of poultry. *Journal of Hygiene, 82*, 133-142.

Miller, R. (1998). Functionality of non-meat ingredients used in enhanced pork. National Pork Board, Des Moines, IA.

Nattress, F. M., & Baker, L. P. (2003). Effects of treatment with lysozyme and nisin on the microflora and sensory properties of commercial pork. *International Journal of Food Microbiology, 85*(3), 259-267. http://dx.doi.org/10.1016/S0168-1605(02)00545-7

Nattress, F. M., & Jeremiah, L. E. (2000). Bacterial mediated off-flavours in retail-ready beef after storage in controlled atmospheres. *Food Research International, 33*(9), 743-748. http://dx.doi.org/10.1016/S0963-9969(00)00064-8

Offer, G., & Trinick, J. (1983). On the mechanism of water holding in meat: The swelling and shrinking of myofibrils. *Meat Science, 8*(4), 245-281. http://dx.doi.org/10.1016/0309-1740(83)90013-X

Phebus, R. K., Mardsen, J. L., Thippareddi, H., Sporing, S., & Ortega, T. (2002). Escherichia coli O157:H7 risk assessment for production and cooking of blade tenderized beef steaks. (USDA-FISIS Public meeting on E.

coli O157:H7 policy, February 29, 2000, Arlington, Virginia). 1-4.

Prestat, C., Jensen, J., McKeith, F. K., & Brewer, M. S. (2002a). Cooking method and endpoint temperature effects on sensory and color characteristics of pumped pork loin chops. *Meat Science, 60*(4), 395-400. http://dx.doi.org/10.1016/S0309-1740(01)00150-4

Prestat, C., Jensen, J., Robbins, K., Ryan, K., Zhu, L., McKeith, F. K., & Brewer, M. S. (2002b). Physical and sensory characteristics of precooked, reheated pork chops with enhancement solutions. *Journal of Muscle Foods, 13*(1), 37-51. http://dx.doi.org/10.1111/j.1745-4573.2002.tb00319.x

Prieto, N., Juárez, M., Larsen, I. L., López-Campos, Ó., Zijlstra, R. T., & Aalhus, J. L. (2015). Rapid discrimination of enhanced quality pork by visible and near infrared spectroscopy. *Meat Science, 110*, 76-84. http://dx.doi.org/10.1016/j.meatsci.2015.07.006

SAS. (2009). SAS User's guide: Statistics. SAS for Windows, Version 9.1., SAS Institute Inc., Cary, NC.

Scramlin, S. M., Newman, M. C., Cox, R. B., Sepe, H. A., Alderton, A. L., O'leary, J., & Mikel, W. B. (2010). Effects of oregano oil brine enhancement on quality attributes of beef longissimus dorsi and semimembranosus muscles from various age animals. *Journal of Food Science, 75*, S89-S94. http://dx.doi.org/10.1111/j.1750-3841.2009.01459.x

Seo, H. W., Seo, J. K., & Yang, H. S. (2016). Effects of injection of hydrolysis plasma protein solution on the antioxidant properties in porcine M. Longissimus Lumborum. *Journal of Animal Science and Technology, 58*, 31. http://dx.doi.org /10.1186/s40781-016-0111-7

Sheard, P. R., Nute, G. R., Richardson, R. I., Perry, A., & Taylor, A. A. (1999). Injection of water and polyphosphate into pork to improve juiciness and tenderness after cooking. *Meat Science, 51*(4), 371-376. http://dx.doi.org/10.1016/S0309-1740(98)00136-3

Shults, G. W., & Wierbicki, E. (1973). Effects of sodium chloride and condensed phosphates on the water-holding capacity, ph and swelling of chicken muscle. *Journal of Food Science, 38*(6), 991-994. http://dx.doi.org/10.1111/j.1365-2621.1973.tb02131.x

Smith, L. A., Simmons, S. L., McKeith, F. K., Bechtel, P. J., & Brady, P. L. (1984). Effects of sodium tripolyphosphate on physical and sensory properties of beef and pork roasts. *Journal of Food Science, 49*(6), 1636-1637. 10.1111/j.1365-2621.1984.tb12869.x

Sutton, D. S., Brewer, M. S., & McKeith, F. K. (1997). Effects of sodium lactate and sodium phosphate on the physical and sensory characteristics of pumped pork loins. *Journal of Muscle Foods, 8*(1), 95-104. http://dx.doi.org/10.1111/j.1745-4573.1997.tb00380.x

Vote, D. J., Platter, W. J., Tatum, J. D., Schmidt, G. R., Belk, K. E., Smith, G. C., & Speer, N. C. (2000). Injection of beef strip loins with solutions containing sodium tripolyphosphate, sodium lactate, and sodium chloride to enhance palatability. *Journal of Animal Science, 78*(4), 952-957. http://dx.doi.org/2000.784952x

Voyle, C., Jolley, P., & Offer, G. (1986). Microscopical observations on the structure of bacon. *Food microstructure, 5*, 63-70.

Wen, X., Li, J., & Dickson, J. S. (2014). Generalized linear mixed model analysis of risk factors for contamination of moisture-enhanced pork with Campylobacter jejuni and Salmonella enterica typhimurium. *Foodborne Pathogens and Disease, 11*(10), 808-814. http://dx.doi.org/10.1089/fpd.2014.1762

Wynveen, E. J., Bowker, B. C., Grant, A. L., Lamkey, J. W., Fennewald, K. J., Henson, L., & Gerrard, D. E. (2001). Pork quality is affected by early postmortem phosphate and bicarbonate injection. *Journal of Food Science, 66*(6), 886-891. http://dx.doi.org/10.1111/j.1365-2621.2001.tb15191.x

8

Properties of Extrusion Processed Corn and Corn Coproducts

Jordan J. Rich[1] & Kurt A. Rosentrater[1]

[1]Department of Agricultural and Biosystems Engineering, Iowa State University, 3327 Elings Hall, Ames, IA 50011, USA

Correspondence: Kurt A. Rosentrater, Department of Agricultural and Biosystems Engineering, Iowa State University, 3327 Elings Hall, Ames, IA 50011, USA. Email: karosent@iastate.edu

Abstract

As the world population continues to grow, the demand for human food and animal feed grows exponentially. Aquaculture is the food sector which has been growing at the greatest rate for several years. Because of the expense of fishmeal in aquaculture fees, an inexpensive protein source could be corn-based proteins. Although many studies have focused on the effects of extruding corn-based blends along with other supplement ingredients, few studies have focused on the extrusion of individual corn-based ingredients. This study examined physical effects of extrusion on distillers dried grains with soluble (DDGS) and corn. Specific objectives included determining moisture content, water activity, color, unit density, durability, water stability, floatability, and bulk density for each corn-based extrudate. Blends were prepared with three levels of moisture (15, 25, and 35% db), and extrusion conditions included three screw speeds (50, 75, and 100 rpm) and three barrel temperatures (100, 125, and 150°C). Results showed that as the moisture content increased, the water activity increased in the raw ingredients, and the moisture content of the extrudates increased. As the screw speed increased, the bulk density decreased in the extrudates, and the mass flow rate increased. As the temperature increased, the floatability of the extrudates increased, while the bulk density decreased. The amount of protein and starch content in the corn products affected the physical quality of the pellets, which is important in aquaculture feed development.

Keywords: extrusion, corn, DDGS, aquaculture, feed

1. Introduction

As the world's population continues to grow exponentially, the need for efficient food and feed production increases each day. Snack foods and breakfast cereals are among many food industries that have grown in the United States, and many of these are produced through extrusion (Razzaq et al., 2012).

Along with the human food industry, aquaculture is one of the fastest growing food production activities globally. It plays a major role in many third world countries because it produces relatively higher income, better nutrition, and better employment opportunities (Kannadhason et al., 2008). Nutritional feed for aquaculture is costly, and the feed cost is generally 30% to 60% of total operational costs (Kannadhason et al., 2009, 2010), both due to raw ingredient costs as well as processing costs. Because feed is so expensive, it is difficult to provide the fish appropriate nutrients they need to thrive.

Extrusion has become a popular processing technique in the feed, cereal, and snack food industries (Delgado-Nieblas et al., 2012). Extrusion is one of the most versatile unit operations (Forsido and Ramaswamy, 2011). Extrusion systems consist of barrel housing with one or two rotating screws, a preconditioner, and an accompanying machine control system (Sorensen, 2012). Extrusion cooking uses high temperatures in a relatively short time to move the ingredients through the barrel and exit through the die. The goals of the extrusion process include cooking, sterilization, expansion, texturization, and product shaping. These factors are especially important in aquaculture feeds because porous pellets that float in water are required. Moreover, these are all important factors for human foods as well (Liu and Rosentrater, 2011).

During the extrusion process, starch is gelatinized, which plays an important role in the final extrudate properties (Chevanan et al., 2008a). Gelatinization is crucial because it affects feed digestibility, expansion, and contributes to water solubility and particle binding (Rosentrater et al., 2009a,b). Along with starch gelatinization, protein denaturation and destruction of microbes and other toxic compounds are other reactions that may also occur during cooking (Kannadhason et al., 2008). The sterilization process allows for the termination of pathogens, which is important in food products for humans, as well as fish (Liu and Rosentrater, 2011).

In the Midwest U.S., corn and soybean crops are the most prevalent for food production and processing. Both corn and soy are relatively high in protein, which is important for aquaculture feeds. According to a study performed at North Dakota State University, corn contains approximately 55-60% escape or bypass protein (Lardy, 2002). The United Soybean Board found soybeans contain 35-38% protein (The United Soybean Board [USB], 2012). Many types of concentrated and isolated meals have been produced for livestock feed. Using the high-protein feed produced for livestock has potential to be extruded for use in aquaculture feed.

Changes of the extrusion conditions, especially barrel temperature, screw speed, and moisture content can have significant effects on the resulting physical properties of the extrudates (Singh et al., 2012). Physical properties are important because they impact the quality of the extrudates. Moisture content is a determining factor for cohesiveness (Ayadi et al., 2011a). Water activity values indicate shelf life (Ayadi et al., 2011b). Because spoilage microorganisms thrive in water activity levels greater than 0.5, a value less than 0.5 is preferred. The greater the values of water activity, the more likely the microorganisms are to thrive, so a lower water activity value is preferred. Simons found that as extruder screw speed increases, water activity decreases (Simons et al., 2012). Color of the extrudate indicates, to some extent, the nutritional quality of the product (Kannadhason et al., 2010), as darker colors can indicate protein degradation. The mass flow rate quantifies the processing performance of the extruder (Ayadi et al., 2011c). Rosentrater found that an increase in both moisture content and temperature result in a decrease in die pressure (Rosentrater et al., 2009a,b). Chevanan found that increasing the moisture content resulted in an increase in pellet durability (Chevanan et al., 2007a). They also found that increasing the temperature resulted in a decrease in apparent viscosity and pellet durability. Chevanan also found that changes in screw speed had a significant effect on all extrusion parameters except for the temperature of the die (Chevanan et al., 2008b). They also found that increasing the screw speed resulted in a decrease in bulk density because of greater expansion (Chevanan et al., 2007b).

The ingredients used in feed extrusion must generally be high in protein, because aquaculture feeds typically required 26 to 50% protein, depending on the fish type, species, and age (Chevanan et al., 2007c). Because of the recent biofuel revolution, the production of ethanol has grown immensely across the United States. A coproduct of ethanol manufacturing is distillers dried grains with solubles (DDGS), which contains moderate levels of protein, fat, and fairly high levels of fiber. Because of the nutritional value provided by DDGS, it has become a key ingredient in feeds for many species, including ruminants and non-ruminants. Research on using DDGS in fish feeds is also promising (Liu and Rosentrater, 2011).

The purpose of this study was to evaluate the effects of extrusion conditions on DDGS, focusing on the effects of ingredient moisture content, extruder screw speed, and barrel temperature. In order to find optimal extrudate properties, extensive physical property testing was conducted after the extrusion process, and results were compared to corn – based extrudate properties.

2. Materials and Methods

2.1 Sample Preparation

Corn ingredients, including DDGS and corn (for comparison purposes), were ground to <1.0 mm particle size using a laboratory mill (Model 4, THOMAS-Wiley, Swedesboro, NJ, USA). The ingredients were then adjusted to three target moisture contents (15, 25, and 35% db) by adding appropriate amounts of water. The ingredients and water were mixed for approximately 20 min, using a small rotary mixer (Model 043206, Type B, Kobalt, Surrey, BC, Canada). The water was added slowly using a spray bottle to prevent agglomeration. The blends were stored overnight to allow for the distribution and equilibration of moisture at room temperature (25±1 °C).

2.2 Extrusion Processing

A single-screw extruder (Model PL2000, Type 680143, C. W. Brabender, South Hackensack, NJ, USA) was used for the cooking process. The screw compression ratio was 1:1, and the diameter of the die was 3.05 mm. The temperature of the feeding zone was room temperature (25±1°C), the metering zone was adjusted to 75°C, and the die zone was adjusted to one of three temperatures (100, 125, and 150°C) throughout the process. The extruder was capable of a screw speed ranging from 0-100 rpm, so the screw speed was adjusted to three speeds (50, 75, and 100 rpm) for the appropriate treatments. Once the processing was completed, the extrudates were stored for approximately 72 h at room temperature (25±1°C), allowing them to air dry.

2.3 Measurement of Extrudate Properties

The extrudates were subjected to extensive physical property analyses, including moisture content (% db), water activity (a_w, -), color (L*, a*, b*), unit density (g/cm^3), pellet durability (%), water stability (s), floatability (s), and bulk density (kg/m^3). The mass flow rate (g/min) and the temperature of the feeding, metering, and die zones ($^\circ$C) were recorded during the extrusion process.

2.3.1 Moisture Content (MC)

Moisture content was determined using approved American Association of Cereal Chemists [AACC] method 44-19 (2000) with a laboratory oven (HERAtherm, Thermo Scientific, Waltham, MA, USA). The samples were prepared and heated for 2 h at 135°C. Moisture content was determined for both the raw ingredients, as well as the extrudates.

2.3.2 Water Activity (a_w)

The water activity was determined for both the raw ingredients, as well as the extrudates using a water activity meter (Series 3TE, AquaLab, Pullman, WA, USA). The meter was calibrated and a sample cup was filled with each sample and placed inside the reading chamber.

2.3.3 Color

Color (L*, a*, and b*) of the raw ingredients and extrudates was determined using a spectrophotometer (LabScan XE, HunterLab, Reston, VA, USA). L* is the brightness/darkness, a* is the redness/greenness, and b* is the yellowness/blueness.

2.3.4 Unit Density (UD)

Extrudates were cut using a razor blade into pieces approximately 2 cm long. Each extrudate was weighed on an electric balance (Model AV114, OHAUS Adventurer Pro, Pine Brook, NJ, USA), and the length and diameter were determined using a digital calliper (Model 14-648-17, Fisher Scientific, Pittsburgh, PA, USA). The unit density was defined as the ratio between the mass compared to the volume, which was based on the cylindrical shape of the extrudate.

2.3.5 Pellet Durability Index (PDI)

Approximately 200 g of extrudates were broken into pieces approximately 2 cm long, then placed in a pellet durability tester (Model PDT-110, Gamet Automatic Sampling Equipment, St. Paul, MN, USA) for a 10-min tumbling period. After tumbling, the samples were then sieved using a no. 8 sieve (2.36 mm) (Model 122622665, Fisher Scientific, Pittsburgh, PA, USA). The pellet durability index was calculated using the following equation:

$$PDI = (M_{at}/M_{bt}) \times 100 \qquad (1)$$

Where,

PDI = Pellet durability index (%)

M_{at} = Mass of the pellets after tumbling (g)

M_{bt} = Mass of the pellets before tumbling (g)

Bulk Density (BD): Using the method described by the USDA (1999), bulk density was measured using a standard bushel tester (Model 151, Seedburo Equipment Co, Chicago, IL, USA) and a test weight 1-L cup and leveling stick (Model 103, Seedburo Equipment Co, Chicago, IL, USA).

2.3.6 Mass Flow Rate

The mass flow rate was determined by collecting extrudates at the die for 30-s intervals, and then weighing them on an electronic balance (Model EX10201, Ohaus Explorer, Pine Brook, NJ, USA). The extrudates were then dried to determine the moisture content, which gave the dry-basis mass flow rate.

2.3.7 Statistical Analysis

Raw ingredients were adjusted to three moisture content levels (15, 25, and 35% db). The blends were extruded at three die temperatures (100, 125, and 150°C) and three screw speeds (50, 75, and 100 rpm). Each of the factors included a high point and a low point with a central composite point, which resulted in 9 treatments (i.e. 2x2x2=8, plus 1 center point). The experimental protocol is shown in Table 1. Most of the physical properties were measured in triplicate (n=3), except for mass flow rate, which was measured in duplicate (n=2). The results were analyzed via Statistical Analysis System (SAS) software, which determined the means and standard deviations for each treatment, as well as the main effects, interactions, and treatment combination effects, using an α=0.05.

Table 1. Experimental design*

Treatment	Screw Speed (rpm)	Moisture Content (% db)	Temperature (°C)
1	50	15	100
2	50	15	150
3	50	35	100
4	50	35	150
5	100	15	100
6	100	15	150
7	100	35	100
8	100	35	150
9 (Center Point)	75	25	125

*The run value describes the order each sample was extruded. Experimental design was a 2 x 2 x 2 + 1 center point = 9 treatment conditions. The Center Point was defined as Treatment 9.

3. Results and Discussion

With the extrusion conditions used in the study, neither the corn nor the DDGS extrudates were of exceptionally high quality, unfortunately. As shown in Figure 1, the DDGS extrudates did not exhibit extensive particle binding during processing, which led to pellets which were not highly cohesive. (Images of corn extrudates are not provided, but exhibited similar behavior.) It is likely that increasing the moisture of the raw blends could have improved particle binding, and thus pellet quality. It should be noted that in subsequent tables that some of the extrusion conditions did not produce results – this occurred due to starch gelatinization within the extruder barrel, which caused the screw to foul, and thus extrusion had to be discontinued.

Figure 1. Bulk and cross-sectional (100x) images of the DDGS extrudates for each treatment (1-9). Image analysis indicates that the quality of the extrudates was low due to incomplete particle binding which occurred during all extrusion conditions used in this study.

3.1 Raw Ingredient Properties

Table 2 and Table 3 provide the results of the effects of moisture content on the water activity and color (L*, a*, b*) of the DDGS and corn, respectively.

Table 2. Raw DDGS properties*

Treatment	Moisture Content (db, %)	Water Activity (a$_w$)	Color L*	a*	b*
1	15.20 (1.11)	0.61 (0.00)	58.92 (0.95)	14.82 (0.28)	48.48 (0.08)
2	15.20 (1.11)	0.60 (0.00)	58.92 (0.95)	14.82 (0.28)	48.48 (0.08)
3	33.78 (0.71)	0.83 (0.00)	58.92 (0.95)	14.82 (0.28)	48.48 (0.08)
4	33.78 (0.71)	0.83 (0.00)	58.92 (0.95)	14.82 (0.28)	48.48 (0.08)
5	15.20 (1.11)	0.60 (0.00)	58.92 (0.95)	14.82 (0.28)	48.48 (0.08)
6	15.20 (1.11)	0.60 (0.00)	58.92 (0.95)	14.82 (0.28)	48.48 (0.08)
7	33.78 (0.71)	0.83 (0.00)	58.92 (0.95)	14.82 (0.28)	48.48 (0.08)
8	33.78 (0.71)	0.83 (0.00)	58.92 (0.95)	14.82 (0.28)	48.48 (0.08)
9	23.95 (0.61)	0.75 (0.00)	58.92 (0.95)	14.82 (0.28)	48.48 (0.08)

* Averages and standard deviations (shown in parentheses) for various physical properties of raw DDGS samples.

Table 3. Raw corn properties *

Treatment	Moisture Content (db, %)	Water Activity (a$_w$)	Color L*	a*	b*
1	16.16 (0.77)	- -	80.66 (1.18)	5.20 (0.20)	32.41 (1.49)
2	16.16 (0.77)	- -	80.66 (1.18)	5.20 (0.20)	32.41 (1.49)
3	34.26 (0.66)	0.97 (0.00)	80.66 (1.18)	5.20 (0.20)	32.41 (1.49)
4	34.26 (0.66)	- -	80.66 (1.18)	5.20 (0.20)	32.41 (1.49)
5	16.16 (0.77)	0.75 (0.00)	80.66 (1.18)	5.20 (0.20)	32.41 (1.49)
6	16.16 (0.77)	0.76 (0.00)	80.66 (1.18)	5.20 (0.20)	32.41 (1.49)
7	34.26 (0.66)	0.98 (0.00)	80.66 (1.18)	5.20 (0.20)	32.41 (1.49)
8	34.26 (0.66)	0.98 (0.00)	80.66 (1.18)	5.20 (0.20)	32.41 (1.49)
9	23.68 (0.47)	- -	80.66 (1.18)	5.20 (0.20)	32.41 (1.49)

*Averages and standard deviations (shown in parentheses) for various physical properties of raw corn samples.

3.1.1 Moisture Content

According to Table 2, the exact moisture contents were relatively close to the desired moisture content of the raw DDGS. The exact moisture content values were off by 1.34, 4.40, and 3.62% for the targets of 15, 25, and 35%, db, respectively. Because of the low variance, the extrusion process was performed without further adjustment.

3.1.2 Water Activity

The water activity level of the raw ingredients is important because the greater the value, the greater the chance for microbial metabolism and growth. The greater the microbial growth rate, the more likely the product is to spoil. According to Table 2, as the moisture content increased, the water activity level increased in the raw DDGS blends. Similar results are seen in Table 3 for the raw corn blends.

3.2 Extrudate Properties

Table 4 and Table 5 provide the results for the treatment effects of moisture content, screw speed, and temperature on the moisture content, water activity, color (L*, a*, b*), unit density, pellet durability, water stability, floatability, and bulk density of the DDGS and corn extrudates, respectively. The tables display the averages for each property, as well as the standard deviations, which are shown in parentheses. Table 6 displays the means and standard deviations for the main effects on the dependent variables of the DDGS extrudates. Table 7 displays the interaction effects.

Table 4. Treatment effects on properties of DDGS extrudates*

Treatment	Moisture Content (db %)	Water Activity (a_w)	Color			Unit Density (g/cm^3)	Pellet Durability Index (%)	Water Stability (s)	Floatability (s)	Bulk Density (kg/m^3)
			L*	a*	b*					
1	15.52cd (0.29)	0.63c (0.01)	52.56abc (1.72)	12.57b (0.18)	41.10bc (0.67)	0.72ab (0.05)	38.83bc (2.27)	4.00de (0.0)	2.67abc (0.58)	378.80bc (0.60)
2	15.32cd (0.81)	0.63c (0.0)	50.45c (1.36)	14.02a (0.33)	41.13bc (0.13)	0.75ab (0.11)	34.87cd (1.60)	8.00b (1.00)	3.33a (0.58)	372.00c (4.93)
3	18.80b (0.37)	0.70a (0.0)	52.34bc (0.93)	12.70b (0.50)	40.03c (0.95)	0.78ab (0.08)	40.60bc (7.59)	6.00c (1.00)	2.00c (0.0)	399.00a (2.16)
4	16.03c (0.33)	0.62c (0.01)	53.49ab (0.94)	14.23a (0.39)	44.45a (1.24)	0.85a (0.04)	- -	1.67fg (0.58)	3.33a (0.58)	- -
5	15.28cd (0.29)	0.65b (0.02)	52.72abc (1.21)	12.92b (0.16)	41.96b (0.52)	0.62b (0.03)	36.53cd (1.12)	3.67de (0.58)	2.33bc (0.58)	355.00d (7.34)
6	14.58d (0.32)	0.63c (0.01)	52.20bc (2.59)	12.77b (0.16)	40.25c (0.20)	0.86a (0.31)	47.90a (1.71)	5.00cd (1.00)	3.00ab (0.0)	324.73e (6.82)
7	20.53a (2.15)	0.68a (0.03)	55.35a (2.11)	13.82a (0.29)	43.81a (0.94)	0.69ab (0.09)	30.63d (4.68)	3.00ef (1.00)	2.67abc (0.58)	381.80b (0.20)
8	18.82b (0.28)	0.69a (0.01)	55.06ab (1.96)	13.89a (0.15)	44.32a (1.48)	0.84ab (0.13)	17.33e (6.05)	0.40g (0.0)	2.67abc (0.58)	385.53b (1.29)
9	16.28c (0.17)	0.64bc (0.01)	52.43bc (1.65)	12.64b (0.28)	40.77bc (0.18)	0.75ab (0.11)	43.83ab (2.68)	9.67a (1.15)	2.00c (0.0)	359.33d (0.99)

Averages and standard deviations (shown in parentheses) for various physical properties of DDGS extrudates. The treatment effect relationships are shown below. Differing letters indicate significant differences among treatments ($\alpha = 0.05$) for a given dependent variable.

Table 5. Treatment effects on properties of corn extrudates*

Treatment	Moisture Content (db, %)	Water Activity (a_w)	Color			Unit Density (g/cm^3)	Pellet Durability Index (%)	Water Stability (s)	Floatability (s)	Bulk Density (kg/m^3)
			L*	a*	b*					
1	- -	- -	-	-	-	-	- -	- -	- -	- -
2	- -	- -	-	-	-	-	- -	- -	- -	- -
3	11.39 (0.96)	0.65 (0.01)	46.66 (1.96)	10.81 (0.89)	40.93 (1.13)	1.32 (0.03)	99.80 (0.10)	1800.00 (0.0)	<1 (0.0)	620.33 (1.70)
4	- -	- -	-	-	-	-	- -	- -	- -	- -
5	12.69 (0.13)	0.65 (0.0)	62.54 (2.31)	4.90 (0.41)	35.06 (1.28)	1.14 (0.08)	97.73 (0.15)	1729.00 (67.77)	<1 (0.0)	490.07 (5.90)
6	12.89 (0.05)	0.63 (0.0)	66.39 (1.81)	3.57 (0.32)	31.37 (0.95)	0.78 (0.05)	- -	468.00 (44.00)	902.67 (85.80)	- -
7	15.42 (5.35)	0.72 (0.04)	52.47 (0.90)	8.99 (0.33)	40.78 (1.22)	1.31 (0.09)	99.40 (0.70)	1800.00 (0.0)	<1 (0.0)	599.20 (3.14)
8	14.41 (0.24)	0.65 (0.01)	57.99 (2.04)	6.16 (0.88)	31.40 (1.62)	1.12 (0.08)	98.90 (0.10)	1800.00 (0.0)	0.33 (0.0)	446.87 (1.68)
9	- -	- -	-	-	-	-	- -	- -	- -	- -

* Averages and standard deviations (shown in parentheses) for various physical properties of corn extrudates.

Table 6. Main effects on properties of DDGS extrudates*

Parameter	Levels	Moisture Content (% db)	Water Activity (aw)	L*	Color a*	Color b*	Unit Density (g/cm³)	Pellet Durability Index(%)	Water Stability (s)	Floatability (s)	Bulk Density (kg/m³)
Moisture Content (% db)	15	15.18a (0.55)	0.64a (0.01)	51.99a (1.80)	13.07a (0.62)	41.11a (0.74)	0.74a (0.17)	39.53a (5.46)	5.17a (1.90)	2.83a (0.58)	357.63a (22.32)
	25	16.28b (0.17)	0.64a (0.01)	52.43ab (1.65)	12.64b (0.28)	40.77a (0.18)	0.75a (0.11)	43.83a (2.68)	9.67b (1.15)	2.00b (0.0)	359.33a (0.99)
	35	18.55c (1.94)	0.68b (0.03)	54.06b (1.86)	13.66c (0.67)	43.15b (2.15)	0.79a (0.10)	29.52b (11.45)	2.77c (2.28)	2.67c (0.65)	388.78b (7.94)
Temperature (°C)	100	17.54a (2.50)	0.67a (0.03)	53.24a (1.85)	13.00a (0.57)	41.72a (1.60)	0.70a (0.08)	36.65a (5.57)	4.17a (1.34)	2.42a (0.51)	378.65a (16.70)
	125	16.27b (0.17)	0.64b (0.01)	52.43a (1.65)	12.64a (0.28)	40.77a (0.18)	0.75ab (0.11)	43.83a (2.68)	9.67b (1.15)	2.00a (0.0)	359.33b (0.99)
	150	16.19b (1.73)	0.64b (0.03)	52.80a (2.35)	13.73b (0.64)	42.54b (2.13)	0.82b (0.16)	33.37a (13.67)	3.77a (3.17)	3.08b (0.51)	360.76b (27.97)
Screw Speed (rpm)	50	16.42a (1.52)	0.65a (0.03)	52.21a (1.59)	13.38a (0.84)	41.68a (1.88)	0.77a (0.08)	38.10a (4.78)	4.92a (2.54)	2.83a (0.72)	383.27a (12.46)
	75	16.28ab (0.17)	0.64a (0.01)	52.43ab (1.65)	12.64b (0.28)	40.77a (0.18)	0.75a (0.11)	43.83a (2.68)	9.67b (1.15)	2.00b (0.0)	359.33b (0.99)
	100	17.31b (2.74)	0.67b (0.03)	53.84b (2.26)	13.35a (0.56)	42.58b (1.85)	0.75a (0.18)	33.10b (11.99)	3.02c (1.88)	2.67a (0.49)	361.77b (25.86)

*Averages and standard deviations (shown in parentheses) of the main effects of the DDGS extrudates. The relationships amongst the main effects are shown below. Differing letters indicate significant differences among levels of a specific independent variable ($\alpha = 0.05$) for a given dependent variable.

Table 7. Interaction effects for properties of DDGS extrudates*

Parameter	Moisture Content (% db)	Water Activity (aw)	L*	Color a*	b*
Moisture Content	<.0001	<.0001	0.0076	0.0001	<.0001
Temperature	0.0007	0.0001	0.5319	<.0001	0.0285
Screw Speed	0.0152	0.0017	0.0301	0.7940	0.0163
Moisture Content*Temperature	0.0145	0.0173	0.2219	0.5380	0.0001
Moisture Content*Screw Speed	0.0006	0.2090	0.3458	0.0024	0.0154
Temperature*Screw Speed	0.6696	0.0043	0.9582	<.0001	0.0006
Moisture Content*Temperature*Screw Speed	0.2523	<.0001	0.2874	0.7624	0.1321

Parameter	Unit Density (g/cm³)	Pellet Durability Index(%)	Water Stability (s)	Floatability (s)	Bulk Density (kg/m³)
Moisture Content	0.3165	<.0001	<.0001	0.3979	<.0001
Temperature	0.0382	0.0352	0.2557	0.0028	0.0022
Screw Speed	0.6934	0.7084	<.0001	0.3979	<.0001
Moisture Content*Temperature	0.8308	<.0001	<.0001	1.0000	<.0001
Moisture Content*Screw Speed	0.6489	0.1258	0.5022	0.3979	0.1793
Temperature*Screw Speed	0.1904	0.0052	0.5022	0.1004	0.0001
Moisture Content*Temperature*Screw Speed	0.5479	-	0.0047	0.1004	-

* Relationships between the treatment interactions of the DDGS extrudates (p-values). Interactions were deemed significant if $p < 0.05$.

3.2.1 Moisture Content

The moisture content of the extrudates is important to consider because it determines the freshness of the pellets. Moisture also impacts the rate of spoilage of the extrudates. When developing a pellet for aquaculture feed, it is important to consider the amount of time the pellets will be stored, so the optimum amount of moisture present in the extrudates can be determined. According to Table 4, treatment 7, which had a screw speed of 100 rpm, raw moisture content of 35% db, and die temperature of 100°C, had the greatest extrudate moisture content and was significantly different from the other eight treatments. Treatments 3 and 8 had similar moisture contents of 18.80% db and 18.82% db, respectively, and treatments 4 and 9 were similar as well, with moisture contents of 16.03% db and 16.28% db, respectively. Treatment 6, which had a screw speed of 100 rpm, raw moisture content of 15% db, and a die temperature of 150°C, had the lowest extrudate moisture content of 14.58% db. Treatments 1, 2, and 5 had similar results to both treatments 4 and 9 and treatment 6, with extrudate moisture contents of 15.52% db, 15.32% db, and 15.28% db, respectively. According to Table 3, the greater the initial moisture content of the raw DDGS blends, the greater the moisture content in the extrudates. A similar relationship between the corn extrudates was determined as well. According to Table 6, as the temperature of the die increased, the moisture content of the DDGS extrudates decreased, due to increased evaporation. According to Table 6, as the screw speed of the extruder increased from 50 to 100 rpm, the moisture content of the DDGS extrudates increased. According to Table 5, the amount of moisture present in the corn extrudates increased as the screw speed of the extruder increased; this was similar behavior as the DDGS.

3.2.2 Water Activity

Water activity determines the amount of water available for microbial metabolism and growth and is important in the extrudates because it determines the rate of spoilage. When comparing moisture content to water activity, the water activity is more important in the biological sense of spoilage because it is the amount of water available for microbial growth. According to Table 4, treatments 3, 7, and 8 have the greatest water activity levels, with values of 0.70, 0.68, and 0.69, respectively. With a value of 0.65, treatment 5 has a slightly smaller water activity value and is significantly different than treatments 3, 7, and 8. Treatments 1, 2, 4, and 6 have the smallest values for water activity, and treatment 9 has a value of 0.64, which is not significantly different from treatments 1, 2, 4, 5, and 6. According to Table 6, as the moisture content of the raw DDGS ingredients increased, the water activity of the extrudates increased. A similar relationship as seen in the corn extrudates and is shown in Table 5. As the temperature of the die of the extruder increased, the water activity in the DDGS extrudates decreased and is displayed in Table 6. Similar results are shown in Table 5 for the corn extrudates. As the screw speed of the extruder increased, there was an increase in water activity levels from the 50 and 75 rpm to the 100 rpm treatments. According to Table 5, as the screw speed of the extruder increased, the water activity level of the corn extrudates increased.

3.2.3 Color

In order to have consumer acceptance, the appearance of extrudates is important. Color is one of the most important appearance attributes because consumers buy products based on expected color from previous experience. The color of the extrudates is measured based on three parameters: L*, a*, and b*. According to Table 4, the L* value for the DDGS was greatest in treatment 7, with a value of 55.35, while treatment 2 had the smallest L* value at 50.45. Treatments 2, 4, 7, and 8 have higher a* values that are significantly different from treatments 1, 3, 5, 6, and 9 (Table 4). The b* values for treatments 4, 7, and 8 were the greatest and were significantly different from the other treatments. Treatments 3 and 6 have the lowest b* values of 40.03 and 40.25, respectively. As the moisture content of the DDGS extrudates increased, the L* and b* values increased, although there was no relationship with a* values, according to Table 6. A similar relationship was true for the corn extrudates for the b* values (Table 5). As the temperature of the die increased, the a* and b* values of the DDGS extrudates increased, but there was no relationship with the L* values. As the screw speed of the extruder increased, the L* and b* values of the DDGS extrudates increased, but there was not a significant relationship with the a* values.

3.2.4 Unit Density

Unit density is important in the extrudates because the amount of mass per unit volume should contain the proper nutrients, in the proper quantities but be able to float in water (i.e., if UD < 1.0 g/cm^3). According to Table 4, treatment 4 and 6 contained the greatest mass per unit volume, with values of 0.85 and 0.86 g/cm^3, respectively, while treatment 5 produced the smallest unit density with a value of 0.62 g/cm^3. Treatments 1, 2, 3, 7, 8, and 9 were similar to treatments 4, 5, and 6. According to Table 6, there was no significant relationship between the unit density of the DDGS extrudates and the change in moisture content or screw speed. However, as the temperature of the die increased, the unit density of the DDGS extrudates increased. Because all treatments had UD values less than 1.0 g/cm^3, they should have been able to float in water (if they could exhibit water stability, that is).

3.2.5 Pellet Durability Index

Pellet durability index is the amount of breakage the extrudate is able to withstand. It is important in aquaculture feeds because of storage and delivery processes. According to Table 4, the greatest value for the pellet durability index was treatment 6 at 47.90%. Treatments 1 and 3 were relatively similar to both treatment 9, as well as treatments 2 and 5. Treatment 7 was similar to treatments 2 and 5 as well. The lowest pellet durability index was treatment 8, at 17.33%. The pellet durability index for treatment 4 was unable to be determined due to the lack of cohesive of extrudates for this treatment, thus PDI could not be reported. According to Table 6, as the moisture content of the DDGS extrudates increased, the pellet durability index decreased. There was no significant relationship between the temperature of the DDGS extrudates and the pellet durability index values. As the screw speed of the extruder increased, the pellet durability index decreased in the DDGS extrudates.

3.2.6 Water Stability

Water stability is the amount of time the pellet holds together in water before it dissolves. It is important for aquaculture feeds because the fish need pellets to last approximately 30 min during feeding time. According to Table 4, the greatest value for water stability was treatment 9, which was the center point. Treatment 9 was followed by treatment 2, and then treatment 3, at 8 and 6 s, respectively. Treatment 6 was relatively similar to both treatment 3 and treatments 1 and 5. Treatments 1 and 5 were also relatively similar to treatment 3, and treatment 3 was relatively similar to treatment 4. Treatment 8 produced the lowest water stability with an average of 0.40 s. According to Table 6, the moisture content of the raw ingredients, temperature of the die, and screw speed of the extruder did not result in significant relationships for the water stability values of the DDGS extrudates. Because DDGS contains very little starch, the pellets lack a binder typically provided by gelatinization of starch. According to Table 5, the increase in moisture content caused an increase in water stability of the corn extrudates, but corn contains large starch content. An increase in both the screw speed of the extruder and the temperature of the die caused a decrease in the water stability of the corn extrudates. It is clear that in order to be effectively used in aquafeeds, DDGS cannot be used alone; instead it must be processed with other ingredients which, hopefully, will improve particle binding characteristics.

3.2.7 Floatability

Floatability is the amount of time the pellets float on top of the water. It is important in aquaculture feeds because some species only eat the feed if it is floating, and the pellets must float for approximately 30 min for appropriate feeding time. According to Table 4, the greatest floatability was produced by treatments 2 and 4 at 3.33 s, while the lowest floatability was produced by treatments 3 and 9 at 2 s. Treatments 1, 5, 6, 7, and 8 were all relatively similar to one another. There was no direct relationship between the moisture content and screw speed and the floatability of the DDGS extrudates, according to Table 6. As the temperature of the extruder increased, the floatability of the DDGS extrudates increased, due to greater expansion. According to Table 5, as the moisture content of the raw corn blends increased, the floatability of the corn extrudates decreased. As the screw speed and temperature of the extruder increased, the floatability of the corn extrudates decreased (Table 5).

3.2.8 Bulk Density

The bulk density quantifies the amount of mass a specific volume can hold. Bulk density is important in storage of aquaculture feeds because the pellets will be distributed in containers, and the containers need to hold as much mass as possible. According to Table 4, the greatest bulk density was produced by treatment 3 at 399.00 kg/m^3. Treatments 7 and 8 followed treatment 3, having slightly lower bulk density. Treatment 1 was relatively similar to treatments 7 and 8, as well as treatment 2. Treatment 5 had a bulk density of 355.00 kg/m^3, which was slightly lower than treatment 2. The smallest value for bulk density was produced by treatment 6, with a value of 324.73 kg/m^3. According to Table 6, as the moisture content of the raw DDGS blend increased, the bulk density of the DDGS increased. The bulk density of the DDGS extrudates decreased as the temperature and screw speed of the extruder increased, due to greater expansion. According to Table 5, similar results were produced for the corn extrudates.

3.3 Implications

Figure 2 is a matrix plot of all independent and dependent variables, and it shows that the extruded DDGS had much different behavior than the extruded corn for all properties studied. This was true for all screw speeds, all ingredient moistures, and all extruder temperatures. This occurred due to the chemical composition being substantially different. As Liu and Rosentrater (2011) discussed, DDGS has approximately three times the levels of protein, fiber, fat, and minerals vis-à-vis corn, whereas DDGS has substantially lower starch (because alcohol manufacturing utilizes only the starch component from the corn kernels). Use of DDGS vs. corn in aquafeed formulations definitely impacts extrudate quality, as shown in Figure 2.

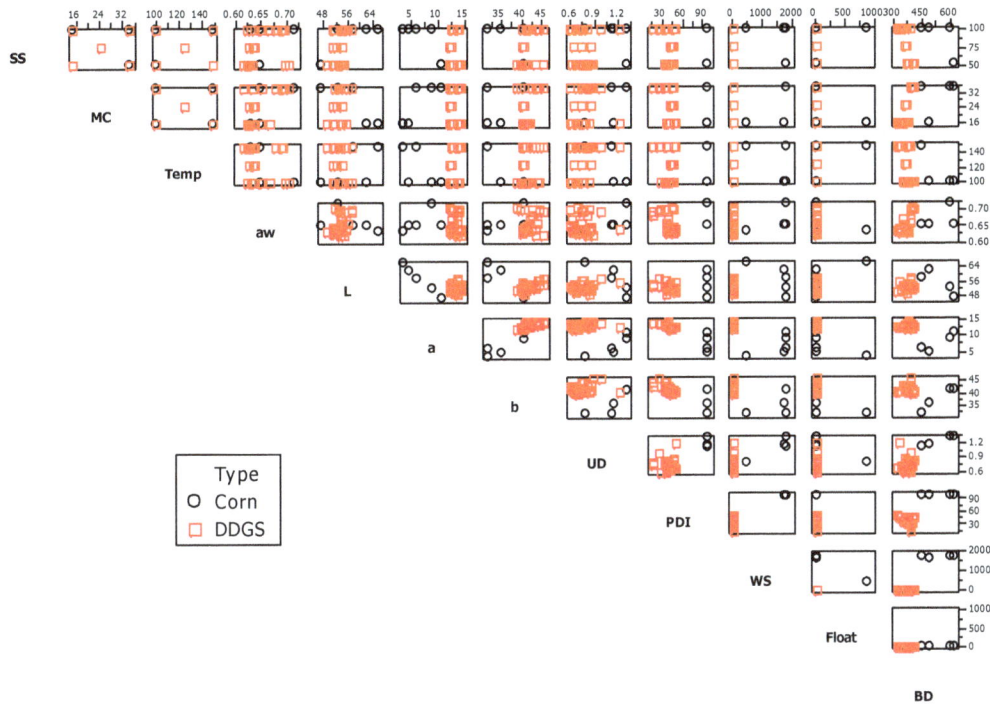

Figure 2. Matrix plot of all independent and dependent variables shows that the extruded DDGS had much different behavior than the extruded corn for all properties studied.

4. Conclusion

In conclusion, the moisture content of the raw ingredients led to an increase in the water activity in the raw ingredients. It also led to an increase in moisture content, water activity level, L* values, b* values, and bulk density of the DDGS extrudates. An increase in moisture content led to a decrease in pellet durability index. There was no significant relationship between the moisture content of the raw ingredients and the a* values, unit density, water stability, or floatability of the DDGS extrudates. The increase in temperature of the extruder die led to an increase in the a* values, b* values, unit density, and floatability of the DDGS extrudates. An increase in the temperature of the die led to a decrease in the moisture content, water activity level, and bulk density of the DDGS extrudates. There was no significant relationship between the temperature of the die and the L* values, pellet durability index, or water stability of the DDGS extrudates. The increase in screw speed caused an increase in the moisture content, water activity level, L* values, and b* values of the DDGS extrudates. An increase in the screw speed of the extruder caused a decrease in the pellet durability index and bulk density of the extrudates. There was no significant relationship between the screw speed of the extruder and the a* values, unit density, water stability, and floatability of the DDGS extrudates.

Ultimately, the composition of the ingredients affects the resulting physical properties of the extrudates. The amount of starch and water present in the ingredients determines the quality of the pellets. Because starch gelatinizes and acts as a binding agent, the more starch present, the better the quality of the pellet. Along with starch content, the amount of fiber and protein in the ingredients affects the quality of the extrudates. It is important to have a high protein diet for aquaculture feed, so the fish reach substantial weight in a reasonable amount of time. The key is to find an appropriate balance between moisture content, screw speed, and temperature, as well as the starch, protein, and fiber content of the ingredients. It is clear that DDGS cannot be processed alone at the extrusion conditions used in this study. DDGS should be used in combination with other ingredients in order to balance nutrient levels as well as improve feed quality characteristics.

References

AACC. (2000). *Method 44-19, moisture-air oven method, drying at 135°C. AACC Approved Methods* (10th ed.). St. Paul, MN: American Association of Cereal Chemists.

Ayadi, F. Y., Muthukumarappan, K., Rosentrater, K. A., & Brown, M. L. (2011a). Single-screw extrusion processing of distillers dried grains with solubles (DDGS)-based yellow perch (*Perca flavescens*) feeds.

Cereal Chemistry, 88(2), 179-188. http://dx.doi.org/10.1094/CCHEM-08-10-0118

Ayadi, F. Y., Muthukumarappan, K., Rosentrater, K. A., & Brown, M. L. (2011b). Twin-screw extrusion processing of rainbow trout (*Oncorhynchus mykiss*) feeds using various levels of corn-based distillers dried grains with solubles (DDGS). *Cereal Chemistry, 88*(4), 363-374. http://dx.doi.org/10.1094/CCHEM-08-10-0120

Ayadi, F. Y., Rosentrater, K. A., Muthukumarappan, K., & Brown M. L. (2011c). Twin-screw extrusion processing of distillers dried grains with solubles (DDGS)-based yellow perch (*Perca flavescens*) feeds. *Food and Bioprocess Techno*logy, *5*(5), 1963-1978.

Chevanan, N., Muthukumarappan, K., & Rosentrater, K. A. (2007a). Extrusion studies of aquaculture feed using distillers dried grains with solubles and whey. *Food and Bioprocess Technology*, *2*, 177-185. http://link.springer.com/content/pdf/10.1007%2Fs11947-007-0036-8.pdf

Chevanan, N., Muthukumarappan, K., Rosentrater, K. A., & Julson, J. L. (2007b). Effect of die dimensions on extrusion processing parameters and properties of DDGS-based aquaculture feeds. *Cereal Chemistry, 84*(4), 389-398. http://dx.doi.org/10.1094/CCHEM-84-4-0389

Chevanan, N., Rosentrater, K. A., & Muthukumarappan, K. (2007c). Twin-screw extrusion processing of feed blends containing distillers dried grains with solubles (DDGS). *Cereal Chemistry, 84*(5), 428-436. http://dx.doi.org/10.1094/CCHEM-84-5-0428

Chevanan, N., Rosentrater, K. A., & Muthukumarappan, K. (2008a). Effect of DDGS, moisture content, and screw speed on physical properties of extrudates in single-screw extrusion. *Cereal Chemistry, 85*(2), 132-139. http://dx.doi.org/10.1094/CCHEM-85-2-0132

Chevanan, N., Rosentrater, K. A., & Muthukumarappan, K. (2008b). Effects of processing conditions on single screw extrusion of feed ingredients containing DDGS. *Food and Bioprocess Technology, 3*(1), 111-120. http://link.springer.com/content/pdf/10.1007%2Fs11947-008-0065-y.pdf

Delgado-Nieblas, C., Aguilar-Palazuelos, E., Gallegos-Infante, A., Rocho-Guzman, N., Zazueta-Morales, J., & Caro-Corrales, J. (2012). Characterization and optimization of extrusion cooking for the manufacture of third-generation snacks with winter squash (*Cucurbita moschata D.*) flour. *Cereal Chemistry, 89*(1), 65-72. http://dx.doi.org/10.1094/CCHEM-02-11-0016

Forsido, S., & Ramaswamy, H. S. (2011). Protein rich extruded products from tef, corn and soy protein isolate blends. *Ethiopian Journal of Applied Sciences and Technology, 2*(2), 75-90. http://www.researchgate.net/publication/224872856_Protein_Rich_Extruded_Products_from_Tef_Corn_and_Soy_Protein_Isolate_Blends

Kannadhason, S., Muthukumarappan, K., & Rosentrater, K. A. (2008). Effect of starch sources and protein content on extruded aquaculture feed containing DDGS. *Food and Bioprocess Technology, 4*, 282-294. http://link.springer.com/content/pdf/10.1007%2Fs11947-008-0177-4.pdf

Kannadhason, S., Muthukumarappan, K., & Rosentrater, K. A. (2009). Effects of ingredients and extrusion parameters on aquafeeds containing DDGS and tapioca starch. *Journal of Aquaculture Feed Science and Nutrition, 1*(1), 6-21. http://www.medwelljournals.com/fulltext/?doi=joafsnu.2009.6.21

Kannadhason, S., Rosentrater. A., Muthukumarappan, K., & Brown, M. L. (2010). Twin-screw extrusion on DDGS-based aquaculture feeds. *Journal of the World Aquaculture Society, 41*(S1), 1-13. http://dx.doi.org/10.1111/j.1749-7345.2009.00328.x

Lardy, G. (2013). *Feeding corn to beef cattle.* Retrieved June 1, 2013, from http://www.ag.ndsu.edu/pubs/ansci/beef/as1238.pdf

Liu, K., & Rosentrater, K. (2011). *Distillers Grains: Production, Properties, and Utilization.* Boca Raton, FL: CRC Press.

Singh, R. K., Majumdar, R. K., & Venkateshwarlu, G. (2012). Optimum extrusion-cooking conditions for improving physical properties of fish-cereal based snacks by response surface methodology. *Journal of Food Science and Technology.* http://link.springer.com/content/pdf/10.1007%2Fs13197-012-0725-9.pdf

Razzaq, M. R., Anjum, F. M., Khan, M. I., Khan, M. R., Nadeem, M., Javed, M. S., & Sajid, M. W. (2012). Effect of temperature, screw speed and moisture variations on extrusion cooking behavior of maize (Zea mays. L). *Pakistan Journal Food Sciences, 22*(1), 12-22. http://www.psfst.com/__jpd_fstr/c2fda8f7cfc6ffb79f9d45ab1cd36929.pdf

Rosentrater, K. A., Muthukumarappan, K., & Kannadhason, S. (2009a). Effects of ingredients and extrusion parameters on properties of aquafeeds containing DDGS and corn starch. *Journal of Aquaculture Feed Science and Nutrition, 1*(2), 44-60. http://www.medwelljournals.com/fulltext/?doi=joafsnu.2009.44.60

Rosentrater, K. A., Muthukumarappan, K., & Kannadhason, S. (2009b). Effects of ingredients and extrusion parameters on aquafeeds containing DDGS and potato starch. *Journal of Aquaculture Feed Science and Nutrition, 1*(1), 22-38.http://docsdrive.com/pdfs/medwelljournals/joafsnu/2009/22-38.pdf

Simons, C. W., Hall, C., & Tulbek, M. (2012). Effects of extruder screw speeds on physical properties and in vitro starch hydrolysis of precooked pinto, navy, red, and black bean extrudates. *Cereal Chemistry, 89*(3), 176-181. http://dx.doi.org/10.1094/CCHEM-08-11-0104

Sorensen, M. (2012). A review of the effects of ingredient composition and processing conditions on the physical qualities of extruded high-energy fish feed as measured by prevailing methods. *Aquaculture Nutrition, 18,* 233-248. http://dx.doi.org/10.1111/j.1365-2095.2011.00924.x

USB. (2012). *Soy Nutritional Content.* Soy Connection. Retrieved June, 1, 2013, from http://www.soyconnection.com/health_nutrition/technical_info/protein_content.php.

USDA. (1999). Practical procedures for grain handlers: inspecting grain. Grain Inspection, Packers, and Stockyards Administration, Washington DC, USA. Retrieved June, 1, 2013, from http://archive.gipsa.usda.gov/pubs/primer.pdf.

9

Stability of Mexican Oregano Essential Oil Double Emulsions Obtained by Ultrasound Formulated With Whey Protein Concentrate and Tween 80

Areli H. Peredo-Luna[1], Aurelio Lopez-Malo[1], Enrique Palou[1] & María Teresa Jiménez-Munguía[1]

[1]School of Engineering, Department of Chemical, Environmental and Food Engineering, Universidad de las Américas Puebla, Puebla, Mexico

Correspondence: María Teresa Jiménez-Munguía, School of Engineering, Department of Chemical, Environmental and Food Engineering, Universidad de las Américas Puebla, Exhacienda Sta. Catarina Mártir S/N, San Andrés Cholula Puebla, 72810, Mexico. E-mail: mariat.jimenez@udlap.mx

Abstract

Water-in-oil-in-water (W/O/W) emulsions have a great potential use for food applications because they can protect sensitive compounds, such as essential oils. The aim of this study was to determine the effect of ultrasonic homogenization parameters: intensity (42 or 54 μm) and time (5 or 7.5 min); and formulation: oil phase proportion (20 or 30%) and emulsifier concentration (0 or 0.3%); on the physical and stability properties of Mexican oregano essential oil (OEO) in double emulsions. The emulsions were made in a two-step process, primary emulsions (W/O) containing OEO in the oil phase and ascorbic acid solution in the water phase, were stabilized with lecithin; while secondary emulsions were stabilized with 6% (w/w) of whey protein concentrate (WPC) and with or without Tween 80 (T80). Creaming, viscosity and droplet size distribution were measured to determine the stability of the W/O/W emulsions; as well as other physical properties like density and pH. The prepared W/O/W emulsions had droplet sizes between 2.89(±0.589) μm and 4.123(±0.964) μm. The most stable emulsions, with no creaming developed after 25 days of storage, were the ones formulated with WPC with T80, and additionally, 30% of the primary emulsion. Besides, higher intensity and longer time of ultrasonic homogenization conditions applied enhances W/O/W stability. Empirical models were developed for viscosity and creaming properties of W/O/W emulsions, with 99.7% of correlation coefficients, finding optimum values for specific homogenization conditions and formulation. Further studies are suggested to evaluate OEO in W/O/W emulsions as controlled release systems in food.

Keywords: double emulsions, ultrasonic homogenization, droplet size, viscosity, creaming

1. Introduction

Essential oils are sensitive compounds and are valuable substances because of their functional properties including, nutritional, organoleptic, antimicrobial and antioxidant properties, among others (Bakkali, Averbeck, Averbeck, & Idaomar, 2008; Burt, 2004; Hammer, Carson, & Riley, 1999). Nevertheless, they are susceptible to oxidative deterioration, chemical interactions and loss of volatile compounds, affecting the final quality of food in which these are added. The nano- or micro-encapsulation by means of emulsification technique has been applied to protect bioactive compounds, like essential oils; the products obtained can be used directly in the liquid state or can be dried (spray- or freeze- drying) to form powders (Bakry et al., 2016). Some important applications of microencapsulated essential oils are in beverages, meat, cheese, and fresh lettuce (Asensio, Grosso & Juliani, 2015; Beirão da Caosta et al., 2013; Bhargava, Conti, da Rocha & Zhang, 2015; Botsoglou, Govaris, Botsoglou, Grigoropoulou & Papageorgiou, 2003; Lakkis, 2007).

The main problem about double emulsions is that they are highly unstable thermodynamic systems. Emulsifiers and stabilizing agents are added to achieve stable systems. The challenge in the food area is to reduce or eliminate synthetic chemical agents and replace them with safe human intake substances. In recent years, biopolymers have been investigated as emulsifiers and stabilizing agents. These biopolymers can be proteins, polysaccharides and phospholipids (Dickinson, 2011). Whey protein concentrated or isolated has been used for the aqueous phases of simple or multiple emulsions and it has been proved its efficiency at concentrations between 1.5% and 15% (Hemar, Cheng, Oliver, Sanguansri, & Augustin, 2010; Mun et al., 2010; Surh, Vladisavljevică, Mun, & MCClements, 2007). Polysaccharides like gums enhance the stability of multiple emulsions due to their thickening properties. Also, stability is reached by the formation of complexes of protein-polysaccharides (Jiménez-Alvarado, Beristain, Medina-Torres, Román-Gerrero, & Vernon-Carter, 2009; Su, Flanagan, & Singh, 2008). In double emulsions of type water-in-oil-in-water (W/O/W) the addition of two types of emulsifiers is suggested, an emulsifier with a low HLB value is to stabilize the first emulsion water-in-oil (W/O), while an emulsifier with high HLB value can stabilize the secondary oil-in-water (O/W) emulsion. The concentration of these must be evaluated to stabilize the double emulsion W/O/W since some interactions may take place if some stabilizers are used (Jiao & Burguess, 2003). A phospholipid that can act as an emulsifying agent is lecithin and is commonly used for food applications since it is considered as GRAS. Phosphatidylcholine-depleted lecithin has been used to stabilize W/O interphases in W/O/W double emulsions (Akhtar & Dickinson, 2001; Knoth, Scherze, & Muschiolik, 2005). Moreover, polyoxyethylene sorbitan esters (Tweens) have minimal toxicity and are widely used in food and pharmaceutical applications forming very stable microemulsions during storage (Jiao & Burguess, 2003; Rukmini, Raharjo, Hastuti & Supriyadi, 2012).

Besides the different compounds used to prepare double emulsions, the homogenization technique has also been studied to optimize the process conditions used. Among the high-energy methods used to prepare emulsions are: high pressure, microfluidization and ultrasound. With these techniques, different emulsion properties (droplet size of the disperse phase, viscosity, density, creaming) are generated according the severity of the treatment and therefore affecting their stability (Cardoso-Ugarte, López-Malo & Jiménez-Munguía, 2016).

Since the process conditions of the homogenization for the double emulsions preparation affects the stability of the final double emulsion, as well as the use of emulsifiers and stabilizers in the formulation, an optimization of the different variables may be approached using a factorial design, in order to evaluate and compare their individual or combined effect on the stability properties of the double emulsions. Therefore, the aim of this study was to determine the effect of the ultrasonic homogenization parameters (intensity and time) and formulation (proportion of oil phase and emulsifier concentration) on the physical and stability properties of Mexican oregano essential oil in double emulsions (W/O/W).

2. Material and Methods

Mexican oregano essential oil (OEO) (*Lippia berlandieri* Schauer) was provided by the Center of Research of Natural Resources (CIRENA) of Chihuahua, Mexico. OEO was mixed with corn oil (La Gloria, Mexico) to adjust the oil phase; while distilled water was used for the preparation of the aqueous phases (inner and outer). The emulsifiers and stabilizing agents were purchased as food grade: soy lecithin (Gelcaps®, Mexico), Tween 80 (sorbitan ester) (Sigma-Aldrich®, México), maltodextrin (DE 10, Globe, CPIngredientes, Mexico) and whey protein concentrate (Fleischmann, Mexico).

2.1 Experimental Design

A factorial design 2^4 (16 experiments) was implemented for this study. Two process variables were studied, ultrasonic intensity adjusted by the wave amplitude (W) and homogenization time (min); and two more variables of formulation were varied, emulsifier concentration (T80) and the proportion of the primary emulsion (E1) used for the double emulsion preparation (E2). The different levels used for each factor are showed in Table 1.

2.1.1 Primary Emulsion Preparation

The primary emulsion W/O was made in a proportion of 20/80 (w/w). The inner aqueous phase was formulated with a solution of 30% (w/w) of maltodextrin (MD) and 0.05 % (total weight of emulsion, tw) of ascorbic acid; while the oil phase was a mixture of corn oil with 0.50% (tw) of Mexican OEO and 2 % (tw) of lecithin. The primary emulsion was prepared with an ultrasonic homogenizer (EW-04711-70 Cole- Parmer, USA). The ultrasonic probe of 1" was introduced in a 150 mL tempering glass beaker, which was connected to a water bath set to 5 °C. The aqueous phase was first added, submerging the probe 1 cm below the liquid surface. The ultrasonic intensity was adjusted to 84 µm or 108 µm of wave amplitude, applied during 5 or 15 min. The dispersed phase was poured into the aqueous phase, using a syringe, when the ultrasonic homogenization process started.

2.2.2 Secondary Emulsion Preparation

The primary emulsion (E1) was dispersed in 70 or 80 % (tw) of the outer aqueous phase (W_2) of the double emulsion. The aqueous phase was formulated with 6% (w/w) of whey protein concentrate (WPC) and with or without 0.3% (tw) of Tween 80 (T80) as emulsifier. The ultrasound homogenization treatment was performed at 42 µm or 54 µm of wave amplitude, for 2.5 or 7.5 min.

2.2 Emulsions Physical Properties

The droplet size distribution of fresh prepared double emulsions, were measured using a dynamic light scattering equipment Nanotrac Wave (Microtrac, USA).

The emulsions densities were determined by the AOAC method 962.37 (1995) using Grease pycnometers, by triplicate, and was reported in g/cm^3.

Since the emulsions prepared were Newtonian systems, a Cannon Fenske viscometer was used to measure the dynamic viscosity of the emulsions (mPa.s); also by triplicate.

The pH values of the emulsions were determined using a potentiometer (Orion Research Inc., model 8005, USA), previously calibrated at pH 4 and pH 7. The measurements were also done by triplicate.

Table 1. Factorial experimental design used for the double emulsions (W/OW) preparation with oregano essential oil using ultrasonic homogenization

Formulation			Ultrasonic Homogenization			
			E1		E2	
System code	Stabilizer-Emulsifier	E1:E2	Wave Amplitude (μm)	Homogenization time (min)	Wave Amplitude (μm)	Homogenization time (min)
A			108	15	54	7.5
B		20:80	108	5	54	2.5
C			84	15	42	7.5
D	WPC		84	5	42	2.5
E			108	15	54	7.5
F		30:70	108	5	54	2.5
G			84	15	42	7.5
H			84	5	42	2.5
I			108	15	54	7.5
J		20:80	108	5	54	2.5
K			84	15	42	7.5
L	WPC-Tween 80		84	5	42	2.5
M			108	15	54	7.5
N		30:70	108	5	54	2.5
O			84	15	42	7.5
P			84	5	42	2.5

E1: Primary emulsion (W/O), E2: Secondary emulsion (W/O/W), WPC: whey protein concentrate

2.3 Emulsions Stability

The creaming of the double emulsions was expressed as percentage and was calculated according to equation 1. An initial fixed volume of the double emulsion was collected in a graduated tube and the creaming phase lecture was registered as proposed by Jiao & Burgess (2003):

$$\text{Creaming (\%)} = 100 \times (V_{\text{total emulsion}} - V_{\text{creaming}}) / V_{\text{total emulsion}} \tag{1}$$

where: $V_{\text{total emulsion}}$ is the initial total double emulsion volume and V_{creaming} is the creaming phase volume, separated at the upper part of the emulsion. Triplicates of the measurements of V_{creaming} were registered every 5 days, during 25 days of storage at room temperature (25 °C).

2.4 Statistical Analysis

The emulsions physical properties were analyzed statistically, applying an ANOVA and Tukey-test to compare among the different experimental systems, using a confidence level of 95%. Besides, the factorial design was also analyzed by ANOVA and two empirical models were obtained, for creaming and viscosity, as functions of the studied variables (Table 1), selecting the main variable effects and the interactions among the variables with significant effect ($p < 0.05$), to include them in the model. Surface response plots were developed from these empirical models. The software used for the statistical analysis was Minitab v.17 (LEAD Technologies Inc., USA).

3. Results and Discussion

3.1 Physical Properties of W/O/W Emulsions

It has been demonstrated that some physical properties of double emulsions are strongly related to their stability (Charoen et al., 2009; Dickinson, 2011), such as particle diameter of the dispersed phase. Small particle diameters are preferred to achieve good stability and this property is dependent of the homogenization parameters and formulation. In this study, the different systems prepared exhibited wide particle distributions; the mean diameter ($D_{4,3}$) is presented in Table 2 in order to compare the different emulsions prepared.

Table 2. Physical and physicochemical properties of double emulsions (W/O/W) prepared with oregano essential oil by ultrasound

System	Mean diameter ($D_{4,3}$) (µm)	Density (g/cm^3)	Viscosity (mPa.s)	pH
A	2.82 ± 0.07^{efg}	1.003 ± 0.012^a	1.57 ± 0.02^g	6.50 ± 0.05^a
B	3.24 ± 0.30^{de}	1.004 ± 0.012^a	1.67 ± 0.05^f	6.52 ± 0.06^a
C	3.23 ± 0.17^{def}	1.003 ± 0.005^a	1.87 ± 0.02^e	6.57 ± 0.01^a
D	2.72 ± 0.06^{fg}	1.005 ± 0.011^a	1.57 ± 0.03^g	6.35 ± 0.05^b
E	4.68 ± 0.48^a	0.992 ± 0.029^a	3.06 ± 0.02^a	5.56 ± 0.01^g
F	4.23 ± 0.42^{ab}	0.995 ± 0.023^a	2.83 ± 0.02^b	5.71 ± 0.01^f
G	3.96 ± 0.22^{bc}	0.996 ± 0.029^a	2.53 ± 0.02^c	5.42 ± 0.01^h
H	4.56 ± 0.60^a	1.009 ± 0.021^a	2.83 ± 0.02^b	5.52 ± 0.03^g
I	2.28 ± 0.18^{gh}	1.009 ± 0.006^a	1.60 ± 0.02^{fg}	6.11 ± 0.03^d
J	2.24 ± 0.11^{gh}	1.013 ± 0.005^a	1.43 ± 0.05^h	6.16 ± 0.01^{cd}
K	1.94 ± 0.11^{hi}	1.008 ± 0.000^a	1.63 ± 0.02^{fg}	6.18 ± 0.00^c
L	1.67 ± 0.00^i	1.007 ± 0.004^a	1.57 ± 0.02^g	6.20 ± 0.01^c
M	3.44 ± 0.42^{cd}	1.008 ± 0.044^a	2.36 ± 0.02^d	5.85 ± 0.01^e
N	3.06 ± 0.10^{def}	0.986 ± 0.033^a	2.55 ± 0.02^c	5.52 ± 0.01^g
O	2.69 ± 0.24^{efg}	0.989 ± 0.031^a	2.48 ± 0.02^c	5.37 ± 0.01^h
P	2.89 ± 0.47^{def}	0.990 ± 0.036^a	2.38 ± 0.00^d	5.51 ± 0.01^g

Values in the same column with different letters are significantly different ($p < 0.05$), analyzed by Tukey-test.

The emulsion systems from A-H were prepared with WPC, while the emulsions systems from I to P were prepared with WPC and the emulsifier Tween 80 (T80). Comparing these two blocks of emulsions, it was demonstrated that the addition of T80 had a significant effect on diminishing the particle size, especially for the emulsions systems I to L which were formulated with 20 % of the primary emulsion (E1). The particle diameter of these latter systems was between 1.67 and 2.28 µm. In contrast, the bigger particle diameters registered were for the emulsions systems prepared with 30 % of E1 and without T80 (emulsions E-H), presenting a particle diameter range between 3.96 and 4.68 µm. The particle size obtained of the emulsions in this research are similar to the ones reported by Jiménez-Alvarado et al. (2009), with particle size diameters around 3.85 ± 0.04 µm; they also used ultrasonic homogenization for the emulsions preparation, adding WPC and arabic gum as stabilizers. Another recent study (Hernández-Marín, Lobato-Caballeros & Vernon-Carter, 2013) of W/O/W emulsions prepared by mechanical homogenization demonstrates that complexes of whey protein and carboximethylcellulose produced stable emulsions with particle diameters in a range of 2.4-3.2 µm. The droplet size range of the disperse phase of the double emulsions presented in this study suggest good stability of the emulsions since the mechanisms of coalescence may occur in a longer period of time, besides, the addition of the emulsifier increases the interfacial film strength (Jiao & Burgess, 2003).

The pH value of the system is an important parameter to control emulsions separation. It has been reported that when using whey protein isolate as emulsifier, it is more effective when the emulsions are prepared at a pH value far from the isoelectric point of these compound, which is 5.2 (Li et al., 2011; Charoen et al., 2011; Djordjevic, Cercaci, Alamed, McClements & Decker, 2008). The pH values obtained for the emulsions systems were among 5.37 – 6.57 (Table 2). However, the emulsions systems M to P which had the lowest pH values from the rest of the prepared emulsions (5.37-5.85) were the most stables ones. It is important to recall the fact that the cavitation phenomenon, occurred during ultrasonic treatments, is able to produce molecular changes of biopolymers and molecules in general, consequently changing some of its properties and functionality (Knorr, Zenker, Heinz & Lee, 2004; Dolatowski, Standnik & Stasiak, 2007).

The values obtained for the emulsions densities are between 0.986 and 1.013 g/cm^3 (Table 2). This is the only physical property that did not presented significant differences (p>0.05) among the different emulsions systems. However, the viscosity of the W/O/W emulsions was clearly affected by the addition of T80 as emulsifier. It is reported that the emulsions tend to be more stable when the viscosity increases (Jiao & Burgess, 2003; Walstra & van Vliet, 2008). Besides, the higher values of viscosity (2.30 – 3.06 mPa.s) for the formulated double emulsions were for the systems prepared with a higher proportion of the oil phase (30 % of E1) which corresponds to the systems E-H (formulated only with WPC) and M-P (formulated with WPC and T80) (Table 2). The effect of the homogenization parameters, intensity and time of homogenization, caused significant differences in viscosity values in the emulsion systems (p < 0.05), however in Table 2 is not clear to appreciate a common trend. Further on in the results of the predictive model for viscosity, trends for these process parameters will be explained.

3.2 Creaming of W/O/W emulsions

Generally, in W/O/W or O/W emulsions creaming occurs when the oil phase droplets in the dispersed phase are accumulated on the top of the aqueous phase. The creaming develops due to the effect of flotation or buoyancy since the oil droplets are less dense. The formation of a phase with cream like appearance is observed as a cloudy separated phase or opaque liquid (Jiao & Burgess, 2003). In this study, it was observed that the W/O/W emulsions prepared with WPC (systems A-H, Table 1) presented creaming values up to 51.7% after 5 days of storage and increased up to 79 % after 25 days of storage at room temperature (Figure 1). These creaming values are considered high, demonstrating lack of emulsions stability during storage when WPC was used as emulsifier for the double emulsions of OEO. In contrast, when WPC and T80 was added to the double emulsions formulation, with 20% of E1, homogenized with ultrasound for 15 min, after 5 days, no phase separation was observed (0% creaming); nevertheless, after 15 days of storage, creaming values for these emulsions was up to 76- 78 % of creaming. Surh et al. (2007) reported 61% of creaming values for a W$_{15\% \text{ WPI}}$/O/W emulsion, only after one day of storage, which was obtained after a simple pass in a high-pressure homogenizer at 7 MPa.

The most stable double emulsions systems of OEO were obtained (M-P) when WPC and T80 were used in the secondary emulsion, with 30% of E1, not mattering the ultrasound homogenization conditions applied (low or high levels of wave amplitude, for short or long time of ultrasound homogenization treatments) (Figure 1).

3.3 Predictive models

Among the studied physical properties of the double emulsions W/O/W with oregano essential oil, the viscosity and creaming were chosen to analyze the influence of the homogenization process parameters and the formulation variables, on these. The factorial design (Table 1) was then submitted to statistical analysis to obtain two empirical models for viscosity and creaming as the main properties mainly related to emulsions stability.

From the statistical analysis and the empirical model obtained for viscosity of the double emulsions (Table 3) it was demonstrated that all the main effects of the factors studied, homogenization wave amplitude (W), homogenization time (t), Tween 80 concentration used (T80) and the primary emulsion concentration in the double emulsion (E1), did had a significant effect (p < 0.05) on this property. The effect of the factors interactions are also showed in Table 3, as well as the correlation coefficient of the model for viscosity, which was 99.9 %. As stated before in other studies (Jiao & Burgess, 2003; Walstra & van Vliet, 2008; Rosa et al., 2016), as the viscosity increases, the emulsions will present better stability. As it can be seen in Figure 2, viscosity values of the double emulsions can be achieved when W and t increases (homogenization process more severe), and with higher values of E1. The effect of the addition of T80 in the W/O/W formulation is also important, however, it seems that when E1 has higher values, the viscosity is even higher for lower values of T80. The presence of T80 in the W/O/W becomes evidently important when creaming is analyzed after 25 days of storage.

The empirical model for the creaming property (Table 3), also presented a significant effect (p < 0.05) of the main factors studied, as well as some factors' interactions. The coefficient correlation was also high (99.9 %), demonstrating a good fitness of the empirical model to the data. As discussed previously, best stability of the emulsions is obtained with lower values of creaming (less phase separation); analyzing the studied factors this was obtained when W and t increases, and when E1 and T80 also are in the highest level tested.

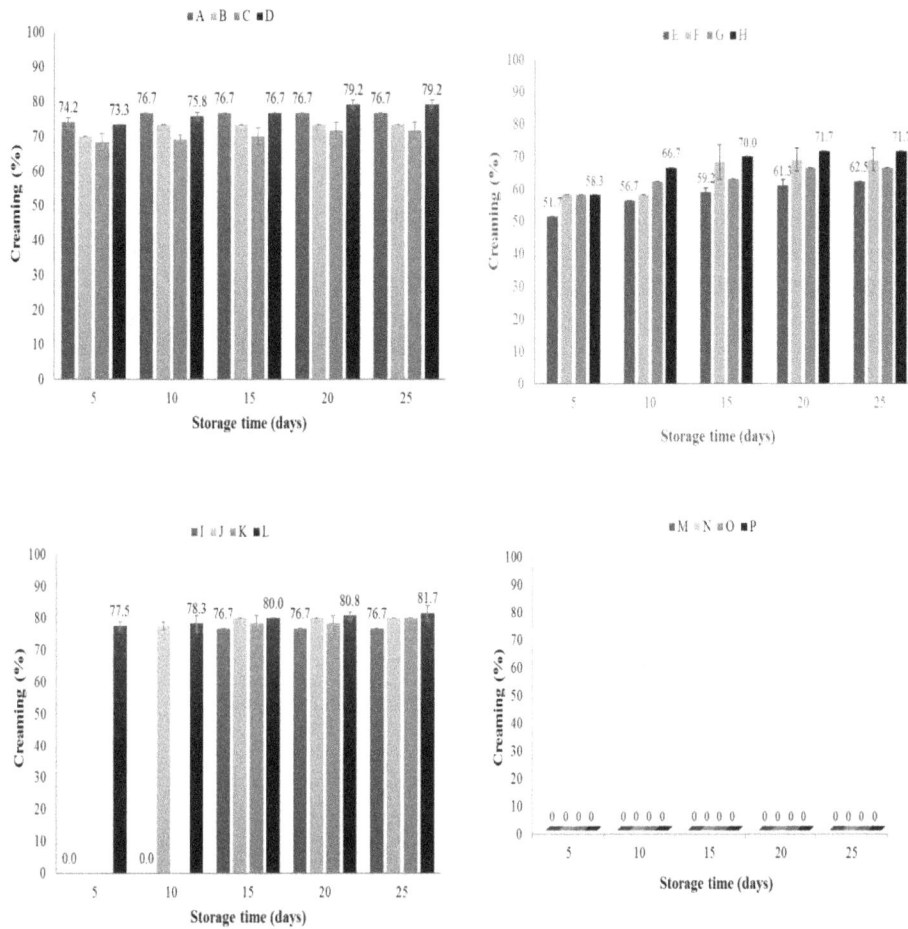

Figure 1. Creaming evolution of W/O/W emulsions prepared with oregano essential oil by ultrasound during storage. See Table 1 for emulsions systems code.

Different combinations of the factors were obtained when both empirical models (viscosity and creaming) were used to optimize the stability of the W/O/W emulsions; considering higher values for the viscosity property and lower values of creaming during storage after 25 days. The combination of the variables for the optimized response was: 54 μm of W, 7.5 min of t, 30% of E1 and 0.28 % of T80. With this combination of variables, the predicted value of viscosity was 2.404 mPas.s, while for creaming was 3.50 % with a prediction desirability average for both empirical models of 0.753.

Table 3. Terms and coefficients of the models obtained for viscosity and creaming of the double emulsions prepared with oregano essential oil by ultrasonic homogenization

Term	Units	Viscosity (mPa.s)		Creaming (%)	
		Coefficient	p-value	Coefficient	p-value
Constant		-6.711		207.50	
Wave amplitude (W)	microns	0.1195	0.000	-2.54	0.003
Homogenization time (t)	min	1.8896	0.000	-27.64	0.000
Tween 80 concentration (T80)	%	33.73	0.000	164.00	0.000
Primary emulsion concentration (E1)	%	0.3542	0.000	-4.23	0.000
W*t		-0.0378	0.048	0.60	0.038
W*T80		-0.673	0.000	-	0.484
W*E1		-0.0047	0.000	-	0.808
t*T80		-	0.449	100.50	0.004
t*E1		-0.0775	0.000	-	0.482
T80*E1		-1.4229	0.000	-14.36	0.000
W*t*T80		0.1682	0.000	-2.27	0.008
W*t*E1		0.0015	0.000	-0.03	0.007
W*T80*E1		0.0278	0.000	-0.24	0.007
t*T80*E1		-	0.919	-3.37	0.003
W*t*T80*E1		-0.0071	0.000	0.08	0.001
Model concordance with data (%)		99.77		99.70	

The terms with significant effect ($p < 0.05$) were included in the respective model.

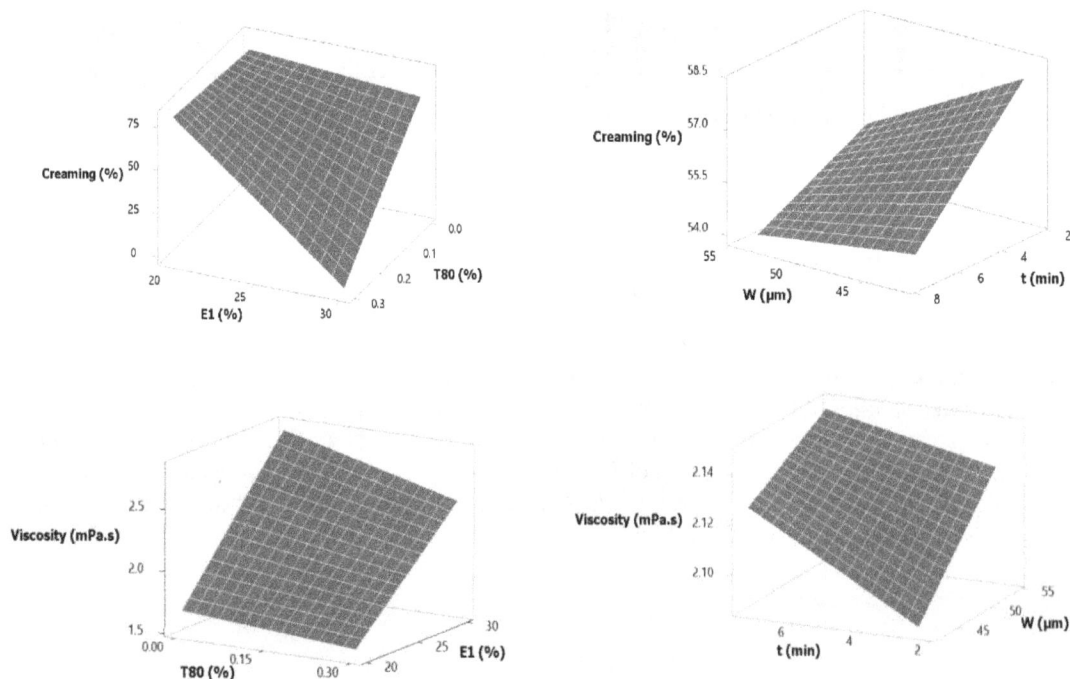

Figure 2. Surface response plots obtained for creaming and viscosity models as a function of ultrasound wave amplitude (W), homogenization time (t), Tween 80 (T80) and primary emulsion concentration (E1) for W/O/W emulsions with oregano essential oil

4. Conclusions

The good stability of double emulsions (W/O/W) with oregano essential oil can be achieved using whey protein concentrate (WPC) in combination with Tween 80, and adjusting the ultrasound homogenization process parameters (intensity and time). With regard to the W/O/W emulsion formulation, WPC demonstrated to be a good food-grade stabilizer since the emulsions containing it, did not showed phase separation (creaming) after 25 days of storage, when Tween 80 was also added, using a proportion 30:70 of the primary emulsion in the double emulsion. Moreover, good stability of the W/O/W emulsions was improved when severe ultrasonic homogenization was applied, highest wave amplitude and longer treatment times, which diminishes the particle size diameter of the dispersed phase and increases their viscosity, which are related to a better stability of the emulsions.

Two empirical models were proposed for viscosity and creaming of the W/O/W emulsions, which resulted to have high correlation coefficients to the experimental data, and included all the main effects of the variables studied and some interactions of these. A combination set of values for the different studied variables was proposed for the optimization of the emulsions stability, maximizing viscosity and minimizing creaming.

Oregano essential oil in W/O/W emulsions may be added in different types of food, in which their antimicrobial or antioxidant properties may be exploited, since these emulsions may serve as controlled release systems.

Acknowledgments

The authors acknowledge the financial support of the National Council of Science and Technology (CONACYT) of Mexico, for the project 168709 CB-2011-01. Author Peredo-Luna gratefully acknowledges the financial support for her PhD studies to CONACyT and the Universidad de las Américas Puebla (UDLAP).

References

A.O.A.C. *Official Methods of Analysis* (1995). 14th ed. Association of Official Analytical Chemists. Inc. Washington, D.C. USA.

Akhtar, M., & Dickinson, E. (2001). Water-in-oil-in-water multiple emulsions stabilized by polymeric and natural emulsifiers. In E. Dickinson, R. Miller (eds), *Food Colloids: Fundamentals of Formulation* (pp 133 - 143). Cambridge, UK: Royal Society of Chemistry.

Asensio, C. M., Grosso, N. R., & Juliani, H. R. (2015). Quality preservation of organic cottage cheese using oregano essential oils. *LWT-Food Science and Technology, 60*(2), 664-671. http://dx.doi.org/10.1016/j.lwt.2014.10.054

Bakkali, F., Averbeck, S., Averbeck, D., & Idaomar, M. (2008). Biological effects of essential oils-A review. *Food and Chemical Toxicology, 46*(2), 446-475. http://dx.doi.org/10.1016/j.fct.2007.09.106

Bakry, A. M., Abbas, S., Ali, B., Majeed, H., Abouelwafa, M. Y., Mousa, A., & Liang, L. (2016). Microencapsulation of oils: a comprehensive review of benefits, techniques, and applications. *Comprehensive Reviews in Food Science and Food Safety, 15*(1), 143-182. http://dx.doi.org/10.1111/1541-4337.12179

Beirão-da-Costa, S., Duarte, C., Bourbon, A. I., Pinheiro, A. C., Januário, M. I. N., Vicente, A. A., Beirão, da Costa, M. L., & Delgadillo, I. (2013). Inulin potential for encapsulaton and controlled delivery of Oregno essential oil. *Food Hydrocolloids, 33,* 199-206. http://dx.doi.org/10.1016/j.foodhyd.2013.03.009

Bhargava, K., Conti, D. S., da Rocha, S. R. P., & Zhang, Y. F. (2015). Application of an oregano oil nanoemulsion to the control of foodborne bacteria on fresh lettuce. *Food Microbiology, 47,* 69-73. http://dx.doi.org/10.1016/j.fm.2014.11.007

Bostsoglou, N. A., Govaris, A., Botsoglou, E. N., Grigoropoulou, S. H., & Papageorgiou, G. (2003). Antioxidant activity of dietary oregano essential oil and alpha-tocopheryl acetate supplementation in logn-term frozen stored turkey meat. *Journal of Agricultural and Food Chemistry, 51*(10), 2930-2936. http://dx.doi.org/10.1021/jf021034o

Burt, S. (2004). Essential oils: their antibacterial properties and potential applications in foods-a review. *International Journal of Food Microbiology, 94,* 223-253. http://dx.doi.org/10.1016/j.ijfoodmicro.2004.03.022

Cardoso-Ugarte, G. A., López-Malo, A., & Jiménez-Munguía, M. T. (2016). Application of nanoemulsions technology for encapsulation and release of lipohilic bioactive compounds in food. In Grumezescu, A. (ed.), *Emulsions* (pp 227-255), Elsevier. ISBN: 978-0-12-804306-6.

Charoen, R., Jangchud, A., Jangchud, K., Harnsilawat, T., Naivikul, O., & David Julian McClements, D.J. (2011). Influence of biopolymer emulsifier type on formation and stability of rice bran oil-in-water emulsions: whey protein, gum arabic, and modified starch. *Journal of Food Science, 76*(1), E165-E172. http://dx.doi.org/10.1111/j.1750-3841.2010.01959.x

Dickinson, E. (2011). Double emulsions stabilized by food biopolimers. *Food Biophysics, 6,* 1-11. http://dx.doi.org/10.1007/s11483-010-9188-6

Djordjevic, D., Cercaci, L., Alamed, J., McClements, D., & Decker, E. A. (2008). Chemical and phisical stabilityof protein- and gum arabic-stabilized oil -in-water emulsions containing Limonene. *Journal of Food Science, 73*(3), C167-C172. http://dx.doi.org/10.1111/j.1750-3841.2007.00659.x

Dolatowski, Z. J., Standnik, J., & Stasiak, D. (2007). Applications of ultrasound in food technology. *Acta Scientiarum Polonorium Technologia Alimentaria, 6*(3), 89-99.

Hammer, K. A., Carson, C. F., & Riley, T. V. (1999). Antimicrobial activity of essential oils and other plant extracts. *Journal of Applied Microbiology, 86,* 985-990. http://dx.doi.org/10.1046/j.1365-2672.1999.00780.x

Hemar, Y., Cheng, L. J., Oliver, C. M., Sanguansri, L., & Augustin, M. (2010). Encapsulation of resveratrol using water-in-oil-in-water double emulsions. *Food Biophysics, 5,* 120-127. http://dx.doi.org/10.1007/s11483-010-9152-5

Hernández-Marín, N. Y., Lobato-Calleros, C., & Vernon-Carter, E. J. (2013). Stability and rheology of water-in-oil-in-water emulsions made with protein-polysaccharide soluble complexes. *Journal of Food Engineering, 119,* 181-187. http://dx.doi.org/10.1016/j.jfoodeng.2013.05.039

Jiao, J., & Burgess, D. J. (2003). Rheology and stability on water-in-oil-in-water emulsions containing Span 83 and Tween 80. *AAPS Pharmaceutical Science, 5*(1), E7-E19. http://dx.doi.org/10.1208/ps050107

Jiménez-Alvarado, R., Beristain, C. I., Medina-Torres, L., Román-Gerrero, A., & Vernon-Carter, E. J. (2009). Ferreus bisglycinate and release in w1/o/w2 multiple emulsions stabilized by protein-polysaccharide complexes. *Food Hydrocolloids, 23,* 2425-2433. http://dx.doi.org/10.1016/j.foodhyd.2009.06.022

Knorr D., Zenker, M., Heinz, V., & Lee, D. U. (2004). Applications and potential of ultrasonics in food processing. *Trends in Food Science and Technology, 15,* 261-266. http://dx.doi.org/10.1016/j.tifs.2003.12.001

Knoth, A., Scherze, I., & Muschiolik, G. (2005.) Stability of water-in-oil-emulsions containing phosphatidylcholine-depleted lecithin. *Food Hydrocolloids, 19,* 635-640. http://dx.doi.org/10.1016/j.foodhyd.2004.10.024

Li, B., Jiang, Y., Liu, F., Chai, Z., Li, Y., & Leng, X. (2011). Study of the encapsulation efficiency and controlled release property of whey protein isolate-polysaccharide complexes in W1/O/W2 double emulsions. *International Journal of Food Engineering, 7*(3), Art. 14. http://dx.doi.org/10.2202/1556-3758.2321

Mun, S., Choi, Y., Rho, S. J., Kang, C. G., Park, C. H., & Kim, Y. R. (2010). Preparation and characterization of water/oil/water emulsions stabilized by polyglycerol polyricinoleate and whey protein isolate. *Journal of Food Science, 75,* E116-E125. http://dx.doi.org/10.1111/j.1750-3841.2009.01487.x

Rosa, T. M. M. G., Silva, E. K., Santos, D. T., Petenate, A. J., Angela, M., & Meireles, A. (2016). Obtaining annatto seed oil miniemulsions by ultrasonication using aqueous extract from Brazilian gingseng roots as a biosurfactant. *Journal of Food Engineering, 168,* 68-78. http://dx.doi.org/10.1016/j.jfoodeng.2015.07.024

Rukmini, A., Raharjo, S., Hastuti, P., & Supriyadi, S. (2012). Formulation and stability of wáter-in-virgin coconut oil microemulsion using ternary food grade nonionic surfactants. *International Food Research Journal, 19*(1), 259-264.

Su, J., Flanagan, J., & Singh, H. (2008). Improving encapsulation efficiency and stability of water-in-oil-in-water emulsions using a modified gum arabic (Acacia (sen) SUPER GUMTM). *Food Hydrocolloids, 22,* 112-120. http://dx.doi.org/10.1016/j.foodhyd.2007.03.005

Surh, J., Vladisavljevicǎ, G. T., Mun, S., & McClements, D. J. (2007). Preparation and characterization of water/oil and water/oil/ water emulsions containing biopolymer-gelled water droplets. *Journal of Agricultural and Food Chemistry, 55,* 175-184. http://dx.doi.org/10.1021/jf061637q

Walstra, P., & van Vliet, T. (2008). Dispersed systems: basic considerations. In S. Damodaran, K. Parking, & O. R. Fennema (Eds.), *Fenema's food chemistry* (pp. 783-847). Boca Raton: CRCPress/Taylor&Fracis.

10

Survival of *Lactobacillus rhamnosus* GG during Simulated Gastrointestinal Conditions Depending on Food Matrix

Riin Karu[1] & Ingrid Sumeri[1]

[1]Competence Center of Food and Fermentation Technologies, Akadeemia tee 15A, 12618 Tallinn, Estonia

Correspondence: Ingrid Sumeri, Competence Center of Food and Fermentation Technologies, Akadeemia tee 15A, 12618 Tallinn, Estonia. E-mail: ingrid@tftak.eu

Abstract

When developing new probiotic foods, their protective properties in maintaining viability of probiotics under gastrointestinal conditions should be evaluated. In the current study, human upper gastrointestinal tract simulator (GITS) was used to compare the effect of different food matrixes on the survival of *Lactobacillus rhamnosus* GG (LGG). pH-auxostat was chosen for the cultivation of LGG cells to obtain culture samples in the same physiological state at maximum growth rate for the GITS experiments. The LGG culture was centrifuged and fast frozen in liquid nitrogen in various liquid food matrixes (commercial UHT milk, soymilk, apple juice, titrated apple juice, whey protein powder drink and M.R.S. Broth as reference) and stored at -40^{0}C. During 3-month storage, reduction of viability was significant only for apple juice. In the GITS experiments, bile had a greater negative impact on LGG than acid conditions, also the effect of food matrix was noted - in the case of milk, soymilk and whey protein powder drink only the highest concentration of bile (0.4%) caused a significant drop in the viability of bacteria when compared to apple juice. To maximize the health benefits of foodstuffs, it should be taken into account that the survival of probiotics during fast freezing, storage and gastrointestinal passage is dependent on the food matrix.

Keywords: bile, food matrix, *in vitro* gastrointestinal simulator, *Lactobacillus rhamnosus* GG, probiotic

1. Introduction

1.1 Food in Human Health

The basic function of a food is to provide vitamins, minerals and energy derived from the proteins, carbohydrates and lipids required for the well-being of the human body (Bigliardi & Galati, 2013). The emphasis on the benefits beyond the basic nutritional functions in the role of foodstuffs has given rise to the term `functional foods´ (Chen, Ma, Liang, Peng, & Zuo, 2011; Eussen *et al.,* 2011).

The most important and the most frequently used functional food compounds can be listed as probiotics, prebiotics, plant antioxidants, vitamins and calcium (Grajek, Olejnik, & Sip, 2005). Probiotics are usually defined as microbial food supplements that have beneficial effects on consumers (Pang, Xie, Chen, & Hu, 2012).

1.2 Probiotics as Food Components

A probiotic must remain viable through the processing and storage of the food product prior to consumption as well as to be resistant to the gastrointestinal conditions during passage (Puupponen-Pimiä *et al.*, 2002). There are variable recommendations for the minimum viable count threshold to provide health benefits (Champagne, Ross, Saarela, Hansen, & Charalampopoulos, 2011; Karimi, Mortazavian, & Amiri-Rigi, 2012). The usual recommendation is that probiotic bacteria be present in products in a minimal viable concentration of 10^6 -10^7 CFU/ml (Cruz, Faria, & Van Dender, 2007; Vasiljevic & Shah, 2008), with the daily intake approximately of 10^8 CFU·g^{-1}, taking into account reductions in numbers during the gastrointestinal passage (Martins *et al.*, 2013).

Food matrix properties have a significant impact on the viability of probiotics, the most influential factors being the pH and buffering capacity, but also the chemical composition of food. The latter includes fat and carbohydrate content, concentration and type of proteins, presence of polyunsaturated fatty acids, plant extracts *etc* (Ranadheera, Baines, & Adams, 2010).

1.3 Different Food Matrixes Used for Incorporation of Probiotics

Traditionally probiotics were mainly incorporated into dairy products but lately there has been increasing interest in manufacturing non-dairy probiotic products, such as fruit juices and soy products. Effect of lactose intolerance and the cholesterol content are listed as the main disadvantages related to consumption of fermented dairy products. Also, there is a rising demand for vegetarian probiotic products (Prado, Parada, Pandey, & Soccol, 2008; Granato, Branco, Nazzaro, Cruz, and Faria, 2010). Soy products are considered to have good nutritional value for being a source for good quality proteins in high content, but also isoflavones, folic acid and some vitamins (Ng, Lye, Easa, and Liong, 2008; Kaur & Das 2011).

The protective effect of milk against low pH and bile in relation to lactic acid bacteria has been well documented (Saarela, 2011). In addition, soy milk has been reported to have protective buffering capacities as well - use of fermented soymilk as a delivery vehicle for probiotic *Lactobacillus casei* Zhang increased the tolerance to simulated gastric and intestinal juice (Wang *et al.*, 2009).

However, in case of fruit juices, the growth and the strain-specific viability of cells depends on pH, oxygen content, final acidity and the concentration of lactic acid and acetic acid of the product (Rivera-Espinoza & Gallardo-Navarro, 2010). In general, *Lb. rhamnosus* has been reported to maintain desirable viable counts when tested on fruity matrix (do Espirito Santo, Perego, Converti, & Oliveira, 2011).

Due to technological advancements high-protein and low-fat functional whey ingredients, such as whey protein concentrates have been developed (Smithers, 2008) and because of the nutritional and functional properties, whey proteins have been successfully formulated into energy and sports drinks (Wright, 2007).

1.4 Hypothesis and Research Design

The aim of the current study was to investigate the effects of various food matrixes on the survival of the probiotic bacterium *Lactobacillus rhamnosus* GG in simulated gastrointestinal conditions, at different bile concentrations.

2. Materials and Methods

2.1 Microorganisms and Media

The probiotic lactobacillus strain *Lactobacillus rhamnosus* GG isolated from Gefilus daily dose drink (Valio Ltd.) was selected for the study.

2.2 pH-auxostat Cultivation

Bacterial cultures were obtained using pH-auxostat cultivation, as described by Sumeri *et al.* (2010). In short, anaerobic cultivation was performed at 37°C and pH = 6.0 with culture volume of 500 ml, using an Applikon 1 L fermenter; controlled by an ADI 1030 biocontroller ("Applikon", The Netherlands) and cultivation control program "BioXpert" ("Applikon"). MRS medium (LAB M, UK) was used as feeding. Upon achieving a stable specific growth rate D = 0.8 h^{-1} sample collection from the fermenter was started.

2.3 Freezing in Liquid Food Matrix

After reaching steady state at the pH-auxostat cultivation, samples of 200 ml (collected in 50ml tubes) were taken from the fermenter, followed by a centrifugation step (10 min 8000 rpm 10°C) and resuspending in different food matrixes in equal volume. Commercial food products were chosen – UHT milk (Tere AS, Tallinn, Estonia), soymilk (Alpro, Gent, Belgium), apple juice (AS Põltsamaa Felix, Tallinn, Estonia) and whey protein powder drink (Func Food Finland OY, Tampere, Finland) with M.R.S. Broth (Lab M Limited, Lancashire, UK) employed as the reference matrix.

Composition of the products is listed in Table 1, with the pH values determined with Mettler Toledo MP 125 pH Meter (Mettler Toledo Inc., Schwarzenbach, Switzerland).

Prior to resuspending, the pH of 200 ml apple juice (originally pH 3.30) was adjusted to pH 6.70 with 2M NAHCO$_3$ titration, to lessen the impact of acidic environment on long-term storage (in comparison, untitrated apple juice was also used). The whey protein powder drink was prepared by mixing 20 g of whey protein powder with 200 ml of distilled water (according to manufacturer's instructions).

The bacterial suspension was fast frozen by dripping into liquid nitrogen. The resulting `probiotic beads` (1-2mm in size) were stored at -40°C until subjected to GITS experiments.

Table 1. Composition of the food products used in the study, according to manufacturers' labels and pH measurements.

Component	Milk	Soymilk	Whey protein drink	M.R.S	Apple juice
Fat	3.2%	1.8%	8.8%	0.1%	0%
Protein	3%	3%	75%	1.4%	0.5%
Lactose	4.7%	0%	3.3%	-	0%
Sugar	-	2.5%	0.5%	2%	10.7%
Fiber	0%	0.5%	0%	-	0%
pH	6.75	6.89	6.46	6.86	3.30

2.4 Gastrointestinal Tract Simulation (GITS) Experiments

The frozen bacterial cultures were subjected to GITS experiments, as previously described by Sumeri, Arike, Adamberg, and Paalme (2008). GITS experiments were carried out with 5 separate food matrixes used for the incorporation of *Lactobacillus rhamnosus* GG and 5 different bile concentrations, tested in 3 parallels for each experimental combination - 75 experiments were carried out in total.

The simulations were carried out in a single fermenter, spatially and temporally mimicking human gastric transit. At the beginning of GITS experiment, the adult fasted stomach conditions were reproduced (100 ml of 10 mmol L^{-1} HCl), then 200 ml of bacterial culture was added, thus reaching the working volume (300 ml) of the vessel.

The next step in the simulation was titration to pH 3.0 with 1 mol L^{-1} HCl at a rate of 20 mmol h^{-1} (`gastric phase`). The ´gastric phase´ was followed by neutralization with 1 mol L^{-1} $NaHCO_3$ to pH 6.0 and subsequent addition of bile salts to simulate the passing into the duodenum, followed by a 30 minute bile incubation phase. Different bile salt concentrations (0.5% to 4%) were applied to obtain the final bile salt concentrations in the range from 0.05% to 0.4%, with a 0% bile concentration serving as a reference control.

The ´ileum phase´ was reproduced with 5,5h dilution (D=0.4 h^{-1}) of the fermenter contents, while the pH was kept constant (= pH 6.5) by $NaHCO_3$ titration.

In preparation for the GITS experiments, 80 g of frozen milk beads were added fresh milk up to 200 ml, mixed till melting and subjected to the GITS experiment. In case of the other food matrixes, titrated apple juice/soymilk/ whey protein powder drink was prepared as previously described and added up to 200 ml to the corresponding beads, mixed till melting and also subjected to GITS experiments.

2.5 Determination of Bacterial Viability in the GITS Experiments

Viability of bacteria was assessed by plate counts. The numbers of colony-forming units (cfu/ml^{-1}) were determined by counting colonies from the serial dilutions plated on M.R.S. agar (LAB M, UK) using a pour-plate method (de Man, Rogosa, & Sharpe, 1960). Viable bacterial counts were determined from the fresh pH auxostat culture, before and after freezing in liquid nitrogen and also from several points during the GIT simulation (inoculum, pH 3, bile, bile 30 min, end of 5,5h dilution phase).

The survival rate of *Lactobacillus rhamnosus* GG was calculated as the difference in the viable bacterial counts (Log (cfu/ml)) at the beginning and the end of the GITS experiments, while the final counts were normalized taking into account the variation of inocula at each experiment.

The viability was calculated as concentration of the bacteria in the fermenter at time t - CFU_t (cfu/ml) as shown in:

$$CFU_t = CFU_{t-1} + (CFU_t * \mu/60) \tag{1}$$

During the dilution GITS phases (after the 30min bile incubation) the correction for the dilution was taken into account according to the specific growth rate μ (1/h^{-1}):

$$\mu = D + ((rate(CFU)/CFU_{t+1}) * 60) \tag{2}$$

where rate(CFU) is calculated according to:

$$rate(CFU) = (CFU_{t+1} - CFU_t)/(t_{t+1} - t_t) \tag{3}$$

The viability calculations were carried out with the BioXpert software followed by normalization of data in Excel.

3. Results

3.1 The Effects of Fast Freezing in Different Food Matrixes

During pH-auxostat cultivation the physiologically stable state of the *Lactobacillus rhamnosus* GG cells was achieved and maintained during continuous culture. The constant specific growth rate (D) was reached after flow through of five working volumes (500ml), at which point sample collecting was started.

The experiments were divided into two series – *Lactobacillus rhamnosus* GG was cultivated in a pH-auxostat (viable count prior to collecting samples $1.31 \pm 0.21 * 10^8$) and incorporated into milk (pH = 6.75) for initial observation of the effects of bile variation and then, a second a pH-auxostat was carried out (viable count prior to collecting samples $1.22 \pm 0.12 * 10^8$) with freezing the culture into different commercially available liquid matrixes - UHT milk, soymilk, apple juice, titrated apple juice, whey protein powder drink and M.R.S. Broth as reference.

Immediately after fast freezing the viable counts of *Lactobacillus rhamnosus* GG remained above 10^8 for all the matrixes tested, as seen in Figure 1. In case of soymilk a 0.2 log unit drop was observed in the viable count after fast freezing, also for untitrated apple juice there was a drop of 0.21 log units. For M.R.S broth, whey protein powder drink and titrated apple juice slightly higher viable counts were detected in freshly frozen samples, which can be related to physical occurrences on the cell chains as noted by Foschino, Fiori, and Galli (1996).

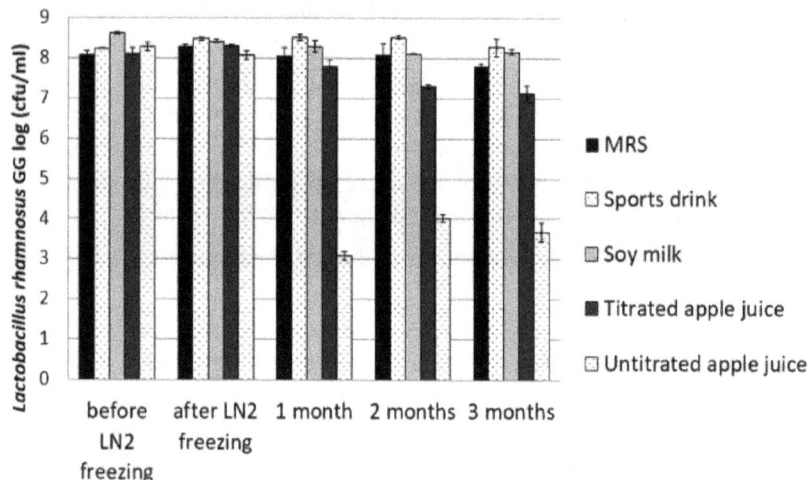

Figure 1. Stability of *Lactobacillus rhamnosus* GG frozen in various matrixes during 3-month storage at -40⁰C

After 3 months of storage at -40 ⁰C the viable counts were $1.48 * 10^8$, $1.33 * 10^8$, $3.85* 10^7$ and $1.05 *10^7$ cfu/ml in frozen whey protein powder drink, soymilk, M.R.S broth and titrated apple juice, respectively, hence remaining at the level between 10^6 -10^8 which is usually recommended for probiotics (Cruz, Antunes, Sousa, Faria, & Saad, 2009). The average loss of viability during storage at -40 ⁰C of *Lactobacillus rhamnosus* GG was 0.99 ± 0.31 log units in titrated apple juice, 0.75 ± 0.12 log units in case of M.R.S broth and only $0.29 \pm 0.0.6$ and 0.11 ± 0.19 in soymilk and whey protein powder drink, demonstrating good viability during storage for the latter matrixes. The storage stability of *Lactobacillus rhamnosus* GG frozen in titrated apple juice was good, remaining above 10^6, however, in untitrated apple juice the viability dropped significantly, to $3.65*10^3$ cfu/ml. The pH was relatively stable during storage (data not shown) for M.R.S broth (remained above pH = 5.50) and soymilk (above pH = 6.30) with a slight drop in case of the whey protein powder drink which had pH 6.45 at the beginning and pH = 6.09 at the end of storage. For untitrated apple juice the pH was low – with an average pH of 3.5. In case of titrated apple juice the pH remained high during storage (about pH = 7), probably due to the titration since the pH of frozen apricot and peach juices and of frozen apple products has been shown to decrease during storage (Elhadad, Alwakdi, Abushita, & Abdulsalam, 2013; Sigita, Krasnova, Seglina, Aboltins & Skrupskis, 2013). Given the low survival during storage untitrated apple juice beads were excluded from further GITS experiments.

3.2 Resistance to Simulated Gastrointestinal Conditions

The GITS experiments with varying bile concentrations were carried out and the results are shown in Figure 2. *Lactobacillus rhamnosus* GG survived well during the `gastric phase` (pH 3) of the experiments, the average loss in viability in M. R. S. Broth was 0.67 ± 0.07 log units; with 0.43 ± 0.08; 0.35 ± 0.09 and 0.25 ± 0.08 log units in soymilk, apple juice and milk, respectively, with the number of bacteria practically unchanged for whey protein powder drink.

'Food' matrixes	Concentration of bile salts solution (vol/vol, %)				
MRS	0%	0.05%	0.10%	0.20%	0.40%
pH 3	-0.59 ± 0.02	-0.75 ± 0.07	-0.7 ± 0.006	-0.59 ± 0.02	-0.71 ± 0.04
Bile	-0.61 ± 0.02	-1.04 ± 0.05	-1.21 ± 0.10	-1.7 ± 0.10	-1.78 ± 0.10
Bile 30min	-0.65 ± 0.08	-1.03 ± 0.02	-2.58 ± 0.20	-3.5 ± 0.33	-5.61 ± 0.34
5,5h dilution	0.45 ± 0.05	-0.89 ± 0.10	-1.71 ± 0.10	-2.63 ± 0.26	-4.3 ± 0.12
Milk	0%	0.05%	0.10%	0.20%	0.40%
pH 3	-0.26 ± 0.01	-0.27 ± 0.01	-0.18 ± 0.01	-0.18 ± 0.02	-0.36 ± 0.03
Bile	-0.26 ± 0.02	-0.28 ± 0.02	-0.27 ± 0.01	-0.2 ± 0.01	-2.28 ± 0.06
Bile 30min	-0.28 ± 0.02	-0.31 ± 0.03	-0.38 ± 0.04	-0.32 ± 0.01	-3.46 ± 0.06
5,5h dilution	0.21 ± 0.02	0.46 ± 0.01	0.38 ± 0.03	0.39 ± 0.04	-2.2 ± 0.20
Apple juice	0%	0.05%	0.10%	0.20%	0.40%
pH 3	-0.27 ± 0.04	-0.32 ± 0.01	-0.36 ± 0.03	-0.51 ± 0.04	-0.31 ± 0.04
Bile	-0.3 ± 0.10	-0.65 ± 0.02	-0.89 ± 0.02	-4.72 ± 0.38	-5.07 ± 0.10
Bile 30min	-0.44 ± 0.16	-0.75 ± 0.06	-1.23 ± 0.11	-4.74 ± 0.40	-5.02 ± 0.08
5,5h dilution	-0.17 ± 0.09	-1.17 ± 0.02	-1.7 ± 0.17	-4.31 ± 0.20	-5.18 ± 0.33
Whey protein powder drink	0%	0.05%	0.10%	0.20%	0.40%
pH 3	0.07 ± 0.006	0.03 ± 0.003	0.05 ± 0.004	0.05 ± 0.004	-0.04 ± 0.002
Bile	0.03 ± 0.002	-0.004 ± 0.0003	-0.04 ± 0.002	-0.02 ± 0.002	-1.93 ± 0.03
Bile 30min	0.01 ± 0.001	-0.05 ± 0.005	-0.05 ± 0.003	-0.07 ± 0.007	-3.85 ± 0.08
5,5h dilution	-0.45 ± 0.01	-0.34 ± 0.02	-0.55 ± 0.03	-0.53 ± 0.03	-3.62 ± 0.10
Soymilk	0%	0.05%	0.10%	0.20%	0.40%
pH 3	-0.51 ± 0.05	-0.46 ± 0.03	-0.39 ± 0.02	-0.31 ± 0.01	-0.47 ± 0.02
Bile	-0.28 ± 0.03	-0.48 ± 0.01	-0.21 ± 0.01	-0.32 ± 0.03	-3.36 ± 0.02
Bile 30min	-0.3 ± 0.01	-0.36 ± 0.03	-0.4 ± 0.04	-0.42 ± 0.03	-3.5 ± 0.05
5,5h dilution	-0.31 ± 0.02	-0.61 ± 0.06	-0.63 ± 0.06	-0.8 ± 0.08	-3.31 ± 0.02

Figure 2. Heatmap representation of survival (Log (cfu/ml)) of *Lactobacillus rhamnosus* GG in various food matrixes during GITS experiments depending on different bile concentrations. Sampling points correspond to the following GITS stages: pH 3 (endpoint of acid addition or `gastric phase`), Bile (directly after bile addition), Bile 30min (after 30min incubation with bile), 5,5h dilution (endpoint of experiment). Color coding added according to the Excel Conditional Formatting on the Green – White – Red Color Scale, with green showing the smallest reduction, red maximum reduction and white median results. Data are means ± SD, calculated from parallel experiments which were averaged and expressed as one data point

During `gastric phase` whey protein powder drink minimalized the effects of the acidic conditions with virtually no detrimental effect on viability. In case of milk the average loss in viability was the smallest, followed by similar losses in soymilk and apple juice, with the highest average loss in viability observed in M. R. S. Broth.

The resistance of *Lactobacillus rhamnosus* GG to bile salts was dependant on the type of liquid food applied for incorporation and the concentration of bile. A similar trend in the effect of bile concentration was observed in the reference matrix M.R.S. broth and in apple juice - the effect of increasing bile concentration was directly proportional to the decrease of viability of *Lactobacillus rhamnosus* GG. In case of 0.1% bile for M.R.S and apple juice the decrease in viability by the end of the experiment was about 2 log units and for 0.4% bile the drop was about 4 and 5 log units for M.R.S and apple juice, respectively.

In case of titrated apple juice the impact of bile was much more pronounced starting with the 0.1% bile, when compared with the effect of acid conditions (pH 3). However the bile damage was observed directly upon addition with no significant added effect during incubation. That was in contrast with M.R.S, which also showed a significant decrease starting from 0.1% bile, but also having an increasingly detrimental influence in relation with the incubation time.

The whey protein powder drink showed the best resistance upon direct bile addition and during the 30 minute bile incubation as seen in Figure 2. In case of milk and soymilk small similar losses were observed. Noticeable decrease in viability for direct bile addition and incubation for milk and soymilk was only caused by the highest bile concentration.

For the highest bile concentration applied (0.4%) the smallest loss in survival by the end of the experiment was observed in case of milk, followed by soymilk and whey protein powder drink. For bile concentrations 0% - 0.2% the decrease by the end of the experiment was less than 0.6 log units for whey protein powder drink and 1 log unit for soymilk.

In case of the control experiment with no bile addition, an increase in viability was noted for M.R.S, however in case of apple juice there was still a small drop in viability by the end of the experiment.

4. Discussion

Traditionally, the beneficial properties of probiotics have been made available by incorporating them into food. Stability of the probiotics therefore depends on the properties of the food matrixes, which also influences the survival of probiotics during gastrointestinal transit. In commercial dairy products the mainly used probiotic strains are *Lactobacillus* spp. and *Bifidobacterium* spp., added as adjunct cultures (Vasiljevic & Shah 2008). In case of non-fermented products probiotics can be added as liquid biomass concentrate after they have been grown in industrial medium and harvested by filtration or centrifugation (Makinen, Berger, Bel-Rhlid, & Ananta, 2012).

The pH-auxostat enables the growth of bacteria at the maximum growth rate in suitable and unstressful conditions. The pH-controlled continuous cultivation was successfully applied in case of *Lb. acidophilus* La-5, *Lb. rhamnosus* GG and *Lb. fermentum* ME-3 with M.R.S broth as growth medium by Sumeri *et al.* (2010). For producing frozen bacterial cultures, the method of making granules by dripping cell suspension into liquid nitrogen (LN$_2$) through a nozzle or disc with bores has been used in commercial applications (Santivarangkna, Kulozik, & Foerst, 2011). So pH-auxostat cultivation and fast freezing of probiotic bacteria in LN$_2$ were applied in the present study as well.

In general, *Lb. rhamnosus* GG is less sensitive to storage in higher pH juices than in low pH juice (Sheehan, Ross, & Fitzgerald, 2007). For *Lb. rhamnosus* R0011 stored in a commercial apple–pear–raspberry juice blend good viability at 4^{0}C was shown (Champagne, Raymond, & Gagnon, 2008). On the other hand, low pH of apple juice (pH 3.4/pH 3.6) has been mentioned to have inhibitory properties on *Lactobacillus rhamnosus* strain E800 (Saarela, Alakomi, Puhakka, & Mättö, 2009), with neutralization suggested to prevent the negative effect of fruit juice on probiotic bacteria (Vinderola, Costa, Regenhardt, & Reinheimer, 2002). Therefore, titration of apple juice was also carried out in the current study.

In our experiments, when incorporated into M.R.S, whey protein powder drink and soy milk the survival of LGG during fast freezing was very good, with a small loss observed in titrated apple juice during storage. However, in case of untitrated apple juice an 1.8-fold average loss in survival over 3 month storage was observed. This significant reduction clearly demonstrates that efficiency of fast freezing is also dependent on the matrix type, pH and chemical composition.

The main gastrointestinal stress factors are considered to be the acidic environment of the stomach and the presence and effect of bile salts in the duodenum (Mills, Stanton, Fitzgerald, & Ross, 2011).

Gorbach and Goldin (1989) showed that *Lactobacillus rhamnosus* GG decreased from 10^8 to 10^6 after 0.5 h at pH 2.5 in normal human gastric juice. During *in vitro* gastric simulation (pH = 3 or pH = 2 for 2h, without pepsin) a drop of 3 and 7 log units, respectively, was observed by Succi *et al.* (2005) when incubation with simulated gastric juice (pH = 2, with pepsin and bile) without added glucose and dilute HCl (pH = 2) for 90 minutes resulted in similar losses (~5.6 log units) in survival (Corcoran, Stanton, Fitzgerald, & Ross, 2005).

However, Papadimitriou *et al.* (2015) have stated that the pH values used for conventional *in vitro* assays for probiotics are unrealistically low, while not taking into consideration the effects of food vehicles and the pre-stressing of probiotics due to food processing and storage. In addition, data from various GIT simulators were found to correlate well with *in vivo* results, with the advantage of employing more realistic pH values. This is well in accordance with the logic behind our simulator profile.

In our GI tract model, the rate of addition for hydrochloric acid was equal to the maximum HCl secretion rate for the human stomach (Ewe & Karbach, 1990). The secretion of human gastric acid is facilitated by gastrin which is stimulated by the presence of food components, *eg* peptides, oligopeptides and aminoacids in the stomach. When pH drops below the pH 3 threshold, the release of gastrin is inhibited and in turn the acid secretion decreases (Ewe & Karbach, 1990). So pH value 3 was chosen as transition point to the small intestine phase. Since the length of `gastric phase` was determined by the time required for lowering the pH of the food matrix, the invidual properties and buffering capacities of different foodstuffs also had an effect. Therefore, in our GIT simulator the human physiological conditions are more closely mimicked, with the pH profile allowing for a more accurate replication of intestinal conditions.

Dietary fat and cholesterol are well known to stimulate bile secretion. In response to increased dietary fat the total concentration of bile acid is also higher (Reddy, Sanders, Owen, & Thompson, 1998). Also, it has been shown that bile salt synthesis rates can vary significantly among people (Bisschop *et al.*, 2004). So, varying bile salt concentrations were used during our gastrointestinal simulations in order to better reflect interpersonal differences and diet influences. A study by He, Zou, Cho, and Ahn (2012) using mixtures with varying concentrations of taurocholic acid, glycocholic acid, taurodeoxycholic acid, and glycodeoxycholate to test probiotic strains, showed that *Lb. rhamnosus* GG was more susceptible to bile acids in comparison to other probiotic strains. Different types of bile have been employed for bile resistance testing. The inhibitory effect of porcine bile has been shown to be greater than bovine or human bile. However, probiotics are usually resistant to human bile, regardless of their response to other types of bile (Dunne *et al.*, 2001; Rakin, Sekulic, & Mojovic, 2012).

In comparison with M.R.S. and titrated apple juice, a smaller reduction of viability in response to bile addition was observed in case of milk and soymilk in our experiments.

In contrast, Champagne and Gardner (2008), found that *Lb. rhamnosus* LB11 stored at 4^{0}C in fruit juice blend was not significantly affected when tested separately with 0.3% bile and pancreatic enzymes.

Fat content in food has been shown to improve probiotic viability in relation to bile, giving protection from the membrane dissolving effect of bile salts (Meira *et al.*, 2015; Ranadheera, Evans, Adams, & Baines, 2012). In case of *Lb. rhamnosus* GG originating from a commercial product, higher viability in gastric and intestinal phase was observed for full fat peanut butter (fat content of 50.10 ± 1.16%) when compared to reduced fat peanut butter (Klu & Chen, 2015). In agreement, bile addition and incubation had virtually no effect on the viability of *Lb. rhamnosus* GG in case of whey protein powder drink, which had the highest fat content of the matrixes in our experiments.

5. Conclusion

In conclusion, the effects of various food matrixes on the survival of the probiotic bacterium *Lactobacillus rhamnosus* GG during the storage following flash freezing in liquid nitrogen and in simulated gastrointestinal conditions were explored. Since it was shown that similarly to milk, soymilk and whey protein powder drink convey protection against gastrointestinal stresses, the latter are suitable candidates for the incorporation of probiotics. Also, due to the stability during flash freezing and storage, a chance for product development arises – for example, probiotic frozen desserts, puddings, sorbets *etc.* can be formulated.

It was demonstrated that lower bile concentrations also caused a decrease in viability for LGG incorporated into apple juice and M.R.S, so it cannot be assumed that foods with low fat content and therefore stimulating less bile secretion would be better for probiotic delivery.

Since apple juice did not prove to be a good candidate for probiotic incorporation, then perhaps protein or fat addition could give improved viability for this matrix, so that could be an avenue for study. Also, a further characterization of specific components responsible for improved viability is required.

To sum up, food matrixes with certain characteristics, such as suitable fat and protein content are better for the formulation of novel probiotic products in the food industry.

References

Bigliardi, B., & Galati, F. (2013). Innovation trends in the food industry: The case of functional foods. *Trends in Food Science and Technology, 31,* 118-129. http://dx.doi.org/10.1016/j.tifs.2013.03.006

Bisschop, P. H., Bandsma, R. H., Stellaard, F., ter Harmsel, A., Meijer, A. J., Sauerwein, H. P., ... Romijn, J. A. (2004). Low-fat, high-carbohydrate and high-fat, low-carbohydrate diets decrease primary bile acid synthesis in humans. *The American Journal of Clinical Nutrition, 79,* 570-576.

Champagne, C. P., Raymond Y. & Gagnon R. (2008). Viability of *Lactobacillus rhamnosus* R0011 in an apple-based fruit juice under simulated storage conditions at the consumer level. *Journal of Food Science, 73*(5), M221-226. http://dx.doi.org/10.1111/j.1750-3841.2008.00775.x

Champagne C. P., & Gardner N. J. (2008). Effect of storage in a fruit drink on subsequent survival of probiotic lactobacilli to gastro-intestinal stresses. *Food Research International, 41,* 539-543. http://dx.doi.org/10.1016/j.foodres.2008.03.003

Champagne C. P., Ross R. P., Saarela M., Hansen K. F., & Charalampopoulos D. (2011). Recommendations for the viability assessment of probiotics as concentrated cultures and in food matrices. *International Journal of Food Microbiology, 149*(3), 185-193. http://dx.doi.org/10.1016/j.ijfoodmicro.2011.07.005

Chen Z.-Y., Ma K. Y., Liang Y., Peng C., & Zuo Y. (2011). Role and classification of cholesterol-lowering functional foods. *Journal of Functional Foods, 3,* 61-69. http://dx.doi.org/10.1016/j.jff.2011.02.003

Corcoran, B. M., Stanton, C., Fitzgerald, G. F., & Ross, R. P. (2005). Survival of probiotic lactobacilli in acidic environments is enhanced in the presence of metabolizable sugars. *Applied and Environmental Microbiology, 71,* 3060-3067. http://dx.doi.org/10.1128/AEM.71.6.3060-3067.2005

Cruz, A. G., Faria, J. A. F., & van Dender, A. G. F. (2007). Packaging system and probiotic dairy foods. *Food Research International, 40,* 951-956. http://dx.doi.org/10.1016/j.foodres.2007.05.003

Cruz A. G., Antunes A. E. C., Sousa A. L. O. P., Faria, J. A. F., & Saad, S. M. I. (2009). Ice-cream as a probiotic food carrier. *Food Research International, 42,* 1233-1239. http://dx.doi.org/10.1016/j.foodres.2009.03.020

De Man, J. C., Rogosa, M., & Sharpe, M. E. (1960). A medium for the cultivation of lactobacilli. *Journal of Applied Bacteriology, 23,* 130-135. http://dx.doi.org/10.1111/j.1365-2672.1960.tb00188.x

do Espirito, S. A. P., Perego, P., Converti, A., & Oliveira, M. N. (2011). Influence of food matrices on probiotic viability - A review focusing on the fruity bases. *Trends in Food Science and Technology, 22,* 377-385. http://dx.doi.org/10.1016/j.tifs.2011.04.008

Dunne, C., O'Mahony, L., Murphy, L., Thornton, G., Morrissey, D., O'Halloran S., ... Collins, J. K. (2001). *In vitro* selection criteria for probiotic bacteria of human origin: correlation with *in vivo* findings. *The American Journal of Clinical Nutrition, 73*(2 Suppl), 386S-392S.

Elhadad, A. S., Alwakdi, O. M., Abushita, A., & Abdulsalam, F. (2013, January). *Influence of some additives on the properties of concentrated apricot and peach juices during freeze storage.* Presented at the 3rd International Conference on Ecological, Environmental and Biological Sciences (ICEEBS'2013), Hong Kong, China. Paper retrieved from http://psrcentre.org/images/extraimages/13.%20ICECEBE%20113849.pdf

Eussen, S. R. B. M., Verhagen, H., Klungel, O. H., Garssen, J., van Loveren, H., van Kranen, H. J., & Rompelberg, C. J. (2011). Functional foods and dietary supplements: Products at the interface between pharma and nutrition. *European Journal of Pharmacology, 668*(Suppl 1), S2-9. http://dx.doi.org/10.1016/j.ejphar.2011.07.008

Foschino, R., Fiori, E., & Galli, A. (1996). Survival and residual activity of *Lactobacillus acidophilus* frozen cultures under different conditions. *Journal of Dairy Research, 63,* 295-303. http://dx.doi.org/10.1017/S0022029900031782

Gorbach, S. L., & Goldin, B. R. (1989). *Lactobacillus strains* and methods of selection. United States Patent No. 4,839,281.

Grajek, W., Olejnik, A., & Sip, A. (2005). Probiotics, prebiotics and antioxidants as functional foods. *Acta Biochimica Polonica, 52,* 665-671.

Granato, D., Branco, G. F., Nazzaro, F., Cruz, A. G., & Faria, J. A. F. (2010). Functional foods and nondairy probiotic food development: trends, concepts, and products. *Comprehensive Reviews in Food Science and Food Safety, 9,* 292-302. http://dx.doi.org/10.1111/j.1541-4337.2010.00110.x

He, X., Zou, Y., Cho, Y., & Ahn, J. (2012). Effects of bile salt deconjugation by probiotic strains on the survival of antibiotic-resistant foodborne pathogens under simulated gastric conditions. *Journal of Food Protection, 75,* 1090-1098. http://dx.doi.org/10.4315/0362-028X.JFP-11-456

Karimi, R., Mortazavian, A. M., & Amiri-Rigi, A. (2012). Selective enumeration of probiotic microorganisms in cheese. *Food Microbiology, 29,* 1-9. http://dx.doi.org/10.1016/j.fm.2011.08.008

Kaur, S., & Das, M. (2011). Functional Foods: An Overview. *Food Science and Biotechnology, 20,* 861-875. http://dx.doi.org/10.1007/s10068-011-0121-7

Klu, Y. A., & Chen, J. (2015). Effect of peanut butter matrices on the fate of probiotics during simulated gastrointestinal passage. *LWT - Food Science and Technology, 62,* 983-988. http://dx.doi.org/10.1016/j.lwt.2015.02.018

Makinen, K., Berger, B., Bel-Rhlid, R., & Ananta, E. (2012). Science and technology for the mastership of probiotic applications in food products. *Journal of Biotechnology, 162,* 356-365.

http://dx.doi.org/10.1016/j.jbiotec.2012.07.006

Martins, E. M. F., Ramos, A. M., Vanzela, E. S. L., Stringheta, P. C., Pinto, C. L. O., & Martins, J. M. (2013). Products of vegetable origin: A new alternative for the consumption of probiotic bacteria. *Food Research International, 51,* 764-770. http://dx.doi.org/10.1016/j.foodres.2013.01.047

Meira, Q. G., Magnani, M., de Medeiros, F. C., do Egito, Q. R., Madruga, M. S., Gullón, B., de Souza, E. L. (2015). Effects of added *Lactobacillus acidophilus* and *Bifidobacterium lactis* probiotics on the quality characteristics of goat ricotta and their survival under simulated gastrointestinal conditions. *Food Research International, 76,* 828-838 http://dx.doi.org/10.1016/j.foodres.2015.08.002

Mills, S., Stanton, C., Fitzgerald, G. F., & Ross, R. P. (2011). Enhancing the stress responses of probiotics for a lifestyle from gut to product and back again. *Microbial Cell Factories, 10*(Suppl 1: S19), 1-15. http://dx.doi.org/10.1186/1475-2859-10-S1-S19

Ng, K. H., Lye, H. S., Easa, A. M., & Liong, M. T. (2008). Growth characteristics and bioactivity of probiotics in tofu-based medium during storage. *Annals of Microbiology, 58,* 477-487. http://dx.doi.org/10.1007/BF03175546

Pang, G., Xie, J., Chen, Q., & Hu, Z. (2012). How functional foods play critical roles in human health. *Food Science and Human Wellness, 1,* 26-60. http://dx.doi.org/10.1016/j.fshw.2012.10.001

Papadimitriou, K., Zoumpopoulou, G., Foligné, B., Alexandraki, V., Kazou, M., Pot, B., & Tsakalidou, E. (2015). Discovering probiotic microorganisms: *in vitro, in vivo,* genetic and omics approaches. *Frontiers in Microbiology, 6*(58), 1-28 http://dx.doi.org/10.3389/fmicb.2015.00058

Prado, F. C., Parada, J. L., Pandey, A., & Soccol, C. R. (2008). Trends in non-dairy probiotic beverages. *Food Research International, 41,* 111-123. http://dx.doi.org/10.1016/j.foodres.2007.10.010

Puupponen-Pimiä, R., Aura A.-M., Oksman-Caldentey, K.-M., Myllaerinen, P., Saarela, M., Mattila-Sandholm, T., & Poutanen, K. (2002). Development of functional ingredients for gut health. *Trends in Food Science and Technology, 13,* 3-11. http://dx.doi.org/10.1016/S0924-2244(02)00020-1

Rakin, M., Sekulic, M. V., & Mojovic, L. (2012). *Handbook of Plant-Based Fermented Food and Beverage Technology.* http://dx.doi.org/10.1201/b12055-26

Ranadheera, R. D. C. S., Baines, S. K., & Adams, M. C. (2010). Importance of food in probiotic efficacy. *Food Research International, 43,* 1-7. http://dx.doi.org/10.1016/j.foodres.2009.09.009

Ranadheera, C. S., Evans, C. A., Adams, M. C., & Baines, S. K. (2012). *In vitro* analysis of gastrointestinal tolerance and intestinal cell adhesion of probiotics in goat's milk ice cream and yogurt. *Food Research International, 49,* 619-625. http://dx.doi.org/10.1016/j.foodres.2012.09.007

Reddy, S., Sanders, T. A., Owen, R. W., & Thompson, M. H. (1998). Faecal pH, bile acid and sterol concentrations in premenopausal Indian and white vegetarians compared with white omnivores. *British Journal of Nutrition, 79,* 495-500. http://dx.doi.org/10.1079/BJN19980087

Rivera-Espinoza, Y., & Gallardo-Navarro, Y. (2010). Non-dairy probiotic products. *Food Microbiology, 27,* 1-11. http://dx.doi.org/10.1016/j.fm.2008.06.008

Saarela, M. H., Alakomi, H. L., Puhakka, A., & Mättö, J. (2009). Effect of the fermentation pH on the storage stability of *Lactobacillus rhamnosus* preparations and suitability of *in vitro* analyses of cell physiological functions to predict it. *Journal of Applied Microbiology, 106,* 1204-1212. http://dx.doi.org/10.1111/j.1365-2672.2008.04089.x

Saarela, M. H. (2011). *Functional foods: Concept to product* (2nd ed.). http://dx.doi.org/10.1533/9780857092557.3.425

Santivarangkna, C., Kulozik, U., & Foerst, P. (2011). *Stress responses of lactic acid bacteria.* http://dx.doi.org/10.1007/978-0-387-92771-8_20

Schmidt, R. F., & Thews, G. (1990). *Physiologie des Menschen* (24th ed.). Berlin: Springer.

Sheehan, V. M., Ross, P., & Fitzgerald, G. F. (2007). Assessing the acid tolerance and the technological robustness of probiotic cultures for fortification in fruit juices. *Innovative Food Science and Emerging Technologies, 8,* 279-284. http://dx.doi.org/10.1016/j.ifset.2007.01.007

Sigita, B., Krasnova, I., Seglina, D., Aboltins, I., & Skrupskis, I. (2013). Changes of pectin quantity in fresh and frozen apple products. *Journal of Chemistry and Chemical Engineering, 7,* 64-69.

Smithers, GW. Whey and whey proteins-From 'gutter-to-gold'. (2008). *International Dairy Journal, 18,* 695- 704. http://dx.doi.org/10.1016/j.idairyj.2008.03.008

Sumeri I., Arike L., Adamberg K., & Paalme T. (2008). Single bioreactor gastrointestinal tract simulator for study of survival of probiotic bacteria. *Applied Microbial and Cell Physiology, 80,* 317-324. http://dx.doi.org/10.1007/s00253-008-1553-8

Sumeri I., Arike L., Stekolstsikova J., Uusna R., Adamberg S., Adamberg K., & Paalme T. (2010).   Effect of stress pretreatment on survival of probiotic bacteria in gastrointestinal tract simulator. *Applied Microbial and Cell Physiology, 86,* 1925-1931. http://dx.doi.org/10.1007/s00253-009-2429-2

Vasiljevic T., & Shah N. P. (2008). Probiotics-From Metchnikoff to bioactives. *International Dairy Journal, 18,* 714-728. http://dx.doi.org/10.1016/j.idairyj.2008.03.004

Vinderola, C. G., Costa, G. A., Regenhardt, S., & Reinheimer, J. A. (2002). Influence of compounds associated with fermented dairy products on the growth of lactic acid starter and probiotic bacteria. *International Dairy Journal, 12,* 579-589. http://dx.doi.org/10.1016/S0958-6946(02)00046-8

Wang, J., Guo Z., Zhang, Q., Yan, L., Chen, W., Liu, X. M., & Zhang, H. P. (2009). Fermentation characteristics and transit tolerance of probiotic *Lactobacillus casei* Zhang in soymilk and bovine milk during storage. *Journal of Dairy Science, 92,* 2468-2476. http://dx.doi.org/10.3168/jds.2008-1849

Wright, S. (2007). A protein punch. Beverage World, *126*(8), 70-71.

11

Nutrient Composition and Adequacy of two Locally Formulated Winged Termite (*Macrotermes Bellicosus*) Enriched Complementary Foods

ADEPOJU, Oladejo Thomas[1] & AJAYI, Kayode[2]

[1] Department of Human Nutrition, Faculty of Public Health, College of Medicine, University of Ibadan, Ibadan, Nigeria

[2] Department of Human Nutrition and Dietetics, College of Medicine and Health Sciences, Afe Babalola University, Ado-Ekiti, Ekiti State, Nigeria

Correspondence: ADEPOJU, Oladejo Thomas, Department of Human Nutrition, Faculty of Public Health, College of Medicine, University of Ibadan, Ibadan, Nigeria. E-mail: tholadejo@yahoo.com

Abstract

The period from birth to two years of age constitute critical window of opportunity for promoting optimal growth and development of a child. Inadequate food intake and poor feeding practices are causes of malnutrition among Nigerian children, as many locally formulated complementary foods are deficient in protein and micronutrients. Roasted *Macrotermes bellicosus* (MB) is nutritious and relished as snack by people living the traditional lifestyle. This study was carried out to investigate possible use of MB in formulating nutrient-dense complementary foods from maize and sorghum. *Macrotermes bellicosus* was collected in Ibadan, Nigeria during their swarming, roasted, de-winged, powdered and added to fermented corn (CF) and sorghum (SF) flour in the ratio 100%flour, 90%flour+10%MB, 85%flour+15%MB, and 80%flour+20%MB to give eight complementary foods, which were analysed for proximate, mineral, vitamin and antinutrient composition using AOAC methods.

Hundred grammes of CF and SF contained 11.7g, 10.6g moisture, 8.9g, 9.7g crude protein, 3.1g, 2.8g fat, 74.3g, 74.8g total carbohydrates, 6.67mg, 26.60mg calcium, 295.50mg, 325.43mg phosphorus, 2.61mg, 7.61mg iron, 3.19mg, 2.41mg zinc, and yielded 353.9kcal, 358.6kcal energy respectively. Significant reduction occurred in moisture and carbohydrate content of MB-incorporated complementary foods while their crude protein, ash, fat, calcium, iron, zinc, vitamins B_3, B_6 B_{12} and β-carotene content increased significantly as the level of inclusion of MB increased ($p<0.05$). Level of atinutritional factors were insignificantly low in the blends, and cannot pose any health risk. *Macrotermes bellicosus* can be used in enriching cereal-based complementary foods as means of reducing infant and young child malnutrition in Nigeria.

Keywords: complementary foods, fermented corn flour, fermented sorghum flour, nutrient adequacy, winged termites

1. Introduction

Undernutrition is a serious medical condition marked by deficiency of energy, essential proteins, fats, vitamins, or minerals in a diet, and it is especially burdensome and dangerous for young, growing children. Malnutrition contributes to one third of the eight million deaths of children under five years of age every year among invisible children, hence, urgent action is required (Black et al., 2010; Rajaratnam, 2010). Most of the damage caused by malnutrition occurs in children before they reach their second birthday. This is the critical window of opportunity when the quality of a child's diet has a profound, sustained impact on his or her health and physical and mental development.

Diets that are lacking in high-quality protein, essential fats, carbohydrates, vitamins, and minerals can impair growth and development, increase the risk of death from common childhood illness, or result in life-long health consequences (Black et al., 2008). Complementary feeding period is the time when malnutrition starts in many infants (Daelmans & Saadeh, 2003), and poor feeding practices coupled with shortfall in food intake are the most important direct factors responsible for under-five malnutrition and illness among children in Nigeria (Solomon, 2005). Traditional complementary foods in the developing countries are of low nutritive value and are characterized by low protein, low energy density and high bulk, because they are usually cereal–based (Shiriki et al., 2015).

Insects link biodiversity conservation and human nutrition in a way that many other food sources do not. They often contain more protein, fat, and carbohydrates than equal amounts of beef or fish, and a higher energy value than soybeans, maize, beef, fish, lentils, or other beans (FAO, 2013). Edible insects consumption has been recently reported in Nigeria (Agbidye et al., 2009; Adeyeye, 2011) among which the termite, (*Macrotermes natalensis*) had the highest mean frequency (Agbidye et al., 2009). Dried *Marcrotermes bellicosus* has been reported to be a good source of dietary protein, fat and micronutrients (Banjo et al., 2006; Ekpo et al., 2009; Adeyeye, 2011; Adepoju & Omotayo, 2014). Adepoju and Omotayo (2014) reported the insect to be low in antinutrients and suggested its possible inclusion in formulating adequate, nutrient-dense complementary foods. This study was therefore carried out to determine the nutrient composition and suitability of *Marcrotermes bellicosus* as source of essential macro and micronutrients in formulating two adequate, nutrient-dense cereal-based complementary foods.

2. Materials and Methods

2.1 Sample Collection and Preparation

Sample of *M. bellicosus* was collected around Alegongo Area, Akobo, Ibadan, Nigeria during their swarming flights. The sample was roasted for ten minutes over a gas cooker, dewinged by rubbing between the palms and then winnowed to remove the wings, labelled as "Roasted sample" and kept in a freezer at - 4°C till when needed for analysis (Adepoju & Omotayo, 2014). About 1.5 kg each of white maize (*Zea mays*) and sorghum (*Sorghum bicolor*) were purchased from Bodija market in Ibadan, Oyo State, Nigeria. The maize and sorghum grains were cleaned and washed thoroughly to remove adhering dirt and dust. The maize and sorghum samples were separately soaked in distilled water for 72 hours, drained and wet-milled. The milled samples were sieved and allowed to settle for 3 hours and then drained. The drained samples were oven-dried for 12 hours at 60°C in plastic trays, packed in cellophane nylon and kept till when needed (Inyang & Idoko, 2006). Various samples of complementary foods were prepared as follows:

Sample A   =   100 g Maize flour
Sample A_1 =   90 g Maize flour + 10 g *M. bellicosus*
Sample A_2 =   85 g Maize flour + 15 g *M. bellicosus*
Sample A_3 =   80 g Maize flour + 20 g *M. bellicosus*

Sample B   =   100 g Sorghum flour
Sample B_1 =   90 g Sorghum flour + 10 g *M. bellicosus*
Sample B_2 =   85 g Sorghum flour + 15 g *M. bellicosus*
Sample B_3 =   80 g Sorghum flour + 20 g *M. bellicosus*

2.2 Proximate Composition Analysis

Moisture content of the samples was determined by air oven at 105°C (Plus 11 Sanyo Gallenkamp PLC UK) for 4 hours. The crude protein of the samples was determined using micro-Kjeldahl method (Method No 978.04, AOAC, 2005). Crude lipid was determined by Soxhlet extraction method (Method No 930.09, AOAC, 2005) and the crude lipid estimated as g/100g of sample. The ash content was determined by weighing 5g of sample in triplicate and heated in a muffle furnace (Gallenkamp, size 3) at 550°C for 4 h (Method No 930.05, AOAC, 2005). The total carbohydrate content was obtained by difference. Gross energy of the samples was determined using ballistic bomb calorimeter (Manufacturer: Cal 2k – Eco, TUV Rheinland Quality Services (Pty) Ltd, South Africa).

2.3 Mineral Analysis

Potassium and sodium content of the samples were determined by digesting the ash of the samples with perchloric acid and nitric acid, and then taking the readings on Jenway digital flame photometer/spectronic20 (AOAC, 2005: (975.11)). Phosphorus was determined by vanado-molybdate colorimetric method (AOAC, 2005: (975.16)). Calcium, magnesium, iron, zinc, manganese, and copper were determined spectrophotometrically by using Buck 200 atomic absorption spectrophotometer (Buck Scientific, Norwalk) and compared with absorption of standards of these minerals (AOAC, 2005: (975.23)).

2.4 Vitamin Analysis

2.4.1 Vitamin A Determination

Vitamin A was determined through ultraviolet absorption measurement at 328 nm after extraction with chloroform. Calibration curve of vitamin A acetate was made and sample vitamin A concentration estimated as microgram (μg) of vitamin A acetate.

2.4.2 Thiamine (Vitamin B_1) Determination

Thiamine content of the sample was determined by weighing 1g of it into 100ml volumetric flask and adding 50ml of 0.1M H_2SO_4 and boiled in a boiling water bath with frequent shaking for 30 minutes. 5ml of 2.5M sodium acetate solution was added and flask set in cold water to cool contents below 50^0C. The flask was stoppered and kept at $45-50^0C$ for 2 hours and thereafter made up to 100ml mark. The mixture was filtered through a No. 42 Whatman filter paper, discarding the first 10ml. 10ml was pipetted from remaining filterate into a 50ml volumetric flask and 5ml of acid potassium chloride solution was added with thorough shaking. Standard thiamine solutions were prepared and treated same way. The absorbance of the sample as well as that of the standards was read on a fluorescent UV Spectrophotometer (Cecil A20 Model) at a wavelength of 285nm.

2.4.3 Riboflavin (Vitamin B_2) Determination

1g of each sample was weighed into a 250ml volumetric flask, 5ml of 1M HCl was added, followed by the addition of 5ml of dichloroethene. The mixture was shaken and 90ml of de-ionized water was added. The whole mixture was thoroughly shaken and was heated on a steam bath for 30 minutes to extract all the riboflavin. The mixture was then cooled and made up to volume with de-ionized water. It was then filtered, discarding the first 20ml of the aliquot. 2ml of the filterate obtained was pipetted into another 250ml volumetric flask and made up to mark with de-ionized water. Sample was read on the fluorescent spectrophotometer at a wavelength of 460nm. Standard solutions of riboflavin were prepared and readings taken at 460nm, and the sample riboflavin obtained through calculation.

2.4.4 Niacin (Vitamin B_3) Determination

5g of sample was extracted with 100ml of distilled water and 5ml of this solution was drawn into 100ml volumetric flask and make up to mark with distilled water. Standard solutions of niacin were prepared and absorbance of sample and standard solutions were measured at a wavelength of 385nm on a spectrophotometer and niacin concentration of the sample estimated.

Pyridoxine (Vitamin B_6) determination

The vitamin B_6 content of the sample was determined by extracting 1g of sample with 0.5g of ammonium chloride, 45 ml of chloroform and 5 ml of absolute ethanol. The mixture was thoroughly mixed in a separating funnel by shaking for 30mins, and 5ml of distilled water added. The chloroform layer containing the pyridoxine was filtered into a 100ml volumetric flask and made up to mark with chloroform. 0-10ppm of vitamin B_6 standard solutions were prepared and treated in a similar way as sample, and their absorbance measured on Cecil 505E spectrophotometer at 415mn. The amount of vitamin B6 in the sample was then calculated.

Cyanocobalamin (Vitamin B_{12}) determination

Cyanocobalamin content of the sample was determined by extracting 1g of sample with distilled water with shaking for 45min followed by filtering the mixture. The first 20 ml of the filterate was rejected, and another 20mls filtrate collected. To the collected filtrate, 5mls of 1% Sodium Dithionite solution was added. Standard cyanocobalamin solutions (0-10μg/ml) were prepared and absorbance of sample as well as standard was read on spectronic21D spectrophotometer at 445nm. The amount of sample cyanocobalamin was then estimated through calculation.

2.4.5 Ascorbic acid Determination

Ascorbic acid in the sample was determined by titrating its aqueous extract with solution of 2, 6-dichlorophenol-indophenol dye to a faint pink end point.

2.5 Antinutrient Analysis

Oxalate was determined by extraction of the samples with water for about three hours and standard solutions of oxalic acid prepared and read on spectrophotometer (Spectronic20) at 420 nm. The absorbance of the samples was also read and amount of oxalate estimated. Phytate was determined by titration with ferric chloride solution (Sudarmadji & Markakis, 1977); while trypsin inhibitory activity was determined on casein and comparing the absorbance with that of trypsin standard solutions read at 280 nm (Makkar & Becker, 1996). The tannin content was determined by extracting the samples with a mixture of acetone and acetic acid for five hours, measuring their absorbance and comparing the absorbance of the sample extracts with the absorbance of standard solutions of tannic acid at 500 nm on spectronic20 (Griffiths & Jones, 1977). Saponin was also determined by comparing the absorbance of the sample extracts with that of the standard at 380 nm (Makkar & Becker, 1996). All determinations were carried out in triplicate.

2.6 Statistical Analysis

The data obtained were subjected to analysis of variance (ANOVA), Fisher's Least Significance Difference (LSD) and Duncan multiple range tests at $p < 0.05$.

3. Results

3.1 Proximate Composition of Roasted M. bellicosus, Cereal Flour and Complementary Foods

The result of proximate composition of roasted *M. bellicosus* and fermented maize and sorghum flour is as shown in Table 1(A). Roasted *M. bellicosus* moisture content was very low while the crude protein, fat, carbohydrate and gross energy content were very high. The moisture and crude fat content of the two cereal flour were low, the crude protein value was moderately high while the ash and total carbohydrates content were very high. The moisture and crude fat values of the maize flour were not significantly different from that of sorghum ($p > 0.05$), while sorghum was significantly higher in crude protein ($p < 0.05$) but insignificantly higher in ash, total carbohydrates and gross energy ($p > 0.05$).

There was a significant reduction in moisture content of the formulated complementary foods, the level of reduction increasing with increasing level of inclusion of MB ($p < 0.05$), Table 1(B). Significant differences also existed between the enriched complementary foods, the level of reduction in moisture content increasing with increase in inclusion level ($p < 0.05$). There were significant increase in values of crude protein and fat, ash, total carbohydrates and gross energy of the enriched maize and sorghum complementary foods ($p < 0.05$), the values increasing with increasing level of inclusion of *M. bellicosus*.

Table 1(A). Proximate composition of roasted *M. bellicosus* and cereal flour (g/100g)

Sample	M. bellicosus	Maize (A)		Sorghum (B)	
		(Wet)	(Dry)	(Wet)	(Dry)
Moisture	4.0±0.04	65.0±0.12	11.7±0.03	64.1±0.50	10.6±0.05
Crude Protein	31.8±0.10	3.3±0.03	8.9±0.09	3.9±0.04	9.7±0.07
Crude Fat	16.4±0.03	1.2±0.01	3.1±0.03	1.1±0.01	2.8±0.02
Ash	3.8±0.03	0.8±0.00	2.0±0.02	0.8±0.00	2.1±0.01
Total Carbohydrates	43.0±0.10	29.7±0.05	74.3±0.11	30.1±0.05	74.8±1.84
Gross Energy (kcal/)	450.7±0.00	-	353.9±0.28	-	358.6±6.81

Table 1(B). Proximate composition of enriched complementary foods (g/100)

Sample	Moisture	C. Protein	C. Fat	Ash	T. Carbohydrates	Gross
A	11.7±0.03[a]	8.9±0.09[a]	3.1±0.03[a]	2.0±0.02[a]	74.3±0.11[a]	353.91±0.28[a]
A$_1$	11.4±0.03[b]	17.8±0.10[b]	5.1±0.02[b]	2.9±0.01[b]	61.7±0.11[b]	362.36±0.64[b]
A$_2$	11.1±0.04[c]	18.6±0.09[c]	5.6±0.03[c]	3.2±0.02[c]	61.6±0.11[b]	364.85±0.09[c]
A$_3$	10.7±0.04[d]	19.7±0.10[d]	5.8±0.04[d]	3.5±0.02[b]	60.5±0.16[c]	367.12±0.12[d]
B	10.6±0.05[e]	9.7±0.07[e]	2.8±0.02[e]	2.1±0.01[e]	73.8±1.84[d]	358.64±6.81[e]
B$_1$	10.3±0.04[f]	18.3±0.07[f]	4.3±0.03[f]	3.8±0.02[f]	61.5±0.12[e]	357.40±0.22[f]
B$_2$	10.1±0.05[g]	19.7±0.11[g]	4.7±0.12[g]	4.1±0.01[g]	60.8±1.84[f]	364.25±0.31[g]
B$_3$	9.8±0.04[h]	21.2±0.12[h]	5.3±0.04[h]	4.4±0.03[h]	56.1±0.02[g]	363.08±0.18[h]
*RV	<5	>15	10-25	<3	64	400-425

Values are means ± standard deviations of triplicate determinations.

Means with different superscripts in a column are significantly different (p<0.05).

*RV = Recommended values (g/100g) *(CODEX CAC/GL 08. 1991): Codex alimentarius: Guidelines on formulated supplementary foods for older infants and young children.

Sample A = 100g Maize flour; Sample A_1 = 90g Maize flour + 10g *M. bellicosus*

Sample A_2 = 85g Maize flour+15g *M. bellicosus;* Sample A_3 = 80g Maize flour+20g *M. bellicosus*

Sample B = 100g Sorghum flour; Sample B_1 = 90g Sorghum flour+10g *M. bellicosus;* Sample B_2 = 85g Sorghum flour+15g *M. bellicosus;* Sample B_3 = 80g Sorghum flour+20g *M. bellicosus*.

3.2 Mineral Composition of Roasted M. bellicosus, Cereal Flour and Complementary Foods

Marcrotermes bellicosus is rich in potassium, calcium, phosphorus, zinc, and copper, moderate in sodium, iron and manganese, but low in magnesium (Table 2(A)). Maize and sorghum flour are high in potassium, phosphorus, and zinc, low in sodium, and very low in calcium, magnesium, and manganese content. Maize flour was significantly higher in potassium, sodium, zinc and copper than sorghum flour (p<0.05), while sorghum flour was significantly higher in calcium, magnesium, phosphorus, iron and manganese than maize flour (p<0.05).

Addition of *M. bellicosus* to maize and sorghum flour (Tables (2B) and (2C)) resulted in significant increase in values of the minerals (p<0.05) in all the formulated complementary diets, the values increasing with increase in inclusion level of *M. bellicosus*. However, the mineral content of the formulated complementary foods were lower than the recommended values by FAO/WHO. The values of the minerals of enriched complementary foods were significantly different (p<0.05) from one another for both maize and sorghum, the 10% *M. bellicosus* incorporated flours having the lowest values while 20% *M. bellicosus* incorporated flours had the highest values.

3.3 Vitamin Composition of Complementary Foods

Addition of *M. bellicosus* to maize and sorghum flour (Table 3(A) and 3(B)) resulted in significant increase in β-carotene, niacin, vitamin B6, and B12 content, with significant reduction in thiamine, riboflavin and ascorbic acid content of enriched complementary foods (p<0.05). For the vitamins with increased content, significant increase was observed as the level of inclusion increased, while there was also significant decrease with increasing level of inclusion for vitamins with decrease in values. Vitamin B_6 content of the formulated diets were higher in value compared with their FAO/WHO recommended value. However, there were significant reduction in thiamine, riboflavin and vitamin C content of the formulated diets (p<0.05). The levels of thiamine, riboflavin and vitamin C content were lower than the value recommended by FAO/WHO.

Table 2(A). Mineral composition of roasted *M. bellicosus* and cereal flour (mg/100)

Parameter	*M. bellicosus*	Maize (A)	Sorghum (B)
Potassium	361.13±0.31	211.83±0.17	205.33±0.25
Sodium	98.40±0.20	52.83±0.25	48.50±0.20
Calcium	227.50±0.20	6.67±0.26	26.60±0.30
Magnesium	24.33±0.15	14.00±0.36	14.73±0.15
Phosphorus	361.30±0.20	295.50±0.30	325.43±0.25
Iron	2.07±0.25	2.61±0.30	7.61±0.03
Zinc	15.03±0.31	3.19±0.04	2.41±0.03
Manganese	2.35±0.25	0.52±0.04	0.62±0.05
Copper	5.07±0.54	1.57±0.25	0.83±0.25

Table 2(B). Mineral composition of formulated maize complementary Foods (mg/100g)

	A	A$_1$	A$_2$	A$_3$	*RV
Potassium	211.83±0.17^a	216.47±0.25^b	219.50±0.30^c	223.57±0.35^d	516
Sodium	52.83±0.25^a	55.60±0.30^b	57.70±0.17^c	59.47±0.25^d	296
Calcium	6.67±0.26^a	7.50±0.20^b	9.27±0.35^c	10.60±0.17^d	500
Magnesium	14.00±0.36^a	15.17±0.25^b	16.73±0.15^c	18.20±0.36^d	76
Phosphorus	295.50±0.30^a	297.30±0.15^b	299.30±0.15^c	301.43±0.15^d	456
Iron	2.61±0.30^a	2.81±0.20^b	3.08±0.03^c	4.13±0.03^d	16
Zinc	3.19±0.04^a	3.49±0.04^b	3.82±0.04^c	4.12±0.03^d	3.2
Manganese	0.52±0.04^a	0.57±0.02^b	0.67±0.02^c	0.82±0.03^d	0.60**
Copper	1.57±0.25^a	2.17±0.35^b	2.40±0.30^c	3.13±0.25^d	0.34**

Values are mean ± standard deviation of triplicate determinations. Mean value with different superscripts in a row are significantly different (p0.05).

*RV = Recommended values (mg/100g) *(CODEX CAC/GL 08. 1991) / **RDA (Sareen, Jack & James, 2009).

Table 2 (C). Mineral composition of formulated sorghum complementary Foods (mg/100g)

	B	B$_1$	B$_2$	B$_3$	*RV
Potassium	205.33±0.25^e	209.50±0.25^f	211.77±0.15^g	214.50±0.20^h	516
Sodium	48.50±0.20^e	49.40±0.30^f	51.43±0.2^g	54.06±0.25^h	296
Calcium	26.60±0.30^e	27.60±0.30^f	29.07±0.35^g	31.50±0.03^h	500
Magnesium	14.73±0.15^e	15.50±0.20^f	16.50±0.17^g	18.63±0.15^h	76
Phosphorus	325.43±0.25^e	327.70±0.17^f	331.37±0.21^g	334.43±0.11^h	456
Iron	7.61±0.03^e	7.91±0.02^f	8.09±0.02^g	9.22±0.02^h	11
Zinc	2.41±0.03^e	2.63±0.05^f	3.31±0.03^g	3.92±0.04^h	3.2
Manganese	0.62±0.05^e	0.66±0.03^f	0.81±0.03^g	0.92±0.03^h	32
Copper	0.83±0.25^e	1.20±0.30^f	1.90±0.25^g	2.53±0.25^h	160

Values are means ± standard deviations of triplicate determinations.

Mean value with different superscripts in a row are significantly different (p<0.05)

Table 3(A). Vitamin composition of maize flour and *M. bellicosus* enriched complementary foods (mg/100 g)

	A	A$_1$	A$_2$	A$_3$	*RV
β-carotene (μg/)	216.23±0.40^a	223.77±0.40^b	225.50±0.30^c	227.60±0.30^d	400
Thiamine	0.35±0.02^a	0.31±0.02^b	0.26±0.02^c	0.21±0.03^d	0.5
Riboflavin	0.12±0.01^a	0.06±0.01^b	0.04±0.01^c	0.02±0.01^d	0.5
Niacin	2.09±0.25^a	2.16±0.03^b	2.24±0.03^c	2.35±0.02^d	6
Vitamin B$_6$	0.72±0.04^a	0.78±0.01^b	0.83±0.02^c	0.91±0.03^d	0.5
Vitamin B$_{12}$ (μg/)	0.18±0.03^a	0.19±0.02^b	0.27±0.03^c	0.31±0.02^d	0.9
Vitamin C	2.83±0.06^a	2.40±0.03^b	2.22±0.04^c	2.00±0.05^d	30

Values are means ± standard deviations of triplicate determinations.

Means with different superscripts in a row are significantly different (p<0.05).

Table 3(B). Vitamin composition of sorghum flour and *M. bellicosus* enriched complementary foods (mg/100 g)

	B	B1	B2	B3	*RV
β-carotene (μg/)	30.23±0.50[e]	34.80±0.30[f]	37.20±0.46[g]	39.10±0.46[h]	400
Thiamine	0.43±0.02[e]	0.37±0.03[f]	0.24±0.01[g]	0.18±0.02[h]	0.5
Riboflavin	0.14±0.01[e]	0.08[j]±0.01[f]	0.05±0.01[g]	0.18±0.02[h]	0.5
Niacin	4.11±0.03[e]	3.63±0.02[f]	3.71±0.04[g]	3.76±0.02[h]	6
Vitamin B_6	0.49±0.02[e]	0.52±0.01[f]	0.56±0.02[g]	0.64±0.02[h]	0.5
Vitamin B_{12} (μg/)	0.27±0.02[e]	0.31±0.02[f]	0.35±0.03[g]	0.64±0.02[h]	0.9
Vitamin C	3.84±0.03[e]	3.01±0.04[f]	2.85±0.04[g]	2.67±0.04[h]	30

Values are means ± standard deviations of triplicate determinations. Means with different.

superscripts in a row are significantly different (p≤0.05).

3.4 Antinutritional Factors of Maize, Sorghum and Enriched Complementary Foods

Maize and sorghum flour were very low in phytate, oxalate, tannin and saponin content, with no detectable level of trypsin inhibitors (Tables 4(A) and (B)). Significant reduction in values resulted on all antinutritional factors studied as the level of inclusion of *M. bellicosus* increased (p<0.05).

Table 4(A) Antinutritional factors in maize flour and *M. bellicosus* enriched complementary foods (mg/100 g)

Parameters	A	A_1	A_2	A_3
Phytate	0.063±0.001[c]	0.051±0.000[b]	0.041±0.000[z]	0.029±0.002[m]
Oxalate	0.030±0.002[w]	0.022±0.001[e]	0.017±0.002[r]	0.012±0.001[t]
Tannin	0.022±0.000[a]	0.013±0.000[h]	0.001±0.000[f]	0.007±0.001[g]
Saponin	0.076±0.002[c]	0.064±0.002[d]	0.051±0.001[e]	0.039±0.003[w]
T. I (TIU/mg)	ND	ND	ND	ND

Mean value with different superscripts in a row are significantly different (p<0.05).

T. I = Trypsin Inhibitors;   ND = Not detected at milligramme level

Table 4(B). Antinutritional factors in sorghum flour and *M. bellicosus* enriched complementary foods (mg/100 g)

Parameters	B	B_1	B_2	B_3
Phytate	0.23±0.00[a]	0.22±0.00[b]	0.20±0.00[c]	0.19±0.00[d]
Oxalate	0.13±0.00[a]	0.12±0.00[b]	0.11±0.00[c]	0.11±0.00[c]
Tannin	0.09±0.00[a]	0.08±0.00[b]	0.07±0.00[c]	0.05±0.00[d]
Saponin	0.10±0.00[a]	0.09±0.00[b]	0.08±0.00[c]	0.07±0.00[d]
T. I (TIU/mg)	ND	ND	ND	ND

4. Discussion

4.1 Proximate Composition

The low moisture content of roasted *M. bellicosus* underscores it high value of dry matter content, and hence high value of macronutrients. The value obtained for the macronutrients of the insect was very similar to those reported for roasted *M. bellicosus* in the literature (Adepoju (2016), and within the range stated for termite (*Trinervitermes germinates*, Afiukwa et al., 2013), while the value for moisture and crude protein content was similar to that of Adepoju & Omotayo (2014). Generally, the insect is rich in crude protein, crude lipid, ash, total carbohydrates and gross energy. The high values of the macronutrients of the insect can contribute significantly to macronutrients of infant complementary foods.

The values of moisture, crude protein, fat total carbohydrate and gross energy of fermented maize and sorghum flour were similar to the ones reported in the literature (Adepoju & Daboh, 2013). The fermented flour were low in moisture and fat content, which is an indication that they can be kept for a period of time before they go bad. They were moderate in protein content compared with that of other plant-based staples used for complementary foods such as yam and rice (Adepoju, 2012; Otegbayo et al., 2001). However, the two fermented flour were high in ash, total carbohydrates and gross energy content. The high values of total carbohydrates and gross energy content underscore the reason why they are used as basic staples for locally made complementary foods.

The reduction in moisture content of the *M. bellicosus* enriched complementary foods was similar to the trend reported by Adepoju and Daboh (2013) for *Cirina forda* powder enriched maize and sorghum complementary foods. The lower moisture content in the enriched complementary foods was an indication that the dried form of the enriched flour can keep long before use.

The increase in protein, fat, ash, total carbohydrate and gross energy content of enriched complementary foods is an indication that addition of *M. bellicosus* resulted in significant improvement in the nutrient content of the formulated foods. The same trend of increase in nutrient content of the enriched flour was observed by Adepoju and Daboh (2013) in *Cirina forda* powder enriched fermented maize and sorghum flour. Inclusion of *M. bellicosus* powder in enriching locally formulated cereal-based complementary foods seems to be a promising way of increasing the protein content of animal origin and gross energy content of the complementary foods.

Feeding infants and young children with the *M. bellicosus* enriched cereal-based complementary foods twice daily will likely provide the greater part of their recommended value of macronutrient needs on daily basis, especially at 20% level of inclusion.

4.2 Mineral Composition

The value obtained for mineral composition of the *M. bellicosus* was closely related to the values reported by Adepoju and Omotayo (2014). *M. bellicosus* is rich in potassium, calcium, phosphorus, zinc, manganese and copper, moderate in sodium and iron content compared with plant sources, and very low in magnesium. Its sodium: potassium ratio is highly desirable as the two are responsible for intra and extra cellular electrolyte balance (Insel et al., 2007; Rolfe et al., 2009)). Calcium is required for development of strong bones and teeth, and participates in muscle contraction, blood clotting and nervous impulses, while phosphorus is required for development of strong bones and teeth, important in genetic material, used in energy transfer and buffer system that maintain acid-base system (Rolfe et al., 2009). Zinc is required for growth, cell replication, fertility and reproduction, and hormonal activities among others (Rolfes et al., 2009; Insel et al., 2007; Roth & Townsend, 2003), while manganese and copper are co-factors in some enzyme activities (Sareen et al., 2009), hence, the inclusion of the insect in complementary foods will be advantageous.

The fermented maize and sorghum flour were high in potassium and phosphorus, moderate in iron but low in sodium, zinc, manganese and copper, and very low in calcium and magnesium in comparison with their daily requirements. The very low values of essential macro and micro minerals explains the inadequacy of these basal staples being used for complementary foods with little or no source(s) of calcium and complete protein, hence they are usually high in energy but low in other essential nutrients. Use of this type of complementary foods will result in undernutrion and growth failure in old infants and young children being fed with this kind of foods.

Addition of *M. bellicosus* powder at various levels of inclusion improved the values of all the minerals in fermented maize and sorghum flour significantly, showing that its inclusion will be beneficial to locally formulated cereal-based complementary foods, especially at 20% w/w level of inclusion. Formulations with 20% inclusion level (A3 and B3) of the insect gave the highest nutrient content for the formulated complementary foods, and hence are the best. A high intake of potassium has been reported to protect against increasing blood pressure and other cardiovascular risk (Insel et al., 2007). Feeding infants and young children the enriched maize and sorghum complementary foods will supply greater percentage of daily mineral needs, except calcium and magnesium. However, milk, which is a rich source of calcium is expected to be fed to the infant and young child on complementary feeding, hence the low value of calcium in this enriched foods should not be a barrier to its adoption.

The recommended daily allowance (RDA) of iron for children aged 1-3years old is 7 mg/day (Faber et al., 2008). According to (Codex Alimentarius, 1991), complementary food which satisfied two third of minerals and/or vitamins RDA is acceptable. The iron content of the formulated diets ranged between 2.81 and 4.13 mg per 100gm which fulfilled the minimum RDA for children aged 1-3yeear, but this may not meet the requirement for infants between ages 6 and 11 months old based on the recommended value. The amount of zinc in the enriched foods meets the requirement for infants at age 6-11 months and 1-3 years.

4.3 Vitamin Composition of Fermented Flour and Enriched Complementary Foods

The fermented maize flour was high in β-carotene content while that of sorghum flour was very low. The fermented flour were however low in all the water soluble vitamins except vitamin B_{12}. The low level of these vitamins is believed to be due to the extent of soaking of the fermented sample, as it has been reported that soaking food samples for a period of time leads to leaching of the water soluble micronutrients into the soaking water (Adepoju et al., 2010, Adepoju, 2012). Addition of *M. bellicosus* powder led to remarkable increase in the β-carotene content of the enriched foods, the increment being due to the vitamin A content of the insect, which has been previously reported to be a good source of vitamin A (330.42 ± 0.12 μg/100g, Adepoju & Omotayo, 2014). Vitamin A is an antioxidant which prevent cells from damage by free radicals, essential for maintaining healthy eyes and skin, needed for normal growth and reproduction, promote healthy immune system and prevention of infections (Rolfe et al., 2009; Roth & Townsend, 2003). Enriched maize complementary foods will supply substantial amount of vitamin A to daily requirements of infants and young children, whereas, the enriched sorghum ones will require a rich source of vitamin A to be able to meet the nutritional needs of the children.

Significant reduction observed in most of the water soluble vitamins of the enriched complementary foods was due to lowering of the fermented maize vitamins by the very low levels of water soluble vitamins of the insect (Adepoju & Omotayo, 2014). Only vitamin B6 daily requirement can be met by the complementary foods, hence, other sources of meeting the other water soluble vitamins are needed to augment the one in the complementary foods.

4.4 Antinutrient Composition of Fermented Flour and Enriched Complementary Foods

The fermented flour were low in all the antinutrients studied. This is believed to be due in part to the fermentation of the grains, as processing has been found to reduce the level of antinutrients in foods (Adepoju & Adeniji, 2008; Adepoju et al., 2010). The flour did not contain trypsin inhibitors at the mg/100g level of detection. The reduction observed in the antinutrient content of formulated complementary foods as the level of inclusion of *M. bellicosus* powder increased was believed to be as a result of very low level of these antinutrients in the insect (Adepoju & Omotayo, 2014). Inclusion of *M. bellicosus* powder is therefore beneficial in reducing the level of antinutrients in complementary foods. The level of the antinutrients were very low and negligible, and hence, cannot constitute any hindrance to nutrient bioavailability in the enriched complementary foods.

5. Conclusion and Recommendation

Marcroterme Bellicosus is high in animal protein, fat, energy, potassium, zinc, iron and vitamin A which are essential for growth and development of infants and young children if bioavailable. Its very low level of antinutrients is suggestive of the bioavailability of these nutrients. Addition of *Marcroterme bellicosus* to maize and sorghum flour improved both macro and micronutrient composition of the resulting complementary foods. There were significant improvement in the crude protein, fat, energy, essential minerals and vitamin content of the enriched formulated diets compared with the unenriched flour.

The levels of antinutrients in the formulated diets were low and cannot interfere with the bioavailability of the nutrients. *Macrotermes bellicosus* can be used to improve the nutrient density of maize - and sorghum - based complementary foods for infants, especially in the rural communities of Nigeria where access to, and affordability of animal-based protein is difficult, thereby reducing the level of malnutrition in these infants. The use of *Marcroterme bellicosus* in enriching infant and young child complementary foods is recommended when the insect is in season.

Acknowledgement

The authors hereby acknowledge and are grateful for the financial support provided by the authority of University of Ibadan through Senate Research Grant for executing this research work.

References

Adepoju, O. T. (2016). Assessment of fatty acid profile, protein and micronutrient bioavailability of Winged Termites (*Marcrotermes bellicosus*) using Albino rats. *Malasian Journal of Nutrition, 22*(1), 153-161.

Adepoju, O. T. (2012). Effects of processing methods on nutrient retention and contribution of white yam (*Dioscorea rotundata*) products to nutrient intake of Nigerians. *African Journal of Food Science, 6*(6), 163-167.

Adepoju, O. T., & Adeniji, P. O. (2008). Nutrient composition, antinutritional factors and contribution of native pear (*Dacryoides edulis*) pulp to nutrient intake of consumers. *Nigerian Journal of Nutritional Sciences, 29*(2), 15 – 23.

Adepoju, O. T., & Omotayo, O. A. (2014). Nutrient Composition and Potential Contribution of Winged Termites (*Marcrotermes bellicosus* Smeathman) to Micronutrient Intake of Consumers in Nigeria. *British Journal of Applied Science & Technology, 4*(7), 1149-1158.

Adepoju, O. T., Adigun, M. O., Lawal, I. M., & Ademiluyi, E. O. (2010). Preliminary investigation of nutrient and antinutrient composition of jams prepared from *Hibiscus sabdariffa* calyx extract. *Nigerian Journal of Nutritional Sciences, 31*(1), 8-11.

Adeyeye, E. I. (2011). Fatty acid composition of *Zonocerus variegatus, Macrotermes bellicosus* and *Anacardium occidentale* kernel. *Intern J Pharma and Bio Sci., 2*(1), 135-144.

Afiukwa, J. N., Okereke, C., & Odo, M. O. (2013). Evaluation of proximate and mineral contents of termite (*Trinervitermes germinatus*) from Abakaliki and Ndieze izzi, Ebonyi state, Nigeria. *Am. J. Food. Nutr., 3*(3), 98-104.

Agbidye, F. S., Ofuya, T. I., & Akindele, S. O. (2009). Some edible insect species consumed by the people of Benue State, Nigeria. *Pakistan J Nutr., 8.*, (7), 946-950.

AOAC. (2005). *Association of Official Analytical Chemists* Official methods of Analysis of AOAC *International, Gaithersburg*, MD. USA.

Banjo, A. D., Lawal, O. A., & Songonuga, E. A. (2006). The nutritional value of fourteen species of edible insects in Southwestern Nigeria. *African J. Biotech., 5*(3), 298-301.

Black, R. E., Cousens, S., Johnson, H. L., Lawn, J. E., Rudan, I., Bassani, D. G., … Mathers, C. (2010). Global, regional, and national causes of child mortality in 2008: a systematic analysis. *The Lancet, 375*(9730), 1969-1987.

Black, R. E., Allen, L. H., Bhutta, Z. A., Caulfield, L. E., de Onis, M., Ezzati, M., … Rivera, J. (2008). Maternal and child undernutrition: global and regional exposures and health consequences. *The Lancet, 371*(9608), 243-260.

Codex Alimentarius. (1991).Guidelines for development of supplementary foods for older infants and young children. (CAC/GL. 08-1991): In report of the 19th session. Rome, Italy (p. 10).

Daelmans, B., & Saadeh, R. (2003). Global initiatives to improve complementary feeding. In SCN Newsletter: Meeting the challenge to improve complementary feeding. United Nations System Standing Committee on Nutrition. Moreira, A.D. Ed. Lavenhem Press, UK (pp. 10-17).

Ekpo, K. E., Onigbinde, A. O. & Asia, I. O. (2009). Pharmaceutical potentials of the oils of some popular insects consumed in southern Nigeria. *African J Pharm and Pharmacol., 3*(2), 051-057.

Faber, M., Laurie, S., & Van Jaarsveld, P. (2008). Nutrient content and consumer acceptability for different cultivars of orange-fleshed sweet potato. South African Sugar Association Project no 202:40.

FAO. (2013). Food and Agricultural Organisation of United Nations. Edible insects: future prospects for food and feed security; 68–71.

Griffiths, D. W., & Jones, D. I. H. (1977). Cellulase inhibition by tannins in the testa of field beans (Vicia faba). *J. Sci. Food Agric., 28*(11), 938-989.

Insel, P., Turner, R. E., Ross, D. (2007). Nutrition, 3rd edn, Jones and Barlett Publishers Inc. USA (pp. 185-186, 424, 472-474, 507).

Makkar, H. P., & Becker, K. (1996). Nutritional value and antinutritional components of whole and ethanol extracted *Moringa oleifera* leaves. *Animal feed Sci Technol., 63*, 211-238.

Otegbayo, B. O., Samuel, F. O., & Fashakin, J. B. (2001). Effect of parboiling on physico-chemical qualities of two local rice varieties in Nigeria. *African J. Food Technol., 6*(4), 130-132.

Rajaratnam, J. K., Marcus, J. R., Flaxman, A. D., Wang, H., Levin-Rector, A., Dwyer, L., … Murray, C. J. (2010). Neonatal, postneonatal, childhood, and under-5 mortality for 187 countries, 1970-2010: a systematic analysis of progress towards Millennium Development Goal 4. *The Lancet, 375*(9730), 1988-2008.

Rolfes, S. R., Pinna, K., & Whitney, E. (2009). Understanding normal and clinical nutrition. Eighth Edn Wadsworth Cengage Learning (pp. 421-423, 455).

Roth, A. R., & Townsend, C. E. (2003). Nutrition and diet therapy, 8th edn. Delmar Learning, Thomson Learning Inc Canada (pp. 150-153).

Sareen, S. G., Jack, L. S, & James, L. G. (2009). Advanced nutrition and human metabolism, 5th ednWadsworth Cengage Learning, Canada. Inside cover pages (pp. 501, 523).

Shiriki, D., Igyor, M. A., & Gernah, D. I. (2015). Nutritional evaluation of complementary food formulations from maize, soybean and peanut fortified with *Moringa oleifera* leaf powder. *Food and Nutrition Sciences, 6,* 494-500.

Solomon, M. (2005) Nutritive value of three potential complementary foods based on cereals and legumes. *African J. Food Agric. Nutr. & Devpt (AJFAND), 5,* 1-14.

Sudarmadji, S., & Markakis, P. (1977). The phytate and phytase of soybean Tempeh. *J. Sci. Food Agric., 28*(4), 381-383.

12

Some Factors Affecting Quality of Crude Palm Oil Sold in Douala, Cameroon

Fabrice F. D. Dongho[1], Inocent Gouado[1], Lambert M. Sameza[1], Raymond S. Mouokeu[2], Adélaïde M. Demasse[1], Florian J. Schweigert[3] & Annie R. N. Ngono[1]

[1]Department of Biochemistry, Faculty of Science, University of Douala, P.O. Box 24157 Douala, Cameroon

[2]Institute of Fisheries and Aquatic Sciences, University of Douala, P.O. Box 2701, Douala, Cameroon

[3]Institute of Nutritional Science, University of Potsdam, Arthur-Scheuert-Allee 114-116, 14558 Bergholz-Rehbrücke, Germany

Correspondence: Annie R. N. Ngono, Department of Biochemistry, Faculty of Science, University of Douala, P.O. Box 24157, Douala-Cameroon. E-mail: angono@yahoo.com

Abstract

Crude palm oil (CPO) is an essential ingredient of Cameroonian recipes. However, its quality is subject to doubt, considering the very often inadequate conditions of extraction, conditioning, storage, and selling in the fast growing small holder sector or in the market. This work aimed to evaluate the influence of seasons and containers on the microbiologic, physicochemical quality and the carotenoids content of CPO sold in Douala. A total of 194 samples of CPO were randomly collected in seven markets among which: 95 during the rainy season and 99 during dry season; 93 from CPO contained in opened containers and 101 in closed containers. In these samples, aerobic count colony (ACC) load, total yeasts and moulds load, peroxide value (PV), free fatty acids content (FFA), impurity level and carotenoids content were assayed.The samples tested had ACC load of $4.48\pm1.86\times10^5$ CFU/ml, total yeasts and moulds load of $0.30\pm0.14\times10^5$ CFU/ml, PV of 1.81 ± 0.74 meqO$_2$/kg, FFA of $4.30\pm1.82\%$, impurity level of $0.34\pm0.16\%$ and carotenoids content of 756.41 ± 110.67 mg/l. Also, none of these parameters had varied according to the market. Moreover, among these parameters, PV and carotenoids content were not varied whatever CPO is sold during rainy or dry season, in open or closed containers while others parameters analysed were significantly ($P<0.05$) higher during dry season or when the CPO was contained in open containers. Consequently, traders should make efforts to avoid CPO contamination during the selling. They could package it first and store it in an adequate space particularly during dry season.

Keywords: carotenoids content, crude palm oil, Douala markets, microbiology quality, physicochemical characteristics, vitamin A deficiency

1. Introduction

"Oils" is a collective term for more or less viscous, generally organic chemical liquids. Palm oil (orange-red to brownish or yellowish-red in colour) is extracted from the fleshy mesocarp of the fruit of *Elaeis guineensis* Jacq (Okechalu, Dashen, Lar, Okechalu, & Gushop, 2011; Olorunfemi et al., 2014). With an annual global production equating to about 39% of world production of vegetable oils, palm oil has outclassed soybean during the last decade to become the most important oil crop in the world (Oil World, 2015). Like other vegetable oil sources, palm oil has found application in food and industries (Chabiri, Hati, Dimari, & Ogugbuaja, 2009; Berger, 2010). Their major applications include biodiesel production, pharmaceutical, cosmetics, polish, detergents, shampoo, lipstick etc. In food industries, it is an ingredient in margarine and confectionaries (Pleanjai, Gheewala, & Garivait, 2007; Berger, 2010; Ohimain, Izah, & Fawar, 2013). Due to long term local eating habits and cheaper cost, palm oil is extensively used in its crude form for food purposes throughout Africa and Asia regions (Ngando, Mpondo, & Ewane, 2013). This can be nutritionally beneficial, as crude palm oil (CPO) is a rich source of some essential nutrients such as vitamin E and carotenoids (Edem, 2002; Berger, 2010). Palm oil vitamin E has been extensively studied for its nutritional and health properties including antioxidant activities, cholesterol lowering, anti-cancer effects and protection against atherosclerosis (Srivastava & Gupta, 2006; Odia, Ofori, & Maduk., 2015). Because of its high content in provitamin A carotenoids, CPO constitutes an important food that could be used to prevent vitamin A deficiency (Zeb & Mehmood, 2004; Edem, 2009; Mukherjee &

Mitra, 2009; Odia et al, 2015).

Worldwide, CPO is extracted by industrial, semi-industrial or traditional methods; non industrial sector representing about 30% of total production. Traditional methods are employed by individuals who have little or no knowledge neither of modern aseptic production techniques nor of the microbiological implication of poor sanitation and storage methods (Ngando, Mpondo, Dikotto, & Koona, 2011; Nkongho, Feintrenie, & Levang, 2014). Therefore, palm oil is prone to contamination by microorganisms found in the environment, raw materials and equipments used for the processing, as well as those used for storage and distribution (Okechalu et al., 2011). The microbial quality of CPO is essential because they play adverse role in food and feed products. Though, CPO used for cooking is subjected to heat which may reduce and/or kill all the microorganisms that could invade the CPO. Unfortunately, in many countries, some individuals still consume CPO raw. Also, in traditional medicine, CPO is also used as ingredient for the cure of ailments. As a result, microorganisms present in oil could cause others diseases. Moreover, microorganisms are known to cause chemical changes in CPO that lead to deterioration in their quality (Okpokwasili & Molokwu, 1996).

In fact, some microorganisms could have a lipolytic activity and therefore lead to increase of oil acidity (Okechalu et al., 2011; Ohimain et al., 2013). Besides, the most effective degradation process of CPO is acidification. Generally, fatty acids are present in oils as part of triacylglycerol molecules. The presence of free fatty acids (FFA) molecules is an indication of the impairment of the quality of oils, as FFA are liberated from the triacylglycerol molecules under the action of lipases and esterases. Acidification is generally assessed through determination of oil acidity or FFA content which is one of the most important criterion for determining the quality of cooking oils (Chabiri et al., 2009). Another degradation reaction of CPO regarding food safety is lipid peroxidation. This process involves unsaturated molecules such as fatty acids and carotenoids which undergo a chain reaction mechanism involving free radicals as intermediates and generating lipid peroxides as end products. The latter undergo additional chain cleavage at the level of the hydroperoxide group to form secondary oxidation products such as short chain aldehydes and products bearing ketone, epoxy or alcohol groups responsible for the rancid smell and taste of the oil. The determination of peroxide value (PV) gives an indication of the level of lipid peroxidation of cooking oils. Alongside oil acidity and PV, impurity level is also one of the most important criterion for determining the quality of cooking oils and fats regarding food safety (Ngando et al., 2011, 2013; Ohimain et al., 2013; Chuks et al., 2016). Contribution in the fight against vitamin A deficiency being one of the most nutritional benefit of CPO, their carotenoids content is an important parameter in the evaluation of its quality.

In Cameroon, palm oil accounts for about 90% of edible oil needs (Ngando et al., 2011). Its great production areas are around Douala. A survey done there showed that CPO is used in 97% of households and that 87% of population regularly consume foods prepared with CPO. In Douala markets, CPO mainly comes from small holders who use traditional methods for the extraction. It is usually marketed in closed containers (barrels, buckets, cans, bottles) although some traders sometimes use open containers (bowls, basins, buckets). Furthermore, environmental conditions of the areas of CPO production and of the markets of its distribution are not always adequate and change enormously from one season to other. In fact, during rainy season, there is mostly mud and stagnant water puddles whereas during dry season there is mainly dust. Considering these observations, one can assume that the quality of CPO available in local markets is subject to doubt. Very little researches carried out on this aspect in Cameroon, particularly in Douala. The few studies on the quality of CPO done by Ngando et al. (2011, 2013) and by Goudoum, Makambeu, Abdou and Mbofung (2015) were only on physicochemical parameters. Besides, they showed that some of these parameters varied according to the production process, to the storage time and to the market of distribution. What about microbiological quality and carotenoids content? Could the season of distribution and the kind of container used influence this quality?

The present work aimed to assess the microbiological quality, the physicochemical characteristics and the carotenoids content of CPO sold in some markets in Douala and to study the influence of season and of the kind of containers used on these parameters.

2. Materials and Methods

2.1 Study Area

This study was carried out in Douala town, Economic Capital of Cameroon, situated near the Atlantic Ocean (Latitude 4°2'51''N, Longitude 9°42'23''E).

2.2 Samples Collection

A total of 194 samples of CPO were randomly collected in seven markets of Douala town particularly

Bonamoussadi (n = 34), *Makepe Missoke* (n = 23), *Dakar* (n = 30), *Ndogpassi* (n = 21), *New-bell* (n = 33), *Grand Hangar Bonabéri* (n = 25) and *Deido* (n = 28). Among these samples, 95 were collected during rainy season (between September and mid-November 2014) and 99 during dry season (between mid-November and December 2014). Moreover, 93 samples were collected from CPO contained in completely open containers (bowls, basins, buckets) and 101 from oils exposed in closed containers (buckets, cans, bottles). For each sample, about 200 ml of CPO was collected in a sterile plastic bottle. Care was taken not to contaminate the bottles before and during collection of the samples. After collection, samples were transported to the laboratory for tests. Once in the laboratory, a part of each sample was used for microbiological assays (aerobic count colony or ACC load, total yeasts and moulds load) before 24 hours. The rest was kept at 4°C until analysis of physicochemical parameters (PV, FFA and impurity level) and of carotenoids content.

2.3 Analysis Methods

2.3.1 Microbiological Analysis

Aerobic count colony and total yeasts and moulds of the samples were enumerated using serial dilution pour plate method of Pepper and Gerba (2004). For each sample, a stock solution was obtained by diluting 10 ml of CPO in 90 ml of sterile Tween 80 (Merck, Germany). Three serial decimal dilutions were made from each stock solution. Aliquots of the two last dilutions (10^{-3} and 10^{-4}) were used to determine the ACC and the total yeast and mould according to AFNOR (*Association Française de la Normalisation*) protocols (AFNOR, 2002, 2013).

For ACC enumeration, plate count agar or PCA (Liofilchem, Italy) was used. Sterile nystatin (250 mg/l) was added to suppress fungi growth. The medium was autoclaved at 121°C for 15 min. For each sample, 1 ml of required dilution was pipetted to appropriate marked duplicate Petri plates. Then, 19 ml of cooled medium was poured into each plate and mixed by rotating and tilting. After solidification, each plate was incubated for 48 to 72 hours at 30°C.

Concerning yeasts and moulds enumeration, Sabouraud CAF agar containing chloramphenicol (Liofilchem, Italy) was used. The medium was autoclaved at 121°C for 15 min. For each sample, 1 ml of required dilution was poured to appropriate marked duplicate Petri plates containing 19 ml of cooled medium. After mixing by rotating and tilting, each plate was directly incubated after solidification at 25°C for 72 hours.

Aseptic conditions were employed in all the procedures. After incubation, the growth colonies were promptly counted and the results were expressed as colony forming units per millilitre (CFU/ml).

2.3.2 Physicochemical Parameters Determination

2.3.2.1 Peroxide Value

Peroxide value was performed by titrimetric method according to Association of Official Analytical Chemists protocol (AOAC, 1990). Briefly, for each sample a mixture of glacial acetic acid and chloroform (Merck, Germany) in the ration 3:2 was added to 2 g of oil sample. Thereafter, one added 0.5 ml of saturated (144 g per 100 mL of distilled water) potassium iodide (Merck, Germany) solution; agitated during 1min and added 15 ml of distilled water and 0.5 ml of starch. Thereafter, this solution was titrated with 0.1N sodium thiosulphate (Merck, Germany) until total disappearance of blue colour. Meanwhile, a blank test without oil was done. Peroxide value was calculated from the equation:

$$PV \ (meqO_2/kg) = 1000 \ x \ (V_2-V_1) \ x \ N \ / \ M \ (1)$$

Where:

M = mass of oil taken

V2 = volume of sodium thiosulphate for essay

V1 = volume of sodium thiosulphate for blank

N = normality of sodium thiosulphate (0.1N)

2.3.2.2 Free Fatty Acids

Free fatty acids level was evaluated by titrimetric method according to AOAC protocol (AOAC, 1990). Briefly, for each sample 25 ml of ethanol (Merck, Germany) was added to about 2 g of oil sample. The mixture was brought to boil in a water bath and then cooled down. After adding 2 drops of phenolphthalein (Merck, Germany) as indicator, 0.1N NaOH (Merck, Germany) was used to titrate the mixture with constant shaking for proper mixing until end-point (appearance of violet colour). The FFA was calculated as followed:

$$FFA = V \ x \ N \ x \ M \ / \ 10xW \ (2)$$

Where:

V = volume of NaOH

N = normality of NaOH (0.1N)

M = molecular weight of palmitic acid (256 g/mol)

W = weight of the sample

2.3.2.3 Impurities Level

The impurities level were determined after dissolving oil in hexane (Merck, Germany), followed by filtration and oven drying to a constant weight as described in Ohimain et al. (2013). Thus, for each sample about 10 g of CPO was weighed into the beaker and 100 ml of hexane was added. After agitation and filtration, the residue obtained was dried until constant weight in an electric oven (Binder FDL 115, Germany). This residue was calculated as the percentage of impurities.

2.3.3 Carotenoids Content Determination

Carotenoids content was evaluated by photometry at 446 nm according to protocol used previously by Dongho, Ngono, Demasse, Schweigert and Gouado (2014). For each sample, 20 µl of oil was diluted with 2 ml of hexane. After vigorous shakeup, the mixture was read with a photometer (iCheck^{TM} Carotene; BioAnalyt GmbH, Germany) which gives the carotenoids content of solution in mg/l. The carotenoids content of CPO was calculated from this value by taking into account the dilution factor.

2.3.4 Statistical Analysis

Data were processed using format designed in Microsoft Excel version 2010. Statistical analyses were performed with Graph Pad Prism package version 5.00 (San Diego California USA). T-Student and one-way ANOVA (analysis of variance) tests were used for multiple comparisons. P-values were used as measure of significance and $P < 0.05$ was considered significant.

3. Results

3.1 Quality of CPO According to the Market

The microbiological and physicochemical quality and carotenoids content of CPO sold in some Douala markets are given in the table 1. The results showed that none of the parameters analysed varied significantly according to the market. For microbiological quality, one noted that ACC load and total yeasts and moulds population are higher than values recommended by standards i.e. $<3 \times 10^5$ UFC/mL for ACC (AFNOR, 2013) and $<10^2$ UFC/mL for total yeasts and moulds (AFNOR, 2002). Concerning physicochemical parameters, contrary to PV and FFA that the values obtained are acceptable according to the recommended standards (i.e. <10 meqO$_2$/kg for PV and $<5\%$ for FFA), the impurity level has the values higher than the recommended values i.e. $<0.05\%$ (Codex Alimentarius Commission/FAO/WHO Food Standards, 2015). As for carotenoids content, their values are ranged in normal values i.e. from 700 and 900 mg/L (Codex Alimentarius Commission/FAO/WHO Food Standards, 2015).

Table 1. Quality and Carotenoids Content of Crude Palm Oil Sold in Douala According to the Market

PARAMETERS	MARKETS							TOTAL	P
	Bonamoussadi (n = 34)	Missoke (n = 23)	Dakar (n = 30)	Ndogpassi (n = 21)	New-Bell (n = 33)	Grand Hangar Bonabéri (n = 25)	Deido (n = 28)		
ACC (CFU/ml x 10⁵)	4.18±1.83ᵃ	4.13±1.92	4.63±1.98	4.70±1.96	4.58±1.74	4.71±1.82	4.48±1.90	4.48±1.86	0.638
	1.47-8.00ᵇ	1.70-8.10	1.30-8.70	1.50-8.00	1.90-9.10	2.80-8.90	2.5-8.7	1.3-9.10	
Total yeasts and moulds (CFU/ml x 10⁵)	0.28±0.12ᵃ	0.32±0.15	0.32±0.15	0.30±0.13	0.25±0.13	0.32±0.14	0.30±0.14	0.30±0.14	0.658
	0.08-0.60ᵇ	0.09-0.60	0.10-0.61	0.12-0.60	0.09-0.61	0.10-0.69	0.1-0.68	0.08-0.69	
PV (meqO₂/kg)	1.82±0.73ᵃ	1.69±0.67	1.70±0.60	1.82±0.79	1.94±0.84	1.80±0.77	1.87±0.82	1.81±0.74	0.8710
	0.90-4.30ᵇ	0.70-3.00	0.8-3.00	0.80-3.10	0.70-3.50	0.70-3.10	0.70-3.10	0.70-4.30	
FFA (%)	4.07±1.64ᵃ	4.87±2.25	3.86±1.33	4.03±1.44	3.97±1.60	4.71±1.90	4.81±2.31	4.30±1.82	0.1355
	2.10-10.00ᵇ	2.00-11.00	1.70-7.80	1.90-7.80	1.50-7.00	1.70-9.00	2.10-10.10	1.50-11.00	
Impurity level (%)	0.32±0.23ᵃ	0.37±0.18	0.33±0.14	0.32±0.16	0.35±0.13	0.36±0.11	0.34±0.14	0.34±0.16	0.8823
	0.128-1.280ᵇ	0.160-0.800	0.144-0.640	0.036-0.640	0.144-0.640	0.224-0.640	0.128-0.640	0.036-1.28	
Carotenoids content (mg/L)	755.82±112.55ᵃ	747.41±110.49	724.87±105.57	779.90±100.36	768.82±124.89	785.32±106.07	739.71±106.72	756.41±110.67	0.3961
	548-960ᵇ	590-915	510-915	619-915	497-1009	597-975	603-954	497-1009	

Note. ACC: aerobic count colony; CFU: colony forming units; PV: peroxide value; FFA: free fatty acids; *a:* mean ± standard deviation; b: range

3.2 Influence of the Season

The results (Table 2) showed that contrary to PV and carotenoids content of CPO which did not vary according to the season of commercialisation, ACC load, total yeasts and moulds, FFA and impurity level were significantly higher in samples collected during dry season than those collected during rainy season.

Table 2. Quality and Carotenoids Content of Crude Palm Oil Sold in Douala According to the Season

PARAMETERS	SEASON		P
	Dry season (n = 99)	Rainy season (n = 95)	
ACC (CFU/ml x 10^5)	4.98±2.01	3.95±1.52	*<0.0001*
Total yeasts and moulds(CFU/ml x 10^5)	0.34±0.15	0.26±0.12	*0.0002*
PV (meqO$_2$/kg)	1.81±0.72	1.81±0.77	*0.9945*
FFA (%)	4.70±1.94	3.88±1.58	*0.0016*
Impurity level (%)	0.37±0.18	0.31±0.12	*0.007*
Carotenoids content (mg/L)	750.22±119.85	762.82±100.51	*0.4295*

Note. ACC: aerobic count colony; CFU: colony forming units; PV: peroxide value; FFA: free fatty acids; values are given as mean ± standard deviation

3.3 Influence of the Kind of Container Used

Table 3 gives the quality and the carotenoids content of CPO sold in Douala according to the kind of container used. As previously, we noted that PV and carotenoids content of CPO did not vary that samples have been collected from CPO exposed in open containers or in closed containers. On the contrary, AAC load, total yeasts and moulds, FFA and impurity level significantly varied from one kind of container to other. In fact, these parameters were significantly higher in samples collected from CPO exposed in open containers compared to those exposed in closed containers.

Table 3. Quality and Carotenoids Content of Crude Palm Oil Sold in Douala According to the Kind of Container Used

PARAMETERS	CONTAINER		P
	Open (n = 93)	Closed (n = 101)	
ACC (CFU/ml x 10^5)	5.48±1.94	3.56±1.19	*<0.0001*
Total yeasts and moulds (CFU/ml x 10^5)	0.37±0.14	0.24±0.10	*<0.0001*
PV (meqO$_2$/kg)	1.85±0.75	1.77±0.74	*0.4891*
FFA (%)	4.60±1.97	4.02±1.63	*0.0262*
Impurity level (%)	0.39±0.18	0.29±0.12	*<0.0001*
Carotenoids content (mg/L)	764.87±115.56	748.58±106.00	*0.3072*

Note. ACC: aerobic count colony; CFU: colony forming units; PV: peroxide value; FFA: free fatty acids; values are given as mean ± standard deviation

4. Discussion

This study was done to assess the microbiological quality, the physicochemical characteristics and the carotenoids content of CPO sold in Douala. The result showed that ACC load and total yeasts and moulds population were higher than values recommended by standards (AFNOR, 2002; 2013). This result agrees with similar studies done in Nigeria by Ekwenye (2006), Okechalu et al. (2011), Enemuor, Adegoke, Haruna and Oguntibeju (2012), and Chuks et al. (2016). The presence of microorganisms in CPO gets us to ask questions about the health of consumer. Indeed, although the majority of Cameroonian dishes CPO-based are prepared hot, there also exist some which are prepared when cold such as yellow sauce. Likewise, CPO is sometimes used as vehicle of traditional medicine. In fact, even if microbial load is an essential marker of the food quality, it is the nature of the microorganisms present which is a sure indicator. Indeed, a food can have a low load, but contain microorganisms that could be harmful for the consumer or that could contribute to the deterioration of that food (Codex Alimentarius Commission/FAO/WHO Food Standards, 2015). Besides, Ekwenye (2006), Okechalu et al. (2011), Enemuor et al. (2012), Izah and Ohimain (2013), Ohimain et al. (2013), and Chuks et al. (2016) revealed the presence in CPO of bacteria and fungi that could be dangerous for the consumers (implication in affections like respiratory tract infection, septicaemia, meningitis, hepatitis and cancers) or increase the acidity of oil (microorganisms with lipase activity). This presence of microorganisms in CPO could be due to the manipulations after production. In fact, according to its extraction process, one will not expect the presence of microorganisms because during the stages of cooking/sterilisation of palm nuts (at about 100°C) and of cooking of oil paste (at about 100°C), all the microorganisms initially present should be death. So, it is probably after production that CPO can be contaminated particularly during the operations of packaging, storage, transport or distribution. This explanation is as much true that the results obtained from our study and from the studies of Ekwenye (2006), Okechalu et al. (2011) and Enemuor et al. (2012) with CPO collected in the markets were in order of 10^4-10^5 for bacteria and of 10^3-10^4 for fungi while the results obtained by Izah and Ohimain (2013) and Ohimain et al. (2013) with CPO newly produced were only in order of 10^3-10^4 for bacteria and of 10^2-10^3 for fungi.

Concerning the physicochemical parameters, we noticed that contrary to PV and FFA the values obtained were acceptable according to the recommended standards, the impurity level has the values higher than the values of recommended standards (Codex Alimentarius Commission/FAO/WHO Food Standards, 2015). These results tally with those obtained in similar studies in Nigeria by Ekwenye (2006), Okechalu et al. (2011), Agbaire (2012), Enemuor et al. (2012), Izah and Ohimain (2013), Ohimain et al.(2013), Olorunfemi et al. (2014), and Chuks et al. (2016); in Egypt by Abd El-Gawad, Hamed,   Zidan and Shain (2015) and in Cameroon by Ngando et al. (2011 ; 2013) and Goudoum et al. (2015).

Peroxide value is a parameter used to assess the quality of cooking oils and fats through the measurement of the amount of lipid peroxides and hydroperoxides formed during the initial stages of oxidative degradation and thus, estimate to which extent spoilage of the oil (expressed by the level of rancidity) has advanced. Beside these visible harmful effects on the sensory quality of the oil, peroxidation also makes the oil dangerous for human health, as the free radicals generated by this process are proven to be carcinogenic (Pignitter & Somoza, 2012). Regarding oxidation, PV does not help us to find out the real oxidation level of oil. In fact, this parameter assesses the peroxides contained in oil. These peroxides are very unstable and can later undergo additional chain cleavage to form secondary oxidation products such as short chain of aldehydes and products bearing ketone, epoxy or alcohol groups responsible for the rancid smell and taste of the oil (Ngando et al. 2011). Unfortunately, PV does not take into account these products. So this parameter just gives the oxidation of oil at a precise moment. Indeed, peroxide formed before the moment of analyse could be already transformed to secondary products. It is the reason why the best method of evaluating of the real oxidation level of oil must take into account not only the peroxides, but also these products. The latter are evaluated by anisidine value (AV) which takes into account aldehydes, ketones, epoxides and alcohols. Total oxidation (TOTOX) of oil can be calculated using the equation: $TOTOX = 2PV + AV$ (Ngando et al., 2011; Pignitter & Somoza, 2012). Thus, even if we obtained a PV respecting the standards in this study, we cannot guarantee that CPO sold in Douala was not oxidised enough after its production.

As for FFA, it is known that their accumulation in CPO is mainly due to the action of an active endogenous lipase presents in the mesocarp of the fruit of the oil palm. This lipase is activated in the fruit at maturity upon wounding and/or bruising and is responsible for the hydrolysis of triglycerides and the liberation of FFA (Ngando et al., 2011). In order to limit the action of this lipase, fresh fruit bunches must be processed rapidly after harvest. The presence of FFA in the CPO could also be explained by the contaminating lipases from microorganisms (Ekwenye, 2006). In fact, although microorganisms present in the palm nuts and in others intrans used during the extraction are almost destroyed during the sterilisation, CPO produced could be contaminated as brought up above by microorganisms (among which those lipase activity) during the operations of packaging, storage, transport or distribution. Moreover, once the fruits are processed, the lipase is no more active, but the FFA of the resulting palm oil may also increase during storage as a result of autocatalytic hydrolysis. In that case, FFA acts as catalysts for the reaction between triacylglycerols and water to produce more FFA (Ngando et al., 2013).

The impurities usually present in oils are all the compounds and/or particles insoluble in oils. We have between others carbohydrate compounds such as gums, metals, metallic particles, metallic ions, metallic complexes and all others solid particles. Their level in CPO is usually associated to production method (Ohimain et al., 2013, Chuks et al., 2016). Besides, Ngando et al. (2011) showed that CPO produced traditionally had an impurity level higher than that of CPO produced industrially. Furthermore, these impurities could come from the environment during the operations of packaging, storage, transport or distribution. The high level of impurities obtained in this study could be explained by the methods of extraction. In fact, CPO sold in Douala mainly comes from small holders who produce oil by traditional method. The important part of CPO produced industrially being generally dedicated to oil refineries and soap factories.

Carotenoids are naturally presented in the mesocarp of palm nuts. Besides, CPO is known as the richest food in carotenoids (700 and 900 mg/L) mainly β-carotene (Codex Alimentarius Commission/FAO/WHO Food Standards, 2015). The results of this study confirm this observation and corroborate with those obtained by Agbaire (2012) and Olorunfemi et al. (2014) in a similar study done in Nigeria.

The results showed that any parameters analysed significantly varied according to the market where samples were collected. These results were not surprising because the major part of CPO sold in Douala markets comes from the same areas of production particularly *Moungo*, *Sanaga Maritime* and *Fako* divisions. Furthermore, sampling was practically the same in all the markets tested. Indeed, in each of these markets, the samples were collected during the rainy season as well as during the dry season. They were collected from CPO contained in open containers as well as from those in closed containers. Our results are not comparable to those of Ngando et al. (2013) which showed a significant difference of the quality of CPO (PV and FFA) from one market to another. The difference noticed between the two studies could be explained by the size of sampling because in this study, we analysed at least 21 samples per market while Ngando et al. (2013) analysed only 04 samples per market.

As for the effect of season, we noticed that contrary to PV and carotenoids content which did not vary according to the season, others parameters analysed (ACC, total yeasts and moulds load, FFA and impurity level) were significantly higher in samples collected during the dry season than those collected during the rainy season. This difference could be attributed to the environment of the areas of production, packaging, transportation and distribution which changes considerably from one season to the other. Indeed, during the rainy season, the environment is most often humid and one usually has water puddles and mud while during the dry season there is usually a dust because the environment is always dry. Contrary to mud, dust is volatile. Therefore, when the wind blows, they can settle on CPO during the operations of packaging, storage, transport or distribution (Nkongho et al., 2014; Bechoff et al., 2015; Goudoum et al., 2015). This dust most often contains microorganisms; hence the direct increase of microbial load of CPO contaminated by dust is evident. Some of these microorganisms could have a lipase activity (Izah & Ohimain, 2013; Ohimain et al., 2013), so it is normal that increasing level of microbial load of a sample automatically lead to its FFA. Beside microorganisms, this dust could contain solid particles insoluble in oil that could explain the increasing of impurity level of samples contaminated by dust. Furthermore, the dust could not contain any factors able to act on oxidation of oil, this could justify why the PV and the carotenoids content did not vary according to the season.

Concerning the effect of the kind of containers used, we observed as previously that PV and carotenoids content did not vary from one kind of container to another while other parameters analysed were significantly higher in samples collected from CPO contained in open containers compared to those collected from CPO contained in closed containers. As previously, this increase could be attributed to the environment of the markets. In fact, whatever the season, CPO samples contained in open containers are more exposed to contamination compared to those contained in closed containers. Indeed, during the dry season, we can have the contamination by dust and in the rainy season, contamination by water or mud spatters. Be it, dust or water/mud spatters, they can contain microorganisms (that could explain the increase of microbial load and of FFA) or solids particles (explaining the increase impurity level).

5. Conclusion

At the end of this study which consisted of evaluating the microbiological quality, the physicochemical characteristics and the carotenoids content of CPO sold in Douala town, we can conclude that with the exception of PV, FFA and carotenoids content which had acceptable values as recommended by standards; impurity level, ACC load, total yeasts and moulds had values higher than the standards. Also, none of these parameters had varied according to the market. Moreover, among these parameters, PV and carotenoids content didn't vary which ever CPO is sold during the rainy or the dry season, in open or closed containers while other parameters analysed were higher during the dry season or when the CPO was contained in open containers. We can recommend traders to make efforts to avoid CPO contamination during the selling. They could for examples keep CPO in closed containers and store it in an adequate space particularly during dry season. Thus, in order to complete this work, we are intent in our next researches to study the effects of these solutions on the quality of CPO.

Conflict of interests

FJS is shareholder of BioAnalyt GmbH, Germany. All other authors declare no conflict of interests regarding the publication of this paper.

Acknowledgements

The authors would like to thank the sellers who provided crude palm oil samples analysed in this study and BioAnalyt GmbH (Teltow, Germany) for gracefully provided us the photometer (iCheck™ carotene) for carotenoid analysis.

References

Abd El-Gawad, I. A., Hamed, E. M., Zidan, M. A., & Shain, A. A. (2015). Fatty acid composition and quality characteristic of some vegetable oils used in making commercial imitation cheese in Egypt. *Journal of Nutrition and Food Sciences, 5*, 380. http://dx.doi.org/10.4172/2155-9600.1000380

Agbaire, P. O. (2012). Quality assessment of palm oil sold in some major markets in Delta State, southern Nigeria. *African Journal of Food Science and Technology, 3*, 223-226.

AFNOR (2002). Food microbiology - Yeasts and moulds count by colony count at 25°C - Routine method. *NF V08-059* November 2002. Retrieved December 12, 2014, from http://www.boutique.afnor.org/norme/nf-v08-059/microbiologie-des-aliments-denombrement-des-levures-et-moisissures-par-comptage-des-colonies-a-25-c-methode-de-routine/article/767939/fa120539?popin=1.

AFNOR (2013). Food microbiology - Horizontal method for microorganisms count - Part 2: colony count at 30°C by surface plating technique. *NF EN ISO 4833-2* October 2013. Retrieved December 12, 2014, from http://www.boutique.afnor.org/norme/nf-en-iso-4833-2/microbiologie-des-aliments-methode-horizontale-pour-le-denombrement-des-micro-organismes-partie-2-comptage-des-colonies-a-/article/798127/fa158338.

AOAC (1990). *Official methods of analysis* (15th ed.). WILLIAM HORWITZ edv., Washington D.C.

Bechoff, A., Chijioke, U., Tomlins, K. I., Govinden, P., Ilona P., Westby, A., & Boy, E. (2015). Carotenoid stability during storage of yellow gari made from biofortified cassava or with palm oil. *Journal of Food Compositionand Analysis, 44*, 36-44. http://dx.doi.org/10.1016/j.jfca.2015.06.002

Berger K. G. (2010). *Quality and functions of palm oil in food applications: a layman's guide.* MALAYSIAN PALM OIL COUNCIL, 2nd Floor, WismaSawit Lot 6, SS6, JalanPerbandaran, 47301 Kelana Jaya, Selangor, Malaysia. ISBN 978-983-9191-09-7.

Chabiri, S. A., Hati, S. S., Dimari, G. A., & Ogugbuaja, V. O. (2009). Comparative quality assessment of branded and unbranded edible vegetable oils in Nigeria. *Pacific Journal of Science and Technology, 10*, 927-934.

Chuks, K. O., Tarfen, Y. A., Ikechukwu, P. O., Paul, E. M., Benedict, T. S., Uche, K. A., ... Amechi, S. N. (2016). Assessment of mold contamination and physicochemical properties of crude palm oil sold in Jos, Nigeria. *Food Science & Nutrition*, published by Wiley Periodicals, Inc. http://dx.doi.org/10.1002/fsn3.393

Codex Alimentarius Commission/FAO/WHO Food Standards (2015). *Joint FAO/WHO food standards programme Codex Alimentarius Commission.* Thirty-eighth Session CICG, REP15/FO, Geneva, Switzerland 6-11 July 2015.

Dongho, D. F. F., Ngono, N. A., Demasse, M.A., Schweigert, F., & Gouado, I. (2014). Effect of heating and of short exposure to sunlight on carotenoids content of crude palm oil. *Journal of Food Processing and Technology, 5(4)*, 314. http://dx.doi.org/10.4172/2157-7110.1000314

Edem, D. O. (2002). Palm oil: biochemical, physiological, nutritional, hematological, and toxicological aspects: a review. *Plant Foods for Human Nutrition, 57*, 319-341. https://doi.org/10.1023/A:1021828132707

Edem, D. O. (2009). Vitamin A: A review. *Asian Journal of Clinical Nutrition, 1*, 65-82. http://dx.doi.org/10.3923/ajcn.2009.65.82

Ekwenye, U.N. (2006). Chemical characteristics of palm oil biodeterioration. *Biokemistri, 18*, 2141-2149.

Enemuor, S. C., Adegoke, S. A., Haruna, A. O., & Oguntibeju, O. O. (2012). Environmental and fungal contamination of palm oil sold in Anyigba Market, Nigeria. *African Journal of Microbiology Research, 6*, 2744-2747. https://doi.org/10.5897/AJMR11.1287

Goudoum, A., Makambeu, N. A., Abdou, B.A., & Mbofung, C.M. (2015). Some physicochemical characteristics and storage stability of crude palm oils (*Elaeis guineensis* Jacq). *American Journal of Food Science and Technolology, 3(4)*, 97-102. http://dx.doi.org/10.12691/ajfst-3-4-1

Izah, S.C., & Ohimain, E.I. (2013). Microbiological quality of crude palm oil produced by smallholder processors in the Niger Delta, Nigeria. *Journal of Microbiology and Biotechnology Research, 3(2)*, 30-36.

Mukherjee, S., & Mitra, A. (2009). Health effects of palm oil. *Journal of Human Ecology, 26*, 197-203.

Ngando, E. G. F., Mpondo, M. E. A., Dikotto, E. E. L., & Koona, P. (2011). Assessment of the quality of crude palm oil from smallholders in Cameroon. *Journal of Stored Products and Postharvest Research, 2*(3), 52-58.

Ngando, E. G. F., Mpondo, M. E. A., & Ewane, M. A. (2013). Some quality parameters of crude palm oil from major markets of Douala, Cameroon. *African Journal of Food Science, 7,* 473-478. http://dx.doi.org/10.5897/AJFS2013.1014

Nkongho, R.N., Feintrenie, L., & Levang, P. (2014). *The non-industrial palm oil sector in Cameroon. Working Paper 139.* Bogor, Indonesia: CIFOR.

Odia, O. J.,Ofori, S., & Maduk, O. (2015). Palm oil and the heart: A review. *World Journal of Cardiology, 7*(3), 144-149. http://dx.doi.org/10.4330/wjc.v7.i3.144

Ohimain, E. I., Izah, S. C., & Fawar, A. D. (2013). Quality assessment of crude palm oil produced by semi-mechanized processor in Bayelsa State, Nigeria. *Discourse Journal of Agricultural and Food Sciences, 1*(11), 171-181.

Oil World (2015). Oil world 2014 annual report. *STA Mielke GmbH Langenberg 25 21077 Hamburg/Germany.* Retrieved June 30, 2015, from http://www.oilworld.biz/annual

Okechalu, J. N., Dashen, M. M., Lar, P. M., Okechalu, B., & Gushop, T. (2011). Microbiological quality and chemical characteristics of palm oil sold within Jos Metropolis, Plateau State, Nigeria. *Journal of Microbiology and Biotechnology Research, 1(2),* 107-112.

Okpokwasili, G. C., & Molokwu, C. N. (1996). Biochemical characteristics of vegetable oil biodeterioration. *Materials and Organisms, 30,* 307-314.

Olorunfemi, M. F., Oyebanji, A. O., Awoite, T. M., Agboola, A. A., Oyelakin, M. O., Alimi, J. P., ... Oyedele, A. O. (2014). Quality assessment of palm oil on sale in major markets of Ibadan, Nigeria. *International Journal of Food Reearch, 1,* 8-15.

Pepper, I. L., & Gerba, C. P. (2004). *Environmental microbiology. A laboratory manual* (2[nd]ed.). Elsevier academic press. 232p.

Pignitter, M., & Somoza, V. (2012). Are vegetable oils always a reliable source of vitamin A? A critical evaluation of analytical methods for the measurement of oxidative rancidity. *Sight and Life Magasine, 26(1),* 18-27.

Pleanjai, S., Gheewala, S. H., & Garivait, S. (2007). Environmental evaluation of biodiesel production from palm oil in a life cycle perspective. *Asian Journal of Energy and Environment, 8,* 15-32.

Srivastava, J.K., & Gupta, S. (2006). Tocotrienol-rich fraction of palm oil induces cell cycle arrest and apoptosis selectively in human prostate cancer cells. *Biochemical and Biophysical Research Communications, 346,* 447-453. http://dx.doi.org/10.1016/j.bbrc.2006.05.147

Zeb, A., & Mehmood, S. (2004). Carotenoids contents from various sources and their potential health applications. *Pakistan Journal of Nutrition, 3,* 199-204. http://dx.doi.org/10.3923/pjn.2004.199.204

13

Ultraviolet-C Light Effect on the Physicochemical and Antioxidant Properties of Blackberry, Blueberry, and Raspberry Nectars

José Fernando Haro-Maza[1] & José Ángel Guerrero-Beltrán[1]

[1]Departamento de Ingeniería Química, Alimentos y Ambiental. Universidad de las Américas Puebla. Ex Hda. Santa Catarina Mártir, San Andrés Cholula, Puebla 72810, Mexico

Correspondance: José Ángel Guerrero-Beltrán, Departamento de Ingeniería Química, Alimentos y Ambiental. Universidad de las Américas Puebla. Ex Hda. Santa Catarina Mártir, San Andrés Cholula, Puebla 72810, Mexico. E-mail: angel.guerrero@udlap.mx; joseangel150@hotmail.com

Abstract

The effect of UV-C light on foodborne microorganisms (mesophilic aerobic bacteria (MAB), molds, and yeasts), physicochemical characteristics (color, total soluble solids, pH, and acidity), and antioxidant properties (ascorbic acid, antioxidant capacity (AC), total monomeric anthocyanins (TMA), and total phenolic compounds (TPC)) in blackberry, blueberry, and raspberry nectars was evaluated. Nectars were UV-C light treated at five times (5, 10, 15, 20, and 25 minutes) at constant flow rate (16.48 mL/s). The best UV-C light treatment for the three nectars, from the microbiological point of view, was 25 min. A statistical difference ($p < 0.05$) in TPC, within treatment times, was observed in nectars; their content was reduced as the treatment time increased, except for the blackberry nectar. The same effect was observed for the antioxidant capacity. The TMA content increased with the UV-C light treatment.

Keywords: UV-C light, blueberry, raspberry, blackberry, antioxidants

1. Introduction

Minimal processed foods is a tendency that has been growing in recent years due to a genuine consumer concern about their health and the effects some food additives might have on them. The production of foods is focusing toward obtaining high nutritional and minimal processed products, a good example of these types of foods are berries nectars, which are high in nutritional value: antioxidant compounds, vitamins and minerals; they can be minimally processed. To gather these criteria of safety and minimal processing some emerging technologies have been implemented, most of them are still being investigated to guarantee their reliability and affectivity. The main objective for using nonthermal processing technologies consists in replacing chemicals and thermal sterilization methods to achieve an adequate microbiological reduction and minimize the effect on the nutritional and sensorial properties of the manufactured foods (Alothman, Bhay, & Karim, 2009).

The UV-C light processing offers a valuable alternative for processing foods. The use of this technology is well established in cases of water, surface and air disinfection; however, its use in food products, including fruit nectars, is still under investigation and its diffusion as a promising alternative to heat treatments is still limited. Thus, UV-C light technology has a wide potential to be developed for its industrial implementation; it can be used in peeled sliced fruits, liquid products, and fresh vegetables (Ribeiro, Canada, & Alvarenga, 2012) because of its high affectivity for inhibiting bacteria and other types of microorganisms.

UV-C light affects microorganisms at DNA level causing mutations due to the separation of the double helix, avoiding in this way their replication (Gardner and Shama, 2000). Another favorable aspect of the UV-C light implementation comes with the approval from the Food and Drug Administration (FDA, 2000); this organization approved its use as an alternative for fruit juices pasteurization. In the same year, the United States Department of Agriculture (USDA) approved this technology and its use in food products.

The objective of this study was to evaluate the effect of UV-C light on microorganisms and on the physicochemical and antioxidant properties of blackberry, blueberry, and raspberry nectars.

2. Method

2.1 Nectar Preparation

Frozen berries (Global premier®), packed in Chicago, Illinois, were homogenized using an immersion food processor (Oster, Mod. 2609 - B2609, Tlalnepantla, Edo de México), filtered through a stainless steel kitchen strainer, and then passed through cheese cloth several times for eliminating coarse fruit particles, resulting in 100% pulp free juices. Juices were analyzed in pH (Orion pH meter, Thermo Scientific, Waltham, MA, USA) and total soluble solids (TSS) (Atago refractometer, Osaka, Japan). Separately, a sucrose syrup (containing the same TSS of fruit juices) was prepared and pH adjusted (similar to the fruit pH). Nectars were prepared mixing 50% of fresh fruit juice and 50% of syrup. Nectars were immediately analyzed and UV-C light processed.

2.2 UV-C Light Processing

Berries nectars were processed in a UV-C light system, similar to a double wall heat exchanger, assembled at the Universidad de las Americas Puebla (Guerrero-Beltrán and Barbosa-Cánovas, 2006; Guerrero-Beltrán, Welti-Chanes, & Barbosa-Cánovas, 2009). The external wall was made of stainless steel with an internal diameter of 4.8 cm. The inner wall was a quartz tube with an external diameter of 2.2 cm. The gap by which liquid flew was 2.6 cm. The UV-C light lamp was hosted in the center of the system, surrounded by the annular part (chamber) by where the liquid product was passed through. The UV-C mercury lamp (Orange, Connecticut, USA), used as the light source, was 303 and 15 mm in length and diameter, respectively, and had an intensity of 17 W that deliver a dosage of 57 $\mu W/cm^2$. The chamber of the system host a volume of 430 mL. In order to process the fluid (600 mL), this was passed through the chamber at 4°C using a recirculating chilling unit model DC50-B12 (Haake instruments, Germany). The nectar was pumped through the UV-C light system and recirculated, using a peristaltic pump model 75553-7 (Master Flex, Vernon, Illinois, USA), at a constant flow rate of 16.48 mL/s (0.989 L/min) (Ochoa-Velasco & Guerrero-Beltrán, 2012). Nectars were UV-C light processed during 5, 10, 15, 20, and 25 min corresponding to doses (D) of 0.171, 0.342, 0.513, 0.684, and 0.860 kJ/m^2, respectively (Guerrero-Beltrán & Barbosa-Cánovas, 2006). Untreated nectars were used as a reference for comparison purposes. The UV-C light treatment was performed in duplicate and each method in triplicate. Control and processed nectars were analyzed in physicochemical, microbiological, and antioxidant characteristics. The residence time (θ = 26.10 s) was obtained dividing the volume of the chamber (V = 430 mL) per the flow rate (q = 16.48 mL/s). The number of passes (NP = t/θ), at each processing time (t), were 11.50, 23.00, 34.49, 45.99, and 57.48, respectively (Guerrero-Beltrán & Barbosa-Cánovas, 2006).

2.3 Physicochemical Analysis

Acidity. The acidity (% p/v as citric acid) was performed according to the 942.15 AOAC (2000) method.

pH. A previously calibrated (buffers 4, 7 and 10) digital pH meter (Jenway, Lansing, USA) was used for measuring pH of nectars.

TSS (°Bx). TSS content was performed using a manual ATAGO refractometer (Osaka, Japan).

2.4 Color

The color was measured using a Colorgard System/05® colorimeter (Gardener, Geretsried, Germany) in the CIELAB (Commission Internationale d'Eclairage L^*, a^*, b^*) scale for measuring L^*, a^* and b^* color parameters in the transmittance mode. For each assay, 4 mL of sample were used. The total change in color (ΔE) was calculated according to McLaren (1986).

2.5 Antioxidant Analysis

Antioxidant capacity. The nectars antioxidant capacity was analyzed according to the Kuskoski, Asuero, Parrilla, Troncoso, & Fett (2004) method with some modifications. The $ABTS^+$ radical was obtained mixing 5 mL of distilled water, 3.3 mg of potassium persulfate, and 19.4 mg of ABTS reagent; the mixture was left to stand in the dark at 25°C for 16 hours. After that, pure ethanol was used to dilute the $ABTS^+$ radical (radical solution) until reaching an absorbance of 0.70 ± 0.02 at 754 nm. Then 80 μL of nectar were mixed with 3920 μL of ABTS radical solution, blended, and the initial absorbance (Ai) measured. After 7 minutes or reaction, the final absorbance (A_f) was measured. The antioxidant capacity was calculated using a standard curve of Trolox (T) with the following equations:

$$UI = \frac{A_i - A_f}{A_i} * 100 \qquad (1)$$

$$UT = \frac{UI - b}{m} * 100 \qquad (2)$$

where UI is the inhibition (%), UT is the amount of Trolox (mg T/mL), b is the intercept, and m is the slope (mL/mg Trolox). The standard curve (0 to 0.2 mg Trolox/mL) was: Abs = 490.40 mL/mg*[mg/mL]+3.16 ($R^2 = 0.981$).

Total phenolic compounds (TPC). They were analyzed according to the Gao, Ohlander, Jeppsson, Bjork, & Trajkovski (2000) method with some modifications. Two milliliters of distilled water were placed in an amber glass tube, then 200 µL of Folin and Ciocalteu reagent (Sigma-Aldrich, Toluca, Mexico), 100 µL of diluted sample and 1 mL of a 20% (w/v) Na_2CO^3 were added and totally mixed. Samples were left in the dark at room temperature for an hour. The absorbance was measured at 675 nm using an UNICO UV-Vis spectrophotometer model 2800 H (NJ, USA). The phenolic compounds calculation was performed using a standard curve of Gallic acid (0 to 0.3 mg/mL) and the next equation:

$$GA = \left(\frac{A - b}{m}\right) * 100 \qquad (3)$$

where GA is the Gallic acid content (mg of Gallic acid/100 mL), A is the sample absorbance, b is the intercept, and m is the slope (mL/mg GA). The standard curve was: Abs = 3.563 mL/mg*[mg/mL]+0.053 ($R^2 = 0.971$).

Total monomeric anthocyanins (TMA). The TMA content was performed according to the Giusti & Wrolstad (2000) method. It is based in the structural transformations of anthocyanins due to the formation of the flavilio cation at pH 1 and its colorless form at pH 4.5.

Ascorbic acid. The ascorbic acid content was analyzed using the 967.21 AOAC (2000) method.

2.6 Microbial Counts

The mesophilic aerobic bacteria (MAB) as well as molds plus yeasts were evaluated using the standard plate count agar and potato dextrose acidified agar (with 10% tartaric acid) methods, respectively. Cultures in Petri dishes were incubated at 35 ± 2 and $25 \pm 2°C$, respectively, for counting colonies forming units per milliliters (CFU/mL) after 2 and 5 days, respectively.

2.7 Statistical Methods

A Microsoft Excel program was used for calculating means and standard deviations. ANOVA and Tukey's test data were calculated using the MINITAB 16 program for determining significant differences among means at a confidence level of 0.05.

3. Results and Discussion

3.1 Physicochemical Analysis

Acidity. Table 1 shows the acidity of control and UV-C light treated nectars. Even though significant differences ($p < 0.05$) were found among some exposition times, they are not directly related to the treatment time. No direct relationship between treatment time and acidity changes was found. In general, it can be said that the treatment time did not affect the acidity of berry nectars. Also, there is no information in the literature that could indicate the opposite. Berries fruits are acid products due to the presence of organic acids; however, the acid content can change depending on factors such as the fruit type and season of production.

Table 1. Effect of UV-C light on acidity of nectars of berries.

	Acidity (g/100 mL)		
Time (min)	Blueberry	Raspberry	Blackberry
0	$0.22 \pm 0.00ab$	$0.69 \pm 0.03b$	$0.62 \pm 0.00a$
5	$0.20 \pm 0.00c$	$0.82 \pm 0.02a$	$0.61 \pm 0.01b$
10	$0.21 \pm 0.01bc$	$0.63 \pm 0.00b$	$0.57 \pm 0.02d$
15	$0.23 \pm 0.01a$	$0.67 \pm 0.01b$	$0.60 \pm 0.00c$
20	$0.22 \pm 0.00ab$	$0.67 \pm 0.01b$	$0.61 \pm 0.00b$
25	$0.23 \pm 0.00a$	$0.65 \pm 0.01b$	$0.60 \pm 0.00c$

Different letters in same column indicate significant differences ($p < 0.05$) within treatment times.

pH. Table 2 shows pH of control and UV-C light treated nectars. It can be observed, that there is not sufficient evidence (p > 0.05) to indicate that the UV-C light treatment affected the pH of nectars. The pH values reported by Purgar, Duralija, Voca, Vokurka, & Ericisli (2012) for blackberry and raspberry are similar to those obtained in this study. They pointed out that berries juices, including blueberry, have a pH between 3.1 and 3.7, depending on the season of production.

Table 2. Effect of UV-C light on pH of nectars of berries.

| Time (min) | pH | | |
	Blueberry	Raspberry	Blackberry
0	3.30 ± 0.00a	3.10 ± 0.05a	3.31 ± 0.00a
5	3.29 ± 0.00a	3.10 ± 0.02a	3.32 ± 0.00a
10	3.30 ± 0.01a	3.14 ± 0.02a	3.32 ± 0.00a
15	3.30 ± 0.00a	3.14 ± 0.02a	3.32 ± 0.00a
20	3.29 ± 0.00a	3.13 ± 0.02a	3.30 ± 0.01a
25	3.29 ± 0.00a	3.15 ± 0.02a	3.30 ± 0.02a

Different letters in same column indicate significant differences (p < 0.05) within treatment times.

Total soluble solids. TSS remained barely constant along the UV-C light treatment (Table 3). Natural berries juices have TSS between 9.0 and 11.8% (Purgar, Duralija, Voca, Vokurka, & Ericisli, 2012), most of them being sugars and organic acids; this is why nectars still have some sour taste. After performing the statistical analysis, significant differences were found (p < 0.05) for total soluble solids within treatment time of nectars; however, an actual tendency of increasing or decreasing of total soluble solids is not observed; therefore, this effect could be associated to experimental measurement.

Table 3. Effect of UV-C light on total soluble solids of nectars of berries.

| Time (min) | Total soluble solids (g/100 g) | | |
	Blueberry	Raspberry	Blackberry
0	10.86 ± 0.05ab	8.90 ± 0.14ab	9.10 ± 0.00a
5	10.96 ± 0.05ab	8.83 ± 0.24ab	9.10 ± 0.00a
10	11.03 ± 0.05a	8.66 ± 0.24b	9.10 ± 0.00a
15	10.83 ± 0.05b	8.60 ± 0.29b	9.10 ± 0.00a
20	10.86 ± 0.12ab	9.16 ± 0.24a	9.10 ± 0.00a
25	10.83 ± 0.12b	9.16 ± 0.24a	9.10 ± 0.00a

Different letters in same column indicate significant differences (p < 0.05) within treatment times.

3.2 Color

In general, nectars were turbid with pale reddish (raspberry), purplish (blackberry), or bluish (blueberry) colors.

$L*$ color parameter. Figure 1 illustrates the UV-C light treatment time effect on the $L*$ color parameter of berries nectars. The $L*$ color parameter refers to the sample lightness; it has values from 0 (perfect black) to 100 (perfect white). Blueberry nectar had an $L*$ value of 38.28 at time 0, after five minutes of UV-C light treatment the $L*$ value increased and this was changing along the treatment time until reaching a value of 38.45 after 25 min; however, those changes are due to other factors such as particles in the nectar because the tendency is not constant; this means that the $L*$ color parameter do not goes down or up constantly. The initial and final $L*$ values were similar. This means that the dark color remained almost constant along the UV-C light treatment. Raspberry nectar was the less dark of the three nectars; it had a similar tendency as blueberry nectar during treatment regarding the $L*$ color parameter; its lightness increased after 5 minutes of treatment from 44.54 to 48.26 and reached a final value of 43.38 after 25 minutes; similar to the initial lightness. Blackberry nectar decreased in lightness after UV-C light treatment; however, the final $L*$ value was barely greater than the initial $L*$ value of nectar without UV-C light treatment, time 0. It should take into account that the scale in Figure 1 is very expanded. According to Guerrero-Beltrán, Welti-Chanes, & Barbosa-Cánovas (2009), lightness in some liquid foods has greater changes when the volumetric flow in the UV-C light process is lower than 1.21 mL/s; however, as flows increases the $L*$ parameter of liquid foods may barely change during UV-C light treatments.

In some cases, if the processing time is prolonged the $L*$ color parameter may increase (fading of color), depending of the color of the product. This is not the case in this study, since only one flow rate was used (16.48 mL/s). According to other UV-C light studies in fruit juices, lightness may change (the $L*$ value may increase) as the treatment time increases (Guerrero-Beltrán, Welti-Chanes, & Barbosa-Cánovas, 2009). These authors also found that after 30 minutes of UV-C light treatment the lightness values of juices were very close to the initial values at time 0 (no treated). According to the statistical analysis, there was no significant differences ($p > 0.05$) for $L*$ values within treatment times (0 to 25 minutes); thus, this shows that the UV-C light treatment does not affect the lightness of these three berries nectars. The three types of nectars have anthocyanins that make them red-purple-blue in color; therefore, UV-C light may affect this types of food pigments (Fulecki & Francis, 1968).

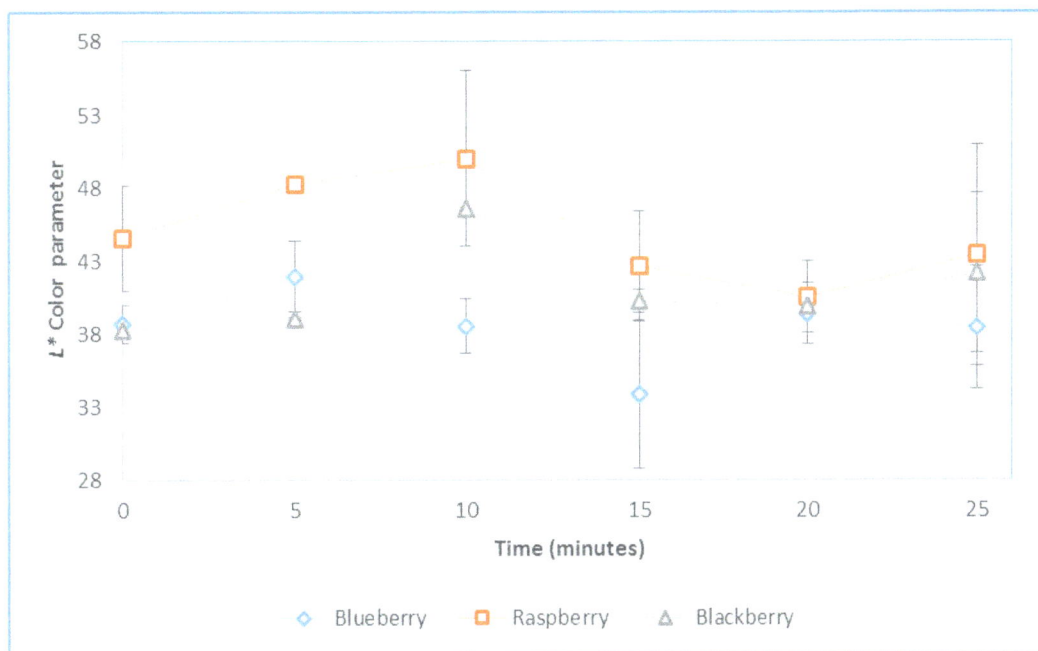

Figure 1. UV-C light effect on the $L*$ color parameter of nectars of berries.

$a*$ color parameter. In Figure 2 is illustrated the effect of the UV-C light on the $a*$ color parameter of berries nectars. The scale of the $a*$ color parameter goes from negative values (green color) to positive values (red color). Raspberry nectar showed a red color. It was observed that the UV-C light treatment had very little effect over the $a*$ value. After 25 minutes of UV-C light treatment, very similar values to time 0 were found (around 5); this value indicates that the color has a tendency to red. Blueberry was in the limit between green and red (an $a*$ value around one was observed) and was maintained along the UV-C light treatment (0 to 25 min). Blackberry nectar had a light change after 25 minutes of UV-C light treatment; showing initially positive $a*$ values (around 0.5) and ending with negative values (around -1). This small change in the $a*$ color parameter may be due to the fact that blackberry had the greatest antioxidant capacity of the three fruits which is due to higher quantities of some phenolic and antioxidant compounds that can cause a light darkening; however, the change in color is barely modified (no detected by the eye). As mentioned above, some berries (blackberry and blueberry) have deep red-blue color due to the anthocyanins content and type. Thus, a change in the product color may mean a change in the anthocyanins content or in their chemical structure which also depend strongly on pH. The statistical analysis indicated no significant differences ($p > 0.05$), within the treatment time, for all nectars treated with UV-C light.

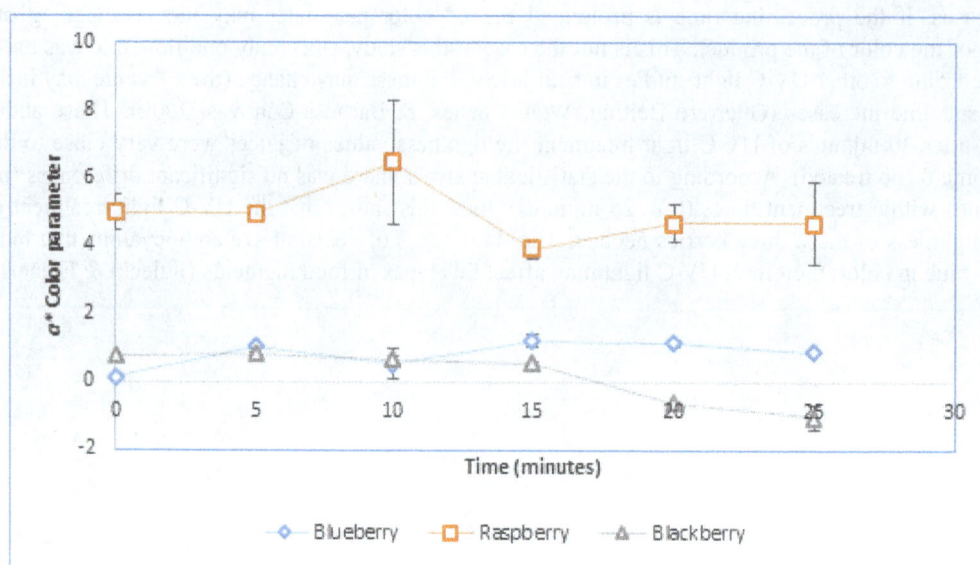

Figure 2. UV-C light effect on the $a*$ color parameter of nectars of berries.

$b*$ color parameter. Figure 3 shows the effect of the UV-C light treatment on the $b*$ color parameter of berries nectars. The $b*$ color parameter on the CIELAB scale corresponds from yellow (positive values) to blue (negative values) hues. In general, the $b*$ color values were between 0.5 and 4 along the UV-C light treatment for the three nectars, being the raspberry nectar barely affected. All $b*$ color values were in the limits of yellow and blue. All nectars were pale in color to the naked eye. The "highest" values (no really high) were for the raspberry nectar, meaning that the product was less bluish. The closer $b*$ values to blue color were for blackberry and blueberry nectars. In general, the $b*$ color parameter was not affected ($p > 0.05$) within the treatment time; therefore, nectars kept their colors similar to those of untreated samples. Results obtained in this study coincide with those reported by Casati, Sánchez, Baeza, Magnani, Evelson, & Zamora (2012); they reported $b*$ color values lower than 5.6 for dark berries such as blackberry and blueberry; $b*$ values were between 3 and 5 for red berries.

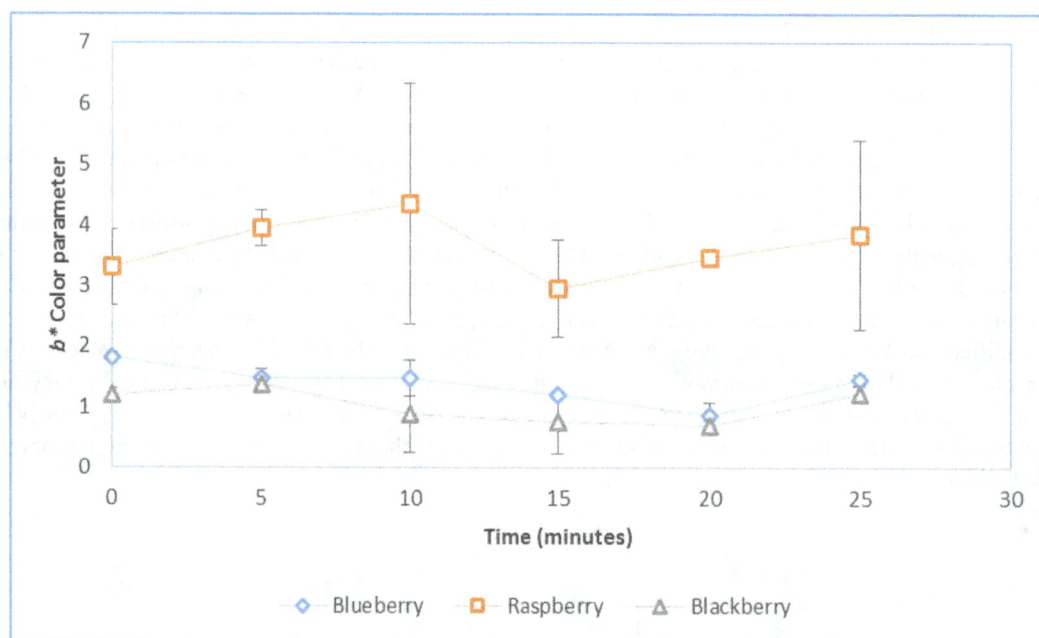

Figure 3. UV-C light effect on the $b*$ color parameter of nectars of berries.

Total change in color (ΔE). Besides of observing low changes in the $L*$, $a*$, and $b*$ color parameters, it is also important to take into account the total change in color of nectars. The ΔE total change in color was 2.2 ± 1.8, 3.3 ± 1.8, and 3.3 ± 2.9 for blueberry, raspberry and blackberry, respectively. The less sensitive nectar to the UV-C light treatment was blueberry because it had the minimum ΔE value. Raspberry and blackberry nectars had similar ΔE values. The penetration of UV-C light depend on the color and turbidity of the liquid fruit product. Guerrero-Beltrán, Welti-Chanes, & Barbosa-Cánovas (2009) pointed out that the UV-C light can generate changes in color, changes that could increase with the exposition time.

3.3 Antioxidant Compounds

Antioxidant Capacity

Figure 4 depicts the effect of UV-C light on the antioxidant capacity of berries nectars. In general, the UV-C light treatment did not affect ($p > 0.05$) the antioxidant capacity within the treatment time. For blackberry, the fresh nectar showed an antioxidant capacity of 216 mg T/100 mL. After 25 minutes of UV-C light treatment, the antioxidant capacity was 218 mg T/100 mL, which means that no UV-C light effect was observed in the antioxidant components. For blueberry something similar happened, the initial and final values were very close to each other, no reduction of antioxidant capacity was observed due to the UV-C light. According to Alothman, Bhay, & Karim (2009), the UV-C light did not modify the antioxidant capacity of fresh-cut tropical fruits (pineapple, banana, and guava); however, those fruits have other types of pigments (carotenoids). López-Rubira, Conesa, Allende, & Artés (2005) found similar results in pomegranate juice which also has anthocyanins as berries have. They reported that the antioxidant capacity of the pomegranate juice did not change after 30 minutes of UV-C light exposure. In raspberry nectar (light reddish), it was observed a decrease in the antioxidant capacity during the UV-C light processing; a decrease from 65 to 15 mg T/100 mL of nectar was observed. Similar results were found by Caminiti, Palgan, Muñoz, Noci, Whyte, & Morgan (2012) who indicated that in clear juices, such as clarified apple juice and raspberry juice, as the exposition time increased, the antioxidant capacity decreased because of a greater penetration of the UV-C light. Another important factor that can contribute to a decrease in the antioxidant capacity of raspberry nectar is the ascorbic acid content. Both, raspberry and blueberry showed a low content of ascorbic acid compared to blackberry; ascorbic acid was lessen during UV-C light treatment (Figure 7). Nevertheless, no significant differences were observed ($p > 0.05$) within the treatment time for the antioxidant capacity in blueberry and raspberry nectars; on the contrary in the case of raspberry nectar where significant differences were observed ($p < 0.05$). Erkan, Wang, & Wang (2008) have pointed out that the UV-C light may increase the antioxidant capacity in strawberries and some other fruits, by increasing the content of phenolic compounds naturally present in fresh fruits which has to be with phytoalexins (antimicrobials and antioxidants).

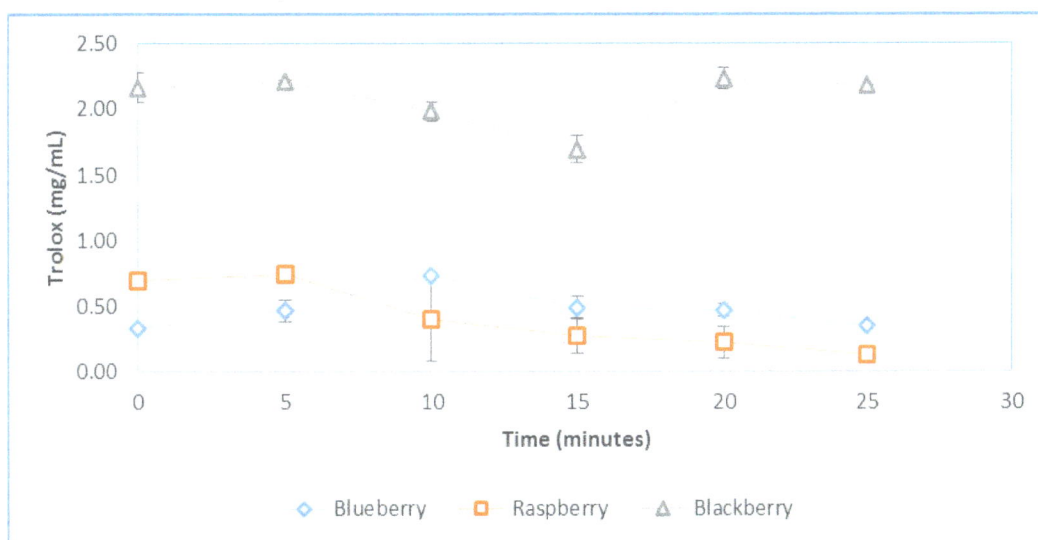

Figure 4. Effect of UV-C light on the antioxidant capacity of nectars of berries.

Total Phenolic Compounds

Figure 5 depicts the effect of the UV-C light on the total phenolic compounds (TPC) of berries nectars. Blackberry and raspberry may contain phenolic compounds such as hydroxybenzoic acids (protocatechuic, Gallic, or p-hydroxybenzoic acids) and blueberry may contain hydroxycinnamic acids (caffeic, chlorogenic, coumaric, ferulic, or sinapic) (Manach, Scalbert, Morand, Remesy, & Jiménez, 2004). TPC slightly decreased as the treatment time increased in blueberry and raspberry nectars. TPC in blackberry nectar remained constant along the treatment, its dark color may minimize the effect of the UV-C light on phenolic compounds. The greater the content of dark colors the lower the penetration of the light; therefore, less effect over some compounds may occur. Ochoa-Velasco & Guerrero-Beltrán (2012) reported that the UV-C light processing remarkably reduced the content of phenolic compounds in pitaya juice as the exposition time increased. The explanation to the reduction of pigments in fruit products is because UV-C light may affect the structures of some compounds. Pala & Toklucu (2011) reported no significant differences ($p > 0.05$) in the content of phenolic compounds in pomegranate juices after UV-C light treatment; the same findings were reported by Caminiti, Palgan, Muñoz, Noci, Whyte, & Morgan (2012) in apple juice. The nectar that was more effected on TPC by the UV-C light was the blueberry nectar; this fact can be attributed to the inactivation of phenolic such as hydroxycinnamic acids and, or flavonols (Manach, Scalbert, Morand, Remesy, & Jiménez, 2004) found in this fruit.

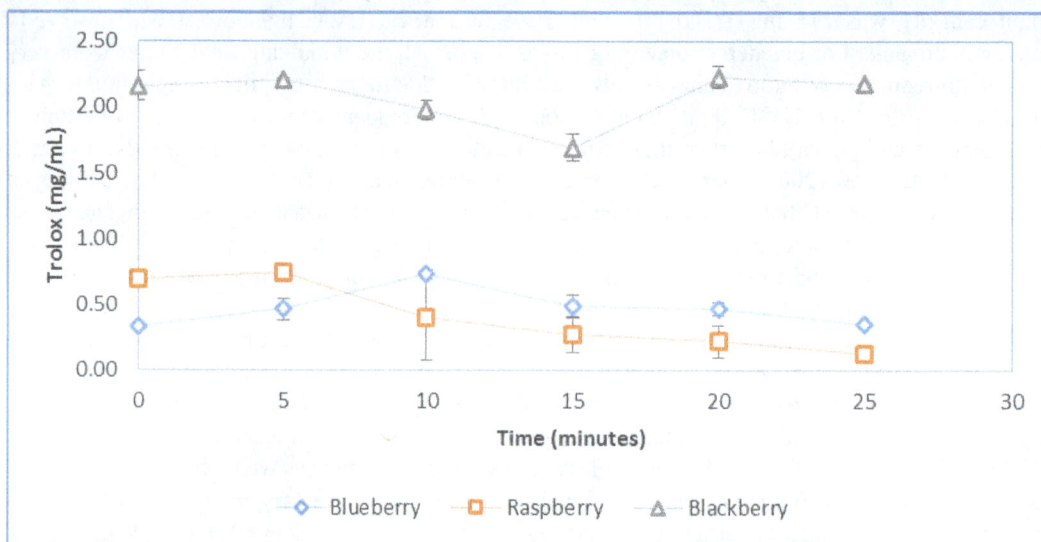

Figure 5. Effect of UV-C light on the total phenolic compounds of nectars of berries.

Total monomeric anthocyanins

Figure 6 shows the effect of the UV-C light on the total monomeric anthocyanins content in berries nectars. All these berries might have anthocyanins such cyanidin, pelargonidin, peonidin, delphinidin or malvidin (Manach, Scalbert, Morand, Remesy, & Jiménez, 2004). The UV-C light increased significantly ($p < 0.05$) the anthocyanins content in blueberry and blackberry nectars during UV-C light processing. A light decrease ($p < 0.05$) of anthocyanins was observed in the raspberry nectar; however, not important changes were observed; actually there is not a decrease, the anthocyanins content was around 35 mg/L in average. Pala & Toklucu (2011) have pointed out that the UV-C light does not generate changes in the anthocyanins content in pomegranate juice.

Ascorbic Acid

Figure 7 illustrates the effect of UV-C light on the ascorbic acid (AA) content in berries nectars. The AA content was significantly reduced ($p < 0.05$) in the three UV-C light treated berries nectars as the exposition time increased.

3.4 Microbial Counts

Mesophilic aerobic bacteria

Figure 8 illustrates the mesophilic aerobic bacteria (MAB) inactivation in berries nectars treated with UV-C light. As stated before, color and turbidity of the liquid food product may affect the UV-C light penetration. Dark and turbid liquid foods are less penetrated by the UV-C light, contrary to the effect on transparent and light colored liquid foods. Blackberry was turbid and had a purplish color. A maximum of 1000 colony forming units per milliliter of MAB were counted in fresh blackberry nectar. A maximum of 0.6 logarithmic cycles of MAB were reduced in blackberry nectar along the treatment time. In blueberry (turbid and bluish) and raspberry (turbid and pale reddish), reductions of 0.97 and 1.3 logarithmic cycles were achieved, respectively, after of 25 min of UV-C light treatment. The raspberry nectar showed the highest MAB reductions, very probably due to its color that favored the penetration of light. The blackberry nectar had the smallest MAB reductions due to its dark color. Lorenzini, Fracchetti, Bolla, Stefanelli, & Rossi (2010) reported 1 - 3 logarithmic cycles reductions of MAB in UV-C light treated grape juice and red wine. Ochoa-Velasco & Guerrero-Beltrán (2012) reported 2.11

logarithmic reductions of MAB in pitaya (Stenocereus griseus) juice after 25 minutes of treatment with UV-C light at the same flow rate (16.48 mL/s) used in this study.

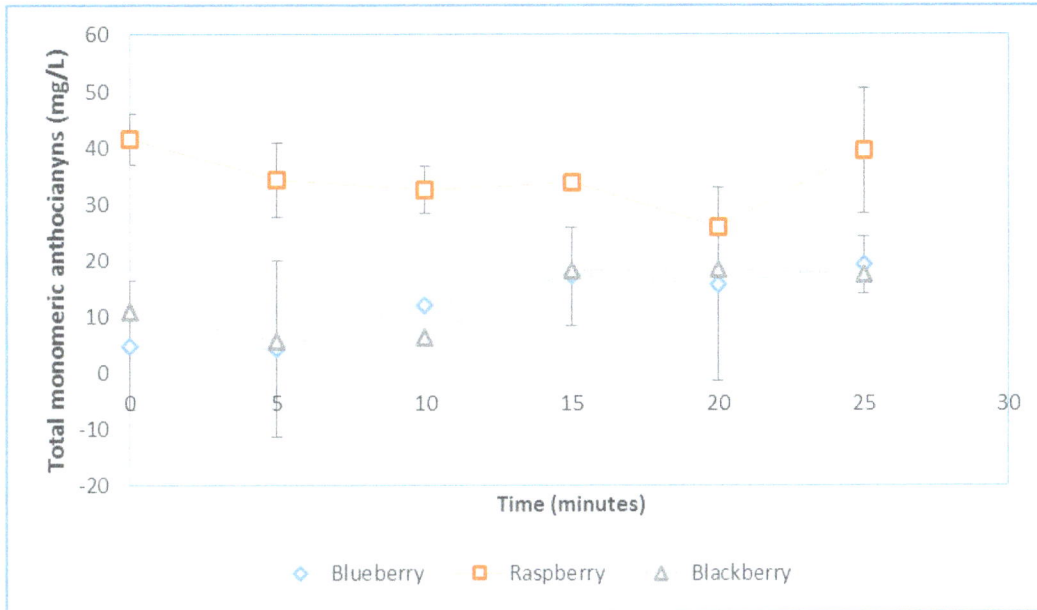

Figure 6. Effect of UV-C light on the total monomeric anthocyanins content in nectars of berries.

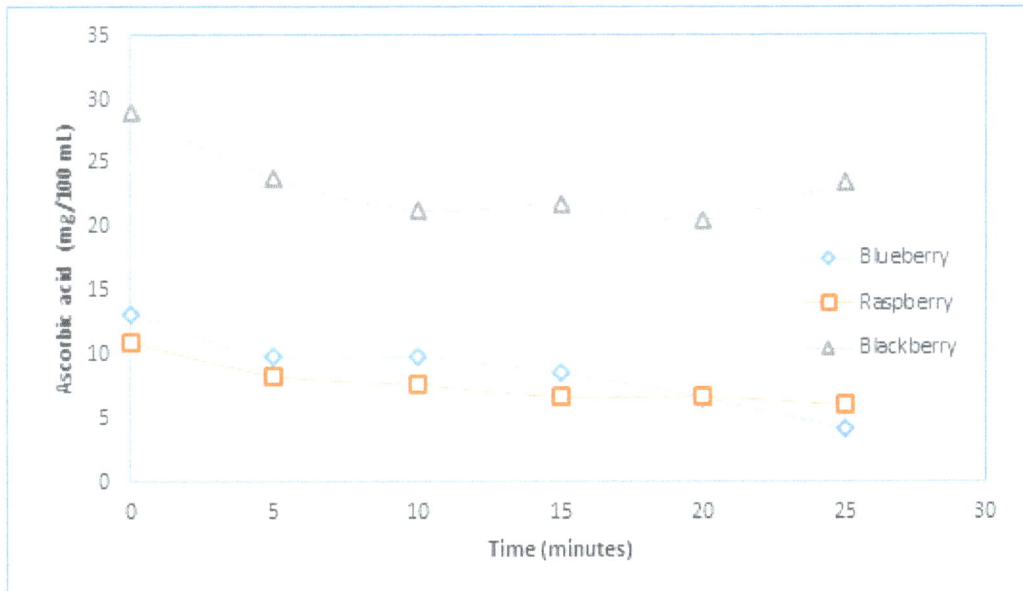

Figure 7. Effect of UV-C light on the ascorbic acid content in nectars of berries.

Molds

Figure 9 shows the effect of UV-C light treatment on the reduction of molds in berries nectars. A maximum of 100 colony forming units per milliliter of molds were counted in fresh raspberry nectar. The highest reduction of molds was achieved in the raspberry nectar; a reduction of 1.62 logarithmic cycles was observed after 25 minutes of UV-C light treatment. In blueberry nectar was observed the smallest reduction (0.75 logarithmic cycles) of molds after 25 min of treatment. In blackberry nectar, a reduction of 0.88 logarithmic cycles was observed. Ochoa-Velasco & Guerrero-Beltrán (2012) reported reductions of 1.14 logarithmic cycles in pitaya juice after 30 min of UV-C light treatment.

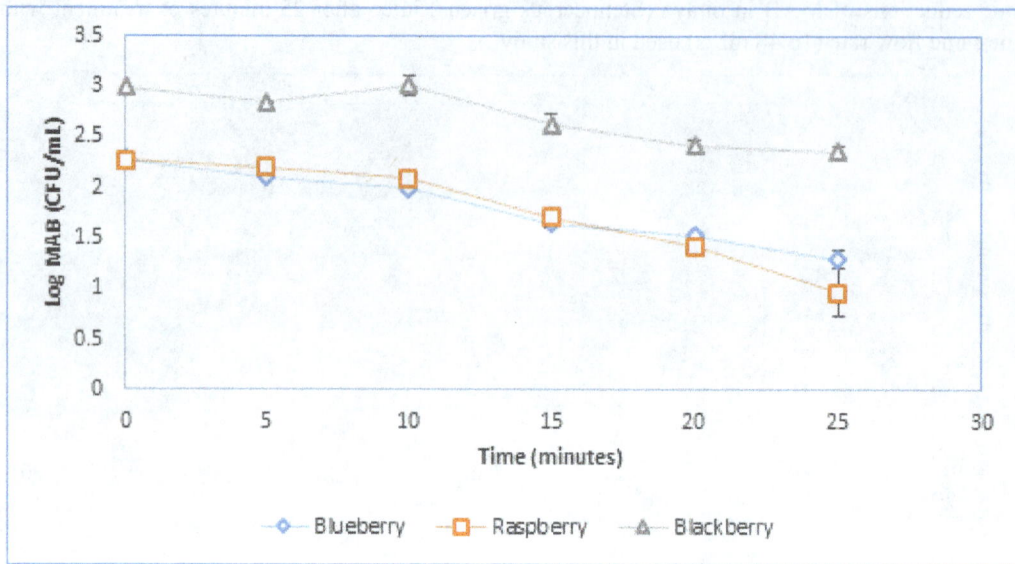

Figure 8. Effect of UV-C light on mesophilic aerobic bacteria reduction in nectars of berries.

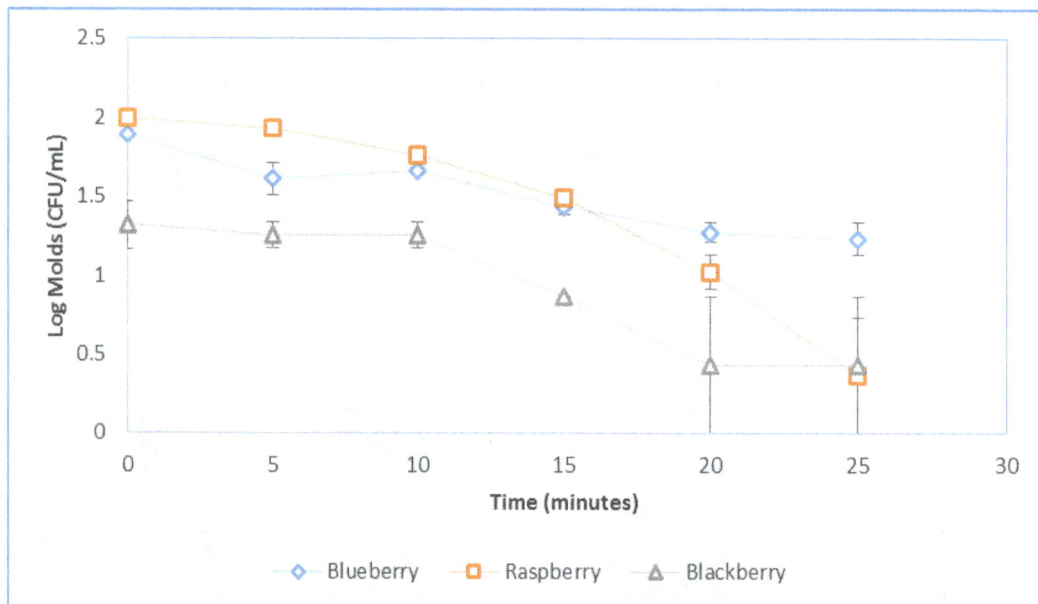

Figure 9. Effect of UV-C light on molds reduction in nectars of berries.

Yeasts

Figure 10 shows the effect of UV-C light on the reduction of yeasts in berries nectars. Reductions of 0.76, 0.80 and 2.02 logarithmic cycles of yeasts were achieved in blackberry, blueberry and raspberry nectars, respectively. Ochoa-Velasco & Guerrero-Beltrán (2013) reported a reductions of 1.8 cycles in UV-C light treated pitaya juice. Guerrero-Beltrán, Welti-Chanes, & Barbosa-Cánovas (2009) observed reductions of 2.5 logarithmic cycles of *S. cerevisiae* in cranberry and grapefruit juices after 30 minutes of UV-C light treatment, being those results similar to results found in this study.

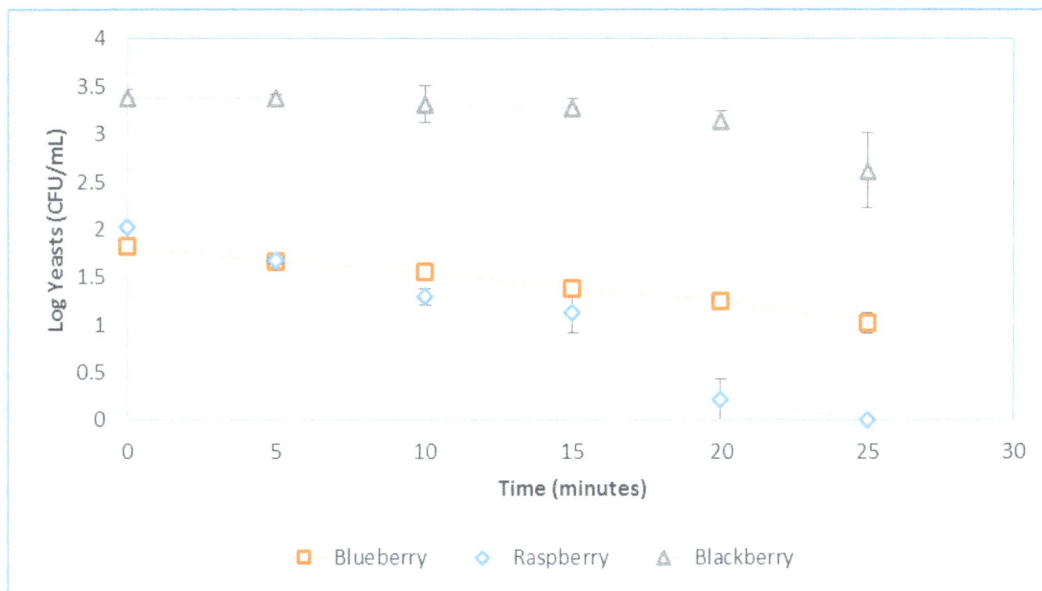

Figure 10. The Effect of UV-C light on yeasts reduction in nectars of berries.

4. Conclusions

According to results found in this study, it was clear that the UV-C light had no effects on pH, acidity or TSS in blueberry, raspberry, and blackberry nectars. Color parameters were affected along the treatment time differently for each type of nectar. According the ΔE total change in color, the less sensitive nectar to the UV-C light was the blueberry nectar; the most sensitive to the UV-C light was the raspberry nectar. The antioxidant capacity was barely affected in the raspberry nectar. Phenolic compounds were decreased lightly in the blueberry and raspberry nectars as the UV-C light exposition time increased. The anthocyanins content increased significantly after 25 minutes of treatment in the blueberry and blackberry nectars. The ascorbic acid content in nectars was lightly lessened due to the UV-C light effect. In average, logarithmic reductions of 1.0, 1.08, and 1.19 were obtained for MAB, molds, and yeasts, respectively, in these types of nectars.

Acknowledgments

J. F. Haro-Maza would like to thanks to the Consejo Nacional de Ciencia y Tecnología (CONACYT) and the Universidad de las Américas Puebla (UDLAP) in Mexico for supporting his Mater Science degree studies.

References

Alothman, M., Bhay, R., & Karim, A. A. (2009). UV radiation-induced changes of antioxidant capacity of fresh-cut tropical fruits. *Innovative Food Science and Emerging Technologies, 10,* 512-516. http://dx.doi.org/10.1016/j.ifset.2009.03.004

AOAC. (2000). Official Methods of Analysis (14th edition). Association of Official Analytical Chemists. USA.

Caminiti, I., Palgan, l., Muñoz, A., Noci, F., Whyte, P., & Morgan, D. (2012). The effect of ultraviolet light on microbial inactivation and quality attributes of apple juice. *Food and Bioprocess Technology, 5*(2), 680-686. http://dx.doi.org/10.1007/s11947-010-0365-x

Casati, C. B., Sánchez, V., Baeza, R., Magnani, N., Evelson, P., & Zamora, C. M. (2012). Relationships between color parameters, phenolic content and sensory changes of processed blueberry, elderberry and blackcurrant commercial juices. *International Journal of Food Science and Technology, 47,* 1728-1736. http://dx.doi.org/10.1111/j.1365-2621.2012.03027.x

Erkan, M., Wang, S. Y., & Wang, C. Y. (2008). Effect of UV treatment on antioxidant capacity, antioxidant enzyme capacity and decay in strawberry fruit. *Postharvest Biology and Technology, 48*(2), 163-171. http://dx.doi.org/10.1016/j.postharvbio.2007.09.028

FDA. (2000). Irradiation in the production, processing and handling of food. 21 CFR Part 179. *Federal Register*, 65(230), 71056-71058.

Fulecki, T., & Francis, F. J. (1968). Quantitative methods for anthocyanins. Extraction and determination of total anthocyanins in cranberries. *Journal of Food Science, 33*, 71-77.

http://dx.doi.org/10.1111/j.1365-2621.1968.tb00887.x

Gao, X., Ohlander, M., Jeppsson, N., Bjork, L., & Trajkovski, V. (2000). Changes in antioxidant effects and their relationship to phytonutrients in fruits of sea Buckthorn (*Hippophae rhamnoides* L.) during maturation. *Journal of Agricultural and Food Chemistry, 48*, 1485-1490. http://dx.doi.org/10.1021/jf991072g

Gardner D. W. M., & Shama, G. (2000). Modeling UV-C induced inactivation of microorganisms on surfaces. *Journal of Food Protection, 63*(1), 63-70.

Giusti, M. M., & Wrolstad, E. R. (2000). Characterization and measurement of anthocyanins by UV-Visible spectroscopy. John Wiley & Sons, Inc. Unit F1.2.

Guerrero-Beltrán, J. A., & Barbosa-Cánovas, G. V. (2006). Inactivation of Saccharomyces cerevisiae and polyphenoloxidase activity in mango nectar treated with ultraviolet light. *Journal of Food Protection, 69*(2), 362-368.

Guerrero-Beltrán, J. A., Welti-Chanes, J., & Barbosa-Cánovas, G. V. (2009). Ultraviolet-C light processing of grape, cranberry and grapefruit juices to inactivate *Saccharomyces cerevisiae*. *Journal of Food Process Engineering, 32*, 916-932. http://dx.doi.org/10.1111/j.1745-4530.2008.00253.x

Kuskoski, M., Asuero, A., Parrilla, M., Troncoso, A., & Fett, R. (2004). Actividad antioxidante de pigmentos antociánicos. *Ciencia y Tecnología de Alimentos (Campinas, Brasil), 24*, 691-693.

López-Rubira, V., Conesa, A., Allende, A., & Artés, F. (2005). Shelf life and overall quality of minimally processed pomegranate arils modified atmosphere packaged and treated with UV-C. *Postharvest Biology and Technology, 37*(2), 174-185. http://dx.doi.org/10.1016/j.postharvbio.2005.04.003

Lorenzini, F., Fracchetti, V., Bolla, E., Stefanelli, F., & Rossi, S. (2010). Ultraviolet light irradiation as an alternative technology for the control of microorganisms in grape juice and wine. Conference: International Organization of Vine and Wine - 33rd World Congress of Vine and Wine, 8th General Assembly of the OIV, V-2010_n° (1240) OR.II.19, 1-8.

Manach, C., Scalbert, A., Morand, C., Remesy, C., & Jiménez, L. (2004). Polyphenols: food source and bioavailability. *American Journal of Clinical Nutrition, 79*, 727-747.

McLaren, K. (1986). The quantification of colour differences. Ch. 10. In *The Colour Science of Dyes and Pigments* (pp. 129-152). Bristol, Great Britain: Adam Hilger Ltd.

Ochoa-Velasco, C. E., & Guerrero-Beltrán, J. A. (2012). Ultraviolet-C light effect on Pitaya (Stenocereus griseus) juice. *Journal of Food Research, 1*(2), 60-70. http://dx.doi.org/http://dx.doi.org/10.5539/jfr.v1n2p60

Ochoa-Velasco, C. E., & Guerrero-Beltrán, J. A. (2013). Short-wave ultraviolet-C light effect on pitaya (Stenocereus griseus) juice inoculated with *Zygosaccharomyces bailii*. *Journal of Food Engineering, 117*(1), 34-41. http://dx.doi.org/10.1016/j.jfoodeng.2013.01.020

Pala, C., & Toklucu, A. (2011). Effect of UV-C light on anthocyanin and other quality parameters of pomegranate juice. *Journal of Food Composition and Analysis, 24*(6), 790-795. http://dx.doi.org/10.1016/j.jfca.2011.01.003

Purgar, D. D., Duralija, B., Voca, S., Vokurka, A., & Ericisli, S. (2012). A comparison of fruit chemical characteristics of two wild grown *Rubus* species from different locations of Croatia. *Molecules Journal, 17*, 10390-10398. http://dx.doi.org/10.3390/molecules170910390

Ribeiro, C., Canada, J., & Alvarenga, B. (2012). Prospects of UV radiation for application in postharvest technology. *Journal of Food Agriculture, 24*(6), 586-597.

14

Enhancing the Nutritional Value of Canola (Brassica napus) Meal Using a Submerged Fungal Incubation Process

Jason R. Croat[1], William R. Gibbons[1], Mark Berhow[2], Bishnu Karki[3], & Kasiviswanathan Muthukumarappan[3]

[1]Biology & Microbiology Department, South Dakota State University, Brookings, SD 57007, USA

[2]USDA, Agricultural Research Service, National Center for Agricultural Utilization Reach; Peoria, IL 61604, USA

[3]Agricultural & Biosystems Engineering Department, South Dakota State University, Brookings, SD 57007, USA

Correspondence: Jason Croat, Biology & Microbiology Department, South Dakota State University, Brookings, SD 57007, USA. E-mail: jason.croat@sdstate.edu

Mention of trade names or commercial products in this paper is solely for the purpose of providing specific information and does not imply endorsement by the U.S. Department of Agriculture. USDA is an equal opportunity provider and employer.

Abstract

The aim of this study was to determine the optimal fungal culture to increase the nutritional value of canola meal so it could be used at higher feed inclusion rates, and for a broad range of monogastrics, including fish. Submerged incubation conditions were used to evaluate the performance of seven fungal cultures in hexane extracted (HE) and cold pressed (CP) canola meal. *Aureobasidium pullulans* (Y-2311-1), *Fusarium venenatum* and *Trichoderma reesei* resulted in the greatest improvements in protein levels in HE canola meal, at 21.0, 23.8, and 34.8 %, respectively. These fungi reduced total glucosinolates (GLS) content to 2.7, 7.4, and 4.9 $\mu M.g^{-1}$, respectively, while residual sugar levels ranged from 0.8 to 1.6 % (w/w). In trials with CP canola meal, the same three fungi increased protein levels by 24.6, 35.2, and 37.3 %, and final GLS levels to 6.5, 4.0, and 4.7 $\mu M.g^{-1}$, respectively. Additionally, residual sugar levels were reduced to 0.3-1.0 % (w/w).

Keywords: canola, fungal incubation, glucosinolates, rapeseed, submerged incubation

1. Introduction

Canola (*Brassica napus*) is grown widely in Canada and the northern United States, and it is the second most abundant source of edible oil in the world (Aider & Barbana, 2011). Canola meal is also the second most abundant protein source for livestock feed, trailing soybean meal (Newkirk, 2009). The abundance and lower price of canola meal have driven interest in replacing soybean meal in ruminant and monogastric feeds (Lomascolo, Uzan-Boukhris, Sigoillot, & Fine, 2012). On a cost per Kg of protein basis, canola protein is typically valued at 80-85 % the value of soybean meal because it contains less gross energy, less protein, and over three times as much fiber. Canola also contains glucosinolates (GLS) that can have anti-nutritional effects on livestock. However, due to its lower cost it may be an economical protein source for animals that do not have high energy or lysine requirements (Bell, 1993).

The presence of GLS in canola meal limits inclusion levels in livestock diets, as they can be toxic when consumed at high levels, dependent on livestock species (Tripathi & Mishra, 2007). GLS and the enzyme myrosinase are compartmentally stored separately in *Brassica* spp. (Rask et al., 2000). Upon mechanical disruption or other stresses on plant tissues, myrosinase cleaves glucose from GLS, which produces toxic compounds such as nitriles, thiocyanates, and isothiocyanates. This self-defense mechanism evolved to reduce animal and insect browsing of the plant (Halkier & Gershenzon, 2006). When consumed, these toxic breakdown products can cause deleterious effects on the thyroid, and ultimately cause goiters from iodine deficiency (Burel et al., 2001). For this reason, canola was bred to contain lower levels of GLS and erucic acid (Newkirk, 2009). However, feed inclusion rates are still limited to 30 %, approximately, and this reduces the value of canola meal

(Newkirk, 2009).

Canadian based MCN Bioproducts Inc. (Saskatoon, SK, Canada) patented a process to fractionate high value protein concentrates from solvent and non-solvent expelled canola meal (Newkirk, Maenz, & Classen, 2006; Newkirk, Maenz, & Classen, 2009). These protein concentrates contained greater than 60 % protein, no detectable phytic acid, and less than 5 $\mu M.g^{-1}$ of total GLS. However, this process utilizes multiple separation steps, which can be expensive and result in a relatively low protein yield in the primary marketed fraction. Bunge licensed this technology in 2012 (All About Food, 2012).

In contrast to mechanical separation to isolate protein, the metabolic diversity of fungi may be exploited to convert canola carbohydrates into protein-rich, single celled protein, and thereby produce a less expensive canola protein concentrate. In addition, fungal bioprocessing has been shown to significantly reduce GLS levels (Croat, Berhow, Karki, Muthukumarappan, & Gibbons, 2015). We hypothesized that this process would generate a more digestible product with enhanced nutritional value to a range of aquaculture and other livestock species.   Fungi selected for initial evaluation included *Aurobasidium pullulans, Trichoderma reesei, Fusarium venenatum, Pichia kudriavzevii,* and *Mucor circinelloides.* Several of these fungi are known to produce cellulose degrading enzymes (Wiebe, 2002; Olempska-Beer, Merker, Ditto, & DiNovi, 2006; Seiboth, Ivanova, & Seidl-Seiboth, 2011; Prajapati, Jani, & Khanda, 2013; Ratledge, 2013). Studies have shown that *F. venenatum* is capable of producing mycotoxins, however their production can be avoided by controlling fermentation conditions (Wiebe, 2002). Both hexane extracted (HE) and cold pressed (CP) canola meals were evaluated with a submerged incubation process, which allowed for better activity of cellulolytic enzymes.

2. Material and Methods

2.1 Feedstocks and Preparation

The HE canola meal was obtained from North Dakota State University (Fargo, ND, USA), while CP canola meal was obtained from Agrisoma Biosciences (Ottawa, Ontario, Canada). Both HE and CP meals were milled through a 2 mm screen via FitzMill model # S-DAS06 knife mill (Elmhurst, IL, USA) prior to use, and were stored at room temperature in sealed bucket throughout the duration of experimentation. Dry weight (dw) analysis was conducted by drying 5 g of canola meal at 80 degrees Celsius (°C) in a drying oven for at least 48 hours (h).

Cultures, Maintenance, and Inoculum Preparation

A. pullulans (NRRL-58522), *A. pullulans* (NRRL-42023), *A. pullulans* (NRRL-Y-2311-1), *T. reesei* (NRRL-3653), and *F. venenatum* (NRRL-26139) were obtained from the National Center for Agricultural Utilization Research (Peoria, IL, USA). *P. kudriavzevii* and *M. circinelloides* were isolated as contaminants from prior trials, and were identified by ARS-USDA (Peoria, IL, USA) using 15 s RNA analysis (O'Donnell, 2000). Short-term maintenance cultures were stored on Potato Dextrose Agar plates and slants at 4 °C. Inocula for all experiments was prepared by transferring isolated colonies or a square section of agar growth (filamentous fungi) into glucose yeast extract (GYE) medium consisting of 5 % glucose and 0.5 % of yeast extract. The pH for *Aureobasidium, Pichia,* and *Mucor* cultures was adjusted to 3.0 ± 0.1 with 5 M sulfuric acid, while a pH of 5.0 to 5.5 was used for *T. reesei* and *F. venenatum.* GYE flasks consisted of 100 milliliter (mL) working volume in 250 mL Erlenmeyer flasks, covered with a foam plug and aluminum foil. Cultures were incubated for 72 h at 30 °C in a New Brunswick Scientific Excella E24 rotary shaker (Hauppauge, NY, USA) at 150 min-1.

2.2 Experimental Procedures

Submerged trials were conducted in 1 L Erlenmeyer flasks with a working volume of 500 mL at a 10 % solid loading rate (SLR) dry weight canola meal. Flasks were covered with foam plugs and aluminum foil. For trials to be subjected to an initial saccharification step, 5 M sulfuric acid was used to adjust the initial pH to 5.0 ± 0.1 (this is the optimal pH level for the commercial cellulase and hemicellulase enzymes used). For trials lacking the saccharification step, the pH was adjusted to the levels indicated previously for specific microbes. Flasks were then autoclaved at 121 °C for 20 min. For saccharification trials, 0.052 mL CTec2 and 0.138 mL HTec2 (Novozymes, Franklinton, NC, USA) were added, and flasks were incubated at 50 °C and 150 min-1 for 24 h. Following saccharification, the pH was adjusted, when necessary, for the specific microbes, and the slurry was cooled to 30 °C. Saccharification and non-saccharification trials were inoculated with 5 mL of a 72 h culture of the appropriate organism and incubated at 30 °C at 150 min-1 during 168 h.   Daily samples of 50 mL were collected and used to monitor pH, cell counts, carbohydrates, protein, fiber, and GLS as described later. Daily samples and remaining slurry at the end of incubation was dried for 2 days at 80 °C using a Fisher Scientific Isotemp oven (Waltham, MA, USA).

2.3 Analytical Methods

2.3.1 Total Protein

The pH of each sample was measured in an Oakton 110 series pH meter (Vernon Hills, IL, USA). Forty-five mL of each sample were dried for 2 days at 80 °C. Approximately 0.5 g of each sample was used for protein analysis in duplicate. Protein was quantified using a LECO model FP528 (St. Joseph, MI, USA) to combust the sample and to measure the total nitrogen gas content in the sample (AOAC Method 990.03). Protein percentage was then calculated from the nitrogen content of the sample using a conversion factor of 6.25. An additional 0.25 g of sample was dried at 80 °C for 48 h to determine the dry matter of protein samples.

2.3.2 Residual Sugars

High Performance Liquid Chromatography (HPLC) was used to measure residual sugars using 5 mL of sample supernatant. Samples were firstly boiled for 10 min to ensure the fungal culture and/or saccharification enzymes were inactivated. Samples were then centrifuged at 10,000 min-1 for 10 min, and the supernatant was poured into 2 mL microcentrifuge tubes and frozen overnight. The supernatant was then thawed and re-centrifuged at 10,000 min-1 for 10 min to remove any precipitants. The final supernatant was then filtered through a 0.2 micrometer (μm) filter and into a HPLC vial and frozen until analysis. A Waters size-exclusion chromatography column (SugarPak column I10 um, 6.5 mm X 300 mm with pre-column module, Waters Corporation, Milford, MA, USA) and a HPLC system (Agilent Technologies, Santa Clara, CA, USA) equipped with refractive index detector (Model G1362A) were used to measure the sugars. The sugars were eluted using a de-ionized water as mobile phase at flow rate of 0.5 mL.min^{-1} and column temperature of 80 °C. Sugars to be quantified included arabinose, galactose, glucose, raffinose, stachyose, and sucrose. All sugar standards were purchase from Sigma-Aldrich (St. Louis, MO, USA) while all standards contained a purity of 99.9 %. The sugar standards were prepared using several concentrations and a calibration curve was constructed using concentration verus HPLC area previously established by Karunanithy, Karuppuchamy, Muthukumarappan, & Gibbons (2012).

2.3.3 Glucosinolates

Approximately 1.5 g of dried 0 h and 168h sample were used for GLS analysis. Individual GLS were confirmed to be present by quadrupole time-of-flight liquid chromatography-mass spectrometry and quantified using reverse phase HPLC (Berhow et al., 2013). For GLS quantitation, a modification of a HPLC method, developed by Betz and Fox (1994), was used. The extract was run on a Shimadzu (Columbia, MD) HPLC System (two LC 20AD pumps; SIL 20A autoinjector; DGU 20As degasser; SPD-20A UV-VIS detector; and a CBM-20A communication BUS module) running under the Shimadzu LC solutions Version 1.25 software. The column was a C_{18} Inertsil reverse phase column (250 mm X 4.6 mm; RP C-18, ODS-3, 5u; with a Metaguard guard column; Varian, Torrance, CA). The glucosinolates were detected by monitoring at 237 nm. The initial mobile phase conditions were 12 % methanol / 88 % aqueous 0.005 M tetrabutylammonium bisulfate (TBS) at a flow rate of 1 mL.min^{-1}. After injection of 15 μl of sample, the initial conditions were held for 2 min, and then up to 35 % methanol over another 20 min, then to 50 % methanol over another 20 min. then up to 100 % methanol over another 10 min.

2.3.4 Fiber

Fiber analysis was completed as Neutral Detergent Fiber (NDF) and Acid Detergent fiber (ADF). NDF is a method commonly used for animal feed analysis to determine the amount of lignin, hemicellulose and cellulose, while ADF represents the least digestible fiber fraction of animal feed including lignin, cellulose, silica but not hemicellulose. NDF and ADF analysis were completed by Midwest Laboratories (Omaha, NE, USA) using ANKOM Technology (Macedon, NY, USA) filter bag methods. Approximately 3 g of dried material was submitted in triplicate for each treatment combination.

3. Results and Discussion

Seven fungal strains were grown on HE vs CP canola meal using a submerged incubation process. Submerged incubation has been defined as processing in the presence of excess water, and has been a proven large-scale process due to easier material handling and process control (Singhania, Sukumaran, Patel, Larroche, & Pandey, 2010). In contrast to solid-state incubation completed in previous work (Croat, Berhow, Karki, Muthukumarappan, & Gibbons, 2016a), submerged incubation has the advantage of being a more homogenous mixture while allowing improved streamlining and standardization of processing (Chicatto, Costa, Nunes, Helm, & Tavares, 2014). The fungi were tested both on raw (non-saccharified) and saccharified meal slurries using commercial cellulases to enhance fiber breakdown. These trials were done in shaker flasks, where mixing and mass transfer were the limiting factors. However, these non-optimized trials were meant to quickly down-select the best microbe for each type of canola meal. Other investigators have previously used a similar submerged incubation process to quickly assess phytase activity of various strains of bacteria, yeasts and fungi when grown on canola and oilseed meals (Nair & Duvnjak, 1991).

3.1 Total Protein

Figures 1 and 2 present the maximum protein levels in HE and CP canola meals, respectively, for raw meal and un-inoculated controls versus the various fungi, both under non-saccharified and saccharified conditions. As expected, protein levels for the un-inoculated controls were similar to the raw meals. In HE meal, protein levels increased from 36.1 % in the raw meal to 39.0-48.7 % after the fungal conversion process (relative improvements of 8.0-34.9 %) (Fig 1). The *M. circinelloides* trial was the only one in which an enzymatic hydrolysis step prior to inoculation proved beneficial. In the case of *T. reesei*, the non-saccharified trial actually resulted in higher protein titers. We had anticipated that saccharification would have a significant positive effect on fiber hydrolysis, and subsequently protein levels. It could be that canola fibers require pretreatment to increase susceptibility to enzymatic hydrolysis (Gattinger, 1990; Yaun, 2014). Proceeding work investigated various pretreatment methods to make canola fibers more susceptible to hydrolysis by the fungal enzymes, thus releasing more sugar for conversion into single celled protein (Croat, Berhow, Iten, Karki, Muthukumarappan, & Gibbons, 2016b). This proceeding work observed pretreatments including extrusion, hot water cook, dilute acid, and dilute alkali compared to non-pretreated canola meal.

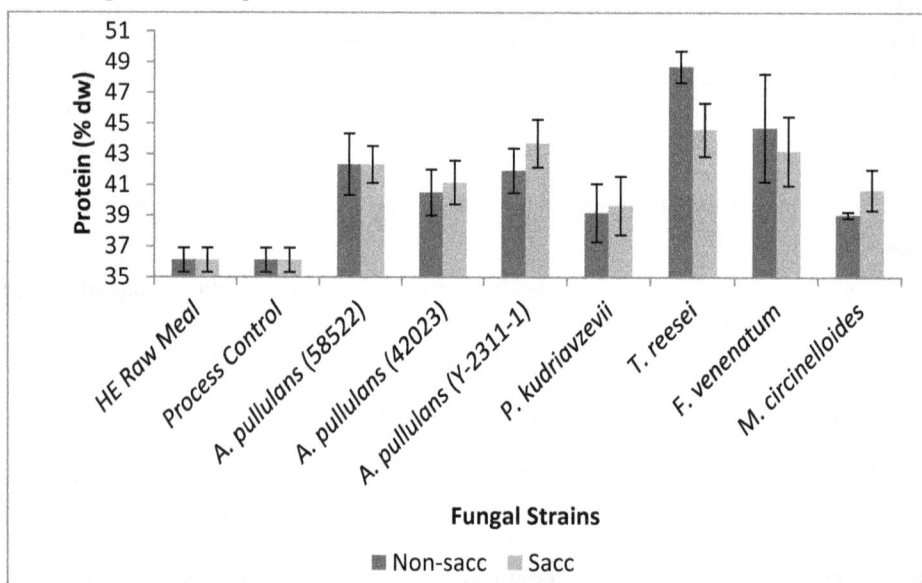

Figure 1. Maximal protein levels ± SD of HE canola meal following submerged fungal incubation

In the CP canola meal (Fig 2) the protein level in the un-inoculated control was 38.6 %, and rose from 40.9 to 53.0 % after microbial conversion, representing relative improvements of 6.0-37.3 %. CP canola meal was about 3 % higher in protein than HE meal and following incubation, protein levels were ~2-8 % higher in CP canola meal trials compared to HE meal for each pair of fungi. HE is a more effective method of removing oil from canola seed, however this process applies significantly higher levels of heat, which may denature or degrade some protein (Spragg & Mailer, 2007). We observed that the enzymatic hydrolysis step prior to inoculation did not significantly affect protein levels for all the fungi tested. Thus for un-pretreated canola meal, there was no benefit to adding cellulolytic enzymes.

T. reesei achieved the highest protein levels for both substrates, while *P. kudriavzevii* exhibited the lowest protein enhancement. *T. reesei* is known to produce many hydrolytic enzymes (Li el al., 2013), and it was expected to provide the greatest conversion of fiber and oligosaccharides into cell mass. As a single-celled yeast, *P. kudriavzevii* does not produce cellulase enzymes and it was therefore anticipated to result the lowest protein improvement. The final protein levels for all other fungal strains were relatively similar, at 40-45 % in HE canola meal and 43-52 % protein in cold pressed canola meal.

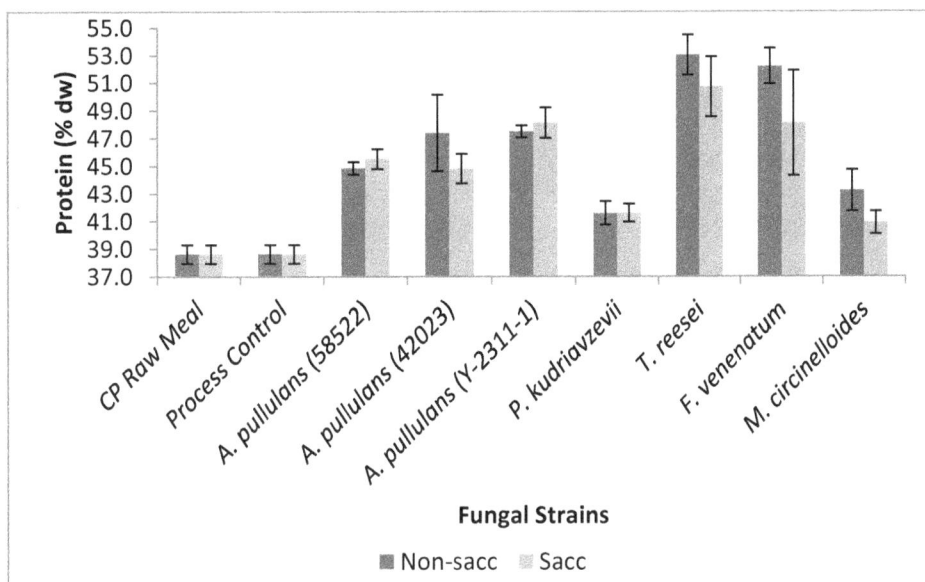

Figure 2. Maximal protein levels ± SD of CP canola meal following submerged fungal incubation

3.2 Residual Sugars

Arabinose, galactose, glucose, raffinose, stachyose, and sucrose were measured throughout incubation via HPLC. For simplicity, the final levels of these sugars were combined and are presented as residual sugars in Figures 3 and 4 for HE and CP canola meal, respectively. The total residual sugar concentrations decreased slightly (2.7-5.5 %) from the raw meals compared to the process controls. Nyombaire, Siddiq, and Dolan (2007) found that a pre-soaking and 80 °C of cooking temperature were sufficient to hydrolyze oligosaccharides such as raffinose and stachyose in red kidney beans. Autoclaving the 10 % SLR canola slurries may have achieved a similar effect, thereby reducing the raffinose and stachyose concentrations.

Between 37.0-94.6 % of sugars present in non-saccharified HE meal (Fig. 3) were used by the fungi during incubation, resulting in residual sugar levels of 0.8-9.4 %. Similarly, 39.0-88.6 % of sugars present in saccharified HE meal was utilized by the fungi, resulting in residual sugar levels of 1.7-9.1 %. *T. reesei* exhibited the lowest residual sugar levels on both non-saccharified and saccharified HE meals, while *M. circinelloides* and *P. kudriavzevii* had the highest final levels in non-saccharified and saccharified trials, respectively. *M. circinelloides* did show a benefit from saccharification, showing a significant drop in residual sugars from 9.4 to 2.7 % w/w when compared to non-saccharification.

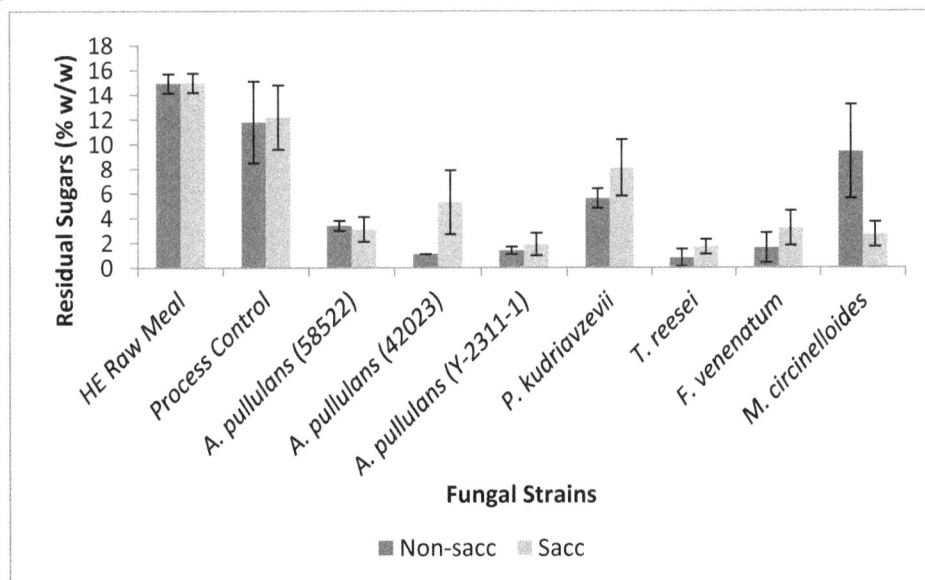

Figure 3. Residual sugar levels ± SD of HE canola meal following submerged fungal incubation

In non-saccharified CP meal (Fig. 4) between 61.0-98.1% of sugars present were metabolized by the fungi during incubation, decreasing residual sugar levels to 0.3-6.3%. Similarly, 40.0-95.0% of sugars present in saccharified CP meal were metabolized by the fungi during incubation, decreasing residual sugar levels to 0.8-9.7%. *F. venenatum* and *T. reesei* exhibited the lowest residual sugar levels on both non-saccharified and saccharified CP meal, while *A. pullulans* (NRRL-42023) and *P. kudriavzevii* had the highest final levels in non-saccharified and saccharified material, respectively. Saccharification significantly reduced residual sugars in trials with *M. circinelloides* and *A. pullulans* (NRRL-42023) when compared to non-saccharification trials.

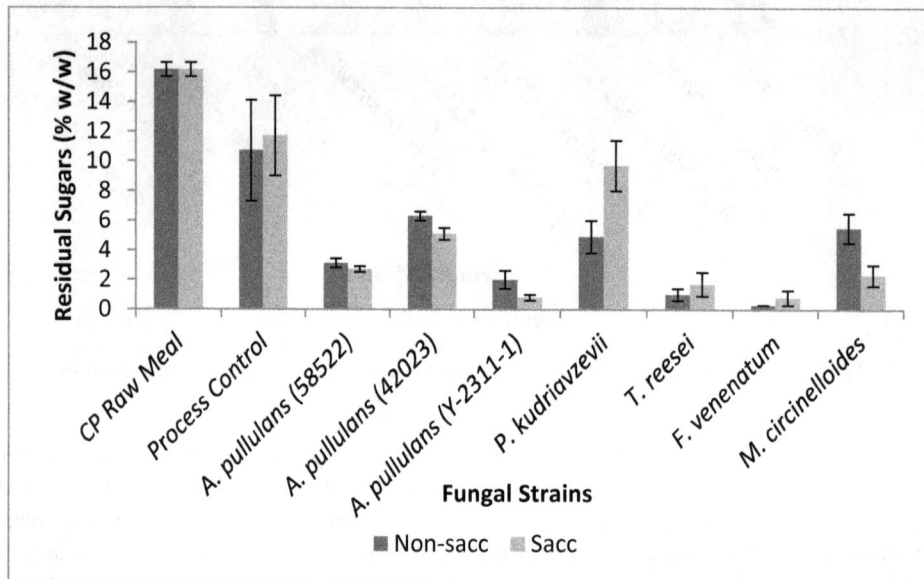

Figure 4. Residual sugar levels ± SD of CP canola meal following submerged fungal incubation

3.3 Glucosinolates

Figures 5 and 6 show GLS levels for the HE and CP canola meal trials, respectively. GLS levels were reduced from 42.8 $\mu M.g^{-1}$ in raw HE meal to 8.7 $\mu M.g^{-1}$ (non-saccharified) and 18.3 $\mu M.g^{-1}$ (saccharified) in the un-inoculated process controls. This represents 79.6 and 57.2 % reductions, respectively, and was presumed due to the conversion of some of the GLS into volatile breakdown products (Halkier & Gershenzon, 2006). Newkirk, Classen, Scott, and Edney (2003) also noted that high processing heat can be used to remove volatile anti-nutritional factors; however this can also denature proteins. Submerged microbial conversion further reduced GLS content to 1.0-14.4 $\mu M.g^{-1}$, representing a total reduction of 66.5-97.8 %.

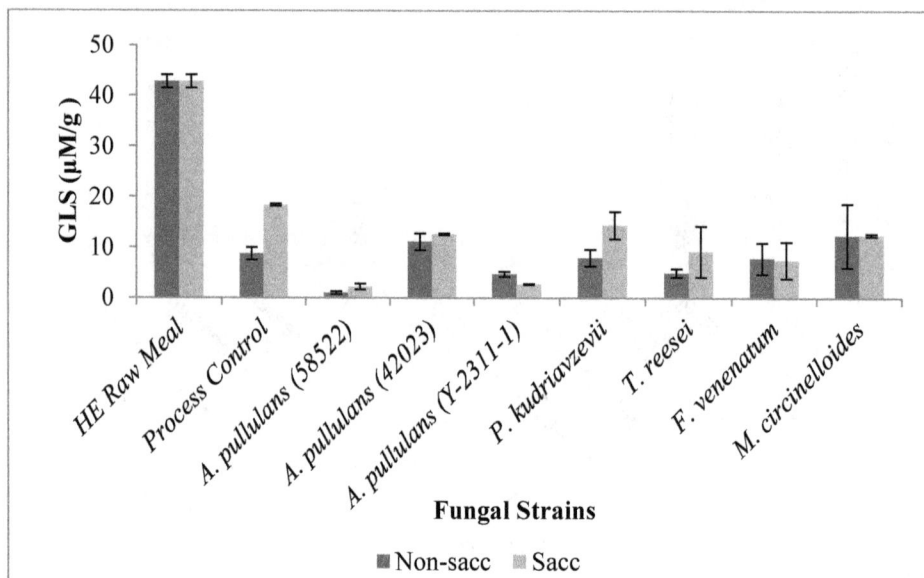

Figure 5. Reduction of total GLS ± SD following sterilization and submerged fungal incubation in HE canola meal

GLS levels in raw CP meal (60.6 $\mu M.g^{-1}$) were higher than in HE meal (42.8 $\mu M.g^{-1}$) since the former does not include the high temperature step to remove the extraction solvent (hexane), which can eliminate GLS. Treatment of the CP meal with the autoclaving and drying steps in the process control reduced GLS levels to 18.6 and 26.2 $\mu M.g^{-1}$, respectively in non-saccharified and saccharified trials (reduction of 69.4 and 56.8 %, respectively). Again, submerged microbial conversion further reduced GLS content to 0.7-23.7 $\mu M.g^{-1}$ (total reduction of 60.8-98.9 %).

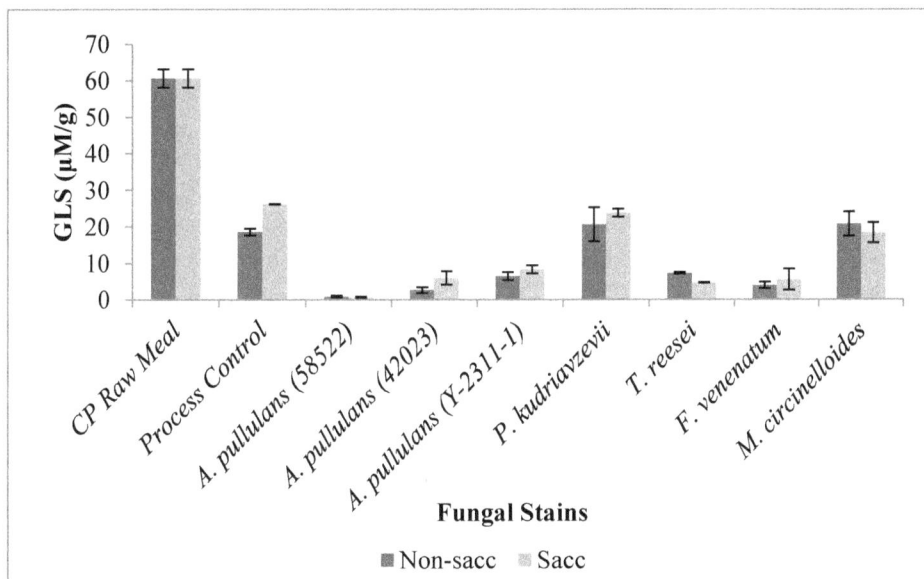

Figure 6. Reduction of total GLS ± SD following sterilization and submerged fungal incubation in CP canola meal

Overall, *A. pullulans* (NRRL-58522) caused the greatest reduction in GLS levels in both HE and CP canola meals (ranging from 94.5-98.9%), likely due to its robust capability for producing extracellular enzymes (Kudanga & Mwenje, 2005). *A. pullulans* (NRRL-Y-2311-1) was also very effective in reducing GLS concentrations (ranging from 86.3-93.7%), followed by *F. venenatum* (81.8-93.5%) and *T. reesei* (78.7-92.2%). Previous studies have shown that various microbes are able to degrade GLS and metabolize the resulting glucose and sulfur moieties. For example, Vig & Walia (2001) observed that *Rhizopus oligosporus* reduced GLS and their byproducts during fungal incubation of *Brassica napus* meal. Similarly, Rakariyatham & Sakorn (2002) reported the complete degradation of GLS after 60-96 h using solid-state fermentation of *Brassica juncea* with *Aspergillus* sp. In the work reported herein, *P. kudriavzevii* and *M. circincelloides* resulted in the least reduction in GLS, as expected due to minimal production of extracellular hydrolytic enzymes when compared to the other fungi tested.

3.4 Fiber

Table 1 provides the ADF and NDF fiber levels of raw, process control, and treated canola meals. In general, most fiber levels were statistically similar to the raw meal, indicating that the conversion process had minimal effects on fiber levels. The only trial to show a statistically significant reduction in ADF in HE meal was *P. kudriavzevii*, while trials with *A. pullulans* (Y-2311-1), *P. kudriavzevii*, *T. reesei*, *F. venenatum*, and *M. circincelloides* all statistically reduced ADF and/or NDF fiber levels in CP canola meal (Table 1). Thus the cellulase producing fungi were effective in hydrolyzing fiber in CP canola meal, however did not show similar results in HE canola meal. A possible explanation for the reduced fiber degradation in HE canola meal is that the heating steps of the hexane extraction process may have reduced the susceptibility of the fibers to subsequent enzymatic hydrolysis. Also, the enzyme cocktail used in the saccharification trials were not optimized for canola fiber, and this may provide a future opportunity to enhance fiber degradation.

In some cases the conversion process actually resulted in a concentration of fibers, caused by the removal of sugars and GLS. Trials with *A. pullulans* (58522), *A. pullulans* (Y-2311-1), *F. venenenatum*, and *M. circinelloides* all increased fiber levels in HE canola meal, while *A. pullulans* (58522) and *A. pullulans* (42023) treatments both increased ADF and/or NDF fiber levels in CP canola meal (Table 1).

We have previously shown that feedstock pretreatment increases the susceptibility of fibers to hydrolysis (Karki, Muthukumarappan, & Gibbons, 2013), and that optimizing the fungal incubation conditions will also enhance cellulase production and activity. This will be evaluated in future studies using extrusion, hot cook, dilute acid, and dilute alkali pretreatments. The resulting sugars would then be available for conversion into additional cell mass and protein

Table 1. Fiber reduction of non-saccharified and saccharified canola meal during submerged fungal incubation

	Hexane Extracted				Cold Pressed			
	Non-Saccharified		Saccharified		Non-Saccharified		Saccharified	
Fungal Culture	ADF (%)	NDF (%)	ADF (%)	NDF (%)	ADF (%)	NDF (%)	ADF (%)	NDF (%)
Raw Meal	19.9±0.2	23.1±0.3	19.9±0.2	23.1±0.3	11.5±0.5	15.0±0.3	11.5±0.5	15.0±0.3
Process Control	18.7±0.3	22.0±0.8	23.0±1.1	29.0±1.8	9.5±0.6	12.4±0.8	14.8±1.4	16.1±1.6
A. pullulans (NRRL-58522)	22.0±1.6[b]	29.1±0.8[b]	20.6±2.4	25.2±4.6	12.1±1.0	16.8±0.7[b]	11.6±1.1	15.9±0.6
A. pullulans (NRRL-42023)	20.4±1.5	24.3±1.3	19.4±2.2	22.6±0.5	12.4±0.4	16.9±0.4[b]	11.2±1.1	15.1±0.3
A. pullulans (NRRL-Y-2311-1)	22.3±0.9[b]	24.5±0.7[b]	21.2±1.4	24.0±0.9	13.6±2.1	14.8±2.4	11.0±0.9	12.6±1.0[a]
P. kudriavzevii	19.7±1.6	23.1±1.9	18.6±0.3[a]	22.7±1.9	11.0±0.4	13.5±0.3[a]	10.4±0.4[a]	12.4±0.5[a]
T. reesei (NRRL-3653)	19.9±3.1	22.5±4.0	19.8±0.8	26.4±3.3	7.6±0.8[a]	10.1±0.6[a]	8.1±1.2[a]	10.8±2.0[a]
F. venenatum (NRRL- 26139)	21.3±2.4	26.7±2.6[b]	20.8±2.2	26.9±0.7[b]	7.6±0.9[a]	10.7±1.2[a]	10.7±2.5	12.9±3.0
M. circinelloides	21.0±0.5[b]	25.9±1.2	19.6±1.2	22.6±1.0	10.2±0.9	12.8±0.9[a]	10.8±0.6	15.6±1.3

[a]Indicates fiber level was statistically lower than raw meal.

[b]Indicates fiber level was statistically higher than raw meal.

4. Conclusions

Submerged incubation with various fungal strains improved the nutritional content of canola meal. *T. reesei* (NRRL-3653), *F. venenatum* (NRRL-26139), and *A. pullulans* (Y-2311-1) resulted in the greatest improvement in protein content in HE canola meal (34.8, 23.8, and 21.0 %), respectively, while reducing total GLS and residual sugar content by 82.6-93.7 % and 89.3-94.6 %. In trials with CP canola meal, the same three fungi increased protein levels to the greatest extent (37.3, 35.2, and 24.6 %), respectively, while reducing total GLS and residual sugar content by 89.3-93.5 % and 93.8-98.1 %.

References

Aider, M., Barbana, C. (2011). Canola proteins: composition, extraction, functional properties, bioactivity, applications as a food ingredient and allergenicity-A practical and critical review. *Trends Food Science & Technology, 22*(1), 21-39.

All About Food. (2012, April 12). Bunge acquires assets of MCN Bioproducts. Retrieved June 24, 2016, from http://www.allaboutfeed.net/Home/General/2012/4/Bunge-acquires-assets-of-MCN-Bioproducts-AAF0130 82W.

Bell, J. M. (1993). Factors affecting the nutritional value of canola meal: A review. *Canadian Journal of Animal Science, 73*(4), 679-697.

Berhow, M. A., Polat, U., Glinski, J. A., Glensk, M., Vaughn, S. F., Isbell, T., ... Gardner, C. (2013). Optimized analysis and quantification of glucosinolates from *Camelina sativa* seeds by reverse-phase liquid chromatography. *Industrial Crops and Products, 43*, 119-125.

Betz, J. M., & Fox, W. D. (1994). High-performance liquid chromatographic determination of glucosinolates in *Brassica* vegetables. In Huang, M. T., Osawa, T., Ho, C. T., Rosen, R. T. (Eds.), Food phytochemicals for cancer prevention I - fruits and vegetables (Vol. 546, pp. 181-196). Washington, D.C.: American Chemical Society.

Burel, C., Boujard, T., Kaushik, S. J., Boeuf, G., Mol, K. A., Van der Geyten, S., ... Ribaillier, D. (2001). Effects of rapeseed meal-glucosinolates on thyroid metabolism and feed utilization in rainbow trout. *General and Comparative Endocrinology, 124*(3), 343-358.

Chicatto, J. A., Costa, A., Nunes, H., Helm, C. V., & Tavares, L. B. B. (2014). Evaluation of hollocelulase production by Lentinula edodes (Berk.) Pegler during the submerged fermentation growth using RSM. *Brazilian Journal of Biology, 74*(1), 243-250.

Croat, J. R., Berhow, M., Karki, B., Muthukumarappan, K., & Gibbons, W. R. (2015). *Reduction of total glucosinolates in canola meal via thermal treatment and fungal bioprocessing*. Poster presented at the 54th Annual Meeting of the Phytochemical Society of North America, Urbana-Champaign, IL. Abstract retrieved from http://www.psna-online.org/PSNA_2015_Full_Program.pdf

Croat, J. R., Berhow, M., Karki, B., Muthukumarappan, K., & Gibbons, W. R. (2016a). Conversion of canola meal into a high protein feed additive via solid-state fungal incubation process. *Journal of the American Oil Chemists' Society, 93*, 499-507.

Croat, J. R., Berhow, M., Iten, L., Karki, B., Muthukumarappan, K., & Gibbons, W. R. (2016b). Utilizing pretreatment and fungal incubation to enhance nutritional value of canola meal (in preparation).

Gattinger, L. D. (1990). *The enzymatic saccharification of canola meal and its utilization for xylanase production by Trichoderma reesei* (Master's thesis, University of Ottawa, Ontario, Canada). Retrieved from http://www.ruor.uottawa.ca/handle/10393/5643

Halkier, B. A., & Gershenzon, J. (2006). Biology and biochemistry of glucosinolates. *Annual Review of Plant Biology, 57*, 303-333.

Karki, B., Muthukumarappan, K., & Gibbons, W. R. (2013). *Optimization of extrusion processing conditions on enzymatic hydrolysis and fermentation of distillers' low oil wet cake*. Poster presented at the 73rd Annual Meeting of the North Central Branch for the American Society for Microbiology, Brookings, SD. Abstract retrieved from https://www.sdstate.edu/biomicro/73rd-Meeting/upload/Abstract-NCB-ASM.pdf

Karunanithy, C., Karuppuchamy, V., Muthukumarappan, K., & Gibbons, W.R. (2012). Selection of enzyme combination, dose, and temperature for hydrolysis of soybean white flakes. *Industrial Biotechnology, 8*(5), 309-317.

Kudanga, T., & Mwenje, E. (2005). Extracellular cellulase production by tropical isolates of *Aureobasidium pullulans*. *Canadian Journal of Microbiology, 51*(9), 773-776.

Li, C., Yang, Z., Zhang, R. H., Zhang, D., Chen, S., & Ma, L. (2013). Effect of pH on cellulase production and morphology of *Trichoderma reesei* and the application in cellulosic material hydrolysis. *Journal of Biotechnology, 168*(4), 470-477.

Lomascolo, A., Uzan-Boukhris, E., Sigoillot, J. C., & Fine, F. (2012). Rapeseed and sunflower meal: a review on biotechnology status and challenges. *Applied Microbiology and Biotechnology, 95*(5), 1105-1114.

Nair, V. C., & Duvnjak, Z. (1991). Phytic acid content reduction in canola meal by various microorganisms in a solid-state fermentation process. *Acta Biotechnologica, 11*(3), 211-218.

Newkirk, R. W., Classen, H. L., Scott, T. A., & Edney, M.J. (2003). The digestibility and content of amino acids in toasted and non-toasted canola meals. *Canadian Journal of Animal Science, 83*, 131-139.

Newkirk, R. W., Maenz, D. D., & Classen, H. L. (2006). Oilseed Processing. United States Patent US7090887 B2, August *15*, 2006.

Newkirk, R. W. (2009). Canola meal feed industry guide. *Canola Council, 4*(2009), 1-48. https://cigi.ca/wp-content/uploads/2011/12/2009-Canola_Guide.pdf

Newkirk, R. W., Maenz, D. D., & Classen, H. L. (2009). Defatted canola flake, protein, and phytate; isoelectric precipitation and/or ultrafiltration. United States Patent US7,629,014 B2, December 8, 2009.

Nyombaire, G., Siddiq, M., & Dolan, K. (2007). Effect of soaking and cooking on the oligosaccharides and lectins in red kidney beans (*Phaseolus vulgaris* L.). *Bean Improvement Cooperative Annual Report, 50*, 31-32.

O'Donnell, K. (2000). Molecular phylogeny of the Nectria haematococca - Fusarium solani species complex. *Mycologia, 92*(5), 919-938.

Olempska-Beer, Z. S., Merker, R. I., Ditto, M. D., & DiNovi, M. J. (2006). Food-processing enzymes from recombinant microorganisms--a review. *Regulatory Toxicology and Pharmacology, 45*(2), 144-158.

Prajapati, V. D., Jani, G. K., & Khanda, S. M. (2013). Pullulan: an exopolysaccharide and its various applications. *Carbohydrate Polymers, 95*(1), 540-549.

Rakariyatham, N., & Sakorn, P. (2002). Biodegradation of glucosinolates in brown mustard seed meal (*Brassica juncea*) by *Aspergillus* sp. NR-4201 in liquid and solid state cultures. *Biodegradation, 13*, 395-399.

Rask, L., Andreasson, E., Ekborn, B., Eriksson, S., Pontoppidan, B., & Meijer, J. (2000). Myrosinase gene family evolution and herbivore defense in *Brassicaceae*. *Plant Molecular Biology, 42*, 93-113.

Ratledge, C. (2013). Microbial production of polyunsaturated fatty acids as nutraceuticals. In B. McNeil, D. Archer, I. Giavasis, & L. Harvey (Eds.), *Microbial production of food ingredients, enzymes and nutraceuticals* (pp. 531-558). Cambridge, England: Woodhead Publishing Limited.

Seiboth, B., Ivanova, C., & Seidl-Seiboth, V. (2011). *Trichoderma reesei*: a fungal enzyme producer for cellulosic biofuels. In: M. A. dos Santos Bernardes (Ed.) *Biofuel production—recent developments and prospects* (pp. 309-340). Rijeka, Croatia: InTech.

Singhania, R. R., Sukumaran, R. K., Patel, A. K., Larroche, C., & Pandey, A. (2010). Advancement and comparative profiles in the production technologies using solid-state and submerged fermentation for microbial cellulases. *Enzyme and Microbial Technology, 46*(7), 541-549.

Spragg, J., & Mailer, R. (2007). Canola Meal Value Chain Quality Improvement. *Australian Oilseeds Federation and Pork CRC*.

Tripathi, M. K., & Mishra, A. S. (2007). Glucosinolates in animal nutrition: A review. *Animal Feed Science and Technology, 132*(1-2), 1-27.

Vig, A. P., & Walia, A. (2001). Beneficial effects of *Rhizopus oligosporus* fermentation on reduction of glucosinolates, fiber and phytic acid in rapeseed (*Brassica napus*) meal. *Bioresource Technologies, 78*(2001), 309-312.

Wiebe, M. G. (2002). Myco-protein from Fusarium venenatum: a well-established product for human consumption. *Applied Microbiology Biotechnology, 58*(4), 421-427.

Yaun, L. (2014). *Pretreatment and enzyme hydrolysis of canola meal (Brassica napus L.) and oriental mustard bran (Brassica juncea)- production of functional oligosaccharides and impact on phenolic content* (Master's thesis, University of Manitoba, Manitoba, Canada). Retrieved from http://link.lib.umanitoba.ca/portal/Pretreatment-and-enzyme-hydrolysis-of-canola-meal/jieE4dVvjJo/

15

Comparison of Post-harvest Practices of the Individual Farmers and the Farmers in Cooperative of Côte d'Ivoire and Statistical Identification of Modalities Responsible of Non-quality

Djedjro C. Akmel[1], Arsène. L. I Nogbou[1], Ibrahima Cissé[1], Kouassi E. Kakou[1], Kisselmina Y. Koné[1], Nogbou E Assidjo[1], & Benjamin Yao[1]

[1]Laboratoire des Procédés Industriels, de Synthèses, de l'Environnement et des Energies Nouvelles, Institut National Polytechnique Félix Houphouet Boigny (LAPISEN/ INP-HB), Yamoussoukro, Côte d'Ivoire

Correspondence: Djedjro C. Akmel, Laboratoire des Procédés Industriels, de Synthèses, de l'Environnement et des Energies Nouvelles, INP-HB, BP 1313 Yamoussoukro, Côte d'Ivoire. E-mail: akmeldc@gmail.com

Abstract

The purposes of this study were to compare the modalities of the post-harvest practices of these two groups and to statistically identify the modalities responsible of non-quality (under grade) on the basis of results of Pareto chart and proportion of successes calculated for each modality. A survey about of modalities of post-harvest processing methods and about the quality of the beans obtained was conducted among producers of the largest producing region of Côte d'Ivoire. The collected data were analyzed by the chi-square test of concordance and the Pareto chart. The results show that there is no correlation between practice of the individual farmers and the farmers in cooperatives. Highly significant differences (p-value < 0.001) were observed in the number of brewing; the fermentation time; the materials of fermentation and drying impacting the quality of merchantable cocoa. Samples collected from farmers into cooperatives have fewer defects than those of individual farmers. Regarding the modalities of the post-harvest practice responsible of the under-grades, the results showed that the samples of farmers in cooperatives had fewer defects than those of individual farmers. Thus, obtaining a good quality cocoa beans must take into account the best modality at each step of post-harvest practices. However certain modality should be avoided. These are: the time breaking pods of one day; the absence of brewing during the fermentation; the time of fermentation less than or equal to three days and the time of drying less than or equal to two days.

Keywords: modalities, post-harvest practices, merchantable cocoa, Pareto chart, chi square test, proportion of successes

1. Introduction

The economy of some countries is largely based on the export of cocoa. This is the case of Côte d'Ivoire, the world's largest cocoa grower, with a production of the order of 1.78 million tons or 40% of world supply (Anonymous, 2015). Its production represents 15% of GDP (Gross Domestic Product) and 30% of export earnings (Koko, 2014).

Although Côte d'Ivoire recorded spectacular results at the quantity level, the results are below expectations at the quality level. Indeed, in recent years, and following liberalization occurred in 1999, we noticed a recurring situation of bad quality cocoa from Côte d'Ivoire on the world market (Anonymous, 2006). The free fatty acid content from Côte d'Ivoire's cocoa exceeds standards allowed 1.75%, which has meant the decline in its quality (Guéhi et al., 2005). In addition, a portion of national production is stored poorly and has a humidity above the standard (more than 8%). This situation is growing and causes the discount of the cocoa from Côte d'Ivoire, what constitutes a financial loss for the producers. The causes for the decreasing of the quality of cocoa are not sufficiently identified. They appear to be multiple and complex and are at all levels of the collection and distribution chain, from producer to foreign markets (Anonymous, 2006). However, some causes have been mentioned for this decreasing. It comes to the inefficient system of quality control of the products; internal marketing system and external difficult to control; unprofessionalism of some producers and the deterioration of the products in farms (Anonymous , 2006; Anonymous , 2016). But of all these causes, the most significant are the post-harvest practices because they are preliminary and decisive step in the transformation of the merchantable cocoa. A bad post-harvest treatment may rot or germinate cocoa beans, which reduces the quality of merchantable cocoa (Janny, Ritchie, & Flood, 2003). According Barel (2013), "Everything is happening in the post-harvest treatment of cocoa! Each cocoa has a potential intrinsic quality which depends on the variety of trees planted, the soil and the know-how of agricultural planter. But this potential can be valorized or deteriorated during the post-harvest pratices."

In Côte d'Ivoire, all post-harvest practices are managed by two groups of producers: farmers organized in cooperatives since the cooperative law of 1997 and the small individual farmers. The farmers' cooperatives receive technical support from the Government, NGOs or Western chocolatiers.

In these cooperatives, there are sometimes large farms whose area may approach 50 ha. The current tendency of producers is to increase acreage to compensate for falling prices (Banzio, 2003; N'Guessan, 2004). Some cooperatives also have many opportunities that can be the direct export of cocoa; guarantees in terms of marketing products; accessibility to additional services such as laboratory for quality control and computer equipment with software for managing. These cooperatives sell about 20 to 30% of national production.

In addition to the cooperatives, there are about 3 to 4 million people working in the cocoa chain and coffee in Côte d'Ivoire. This is usually small independent cocoa farmers and coffee, with mostly family farms ranging in size from 4 to 5 ha on average. These deliver their produce to buyers or subcontractors and trackers. It is estimated that one hectare of cocoa plantation can naturally produce 1.5 tons of cocoa beans. For now in Côte d'Ivoire, one hectare of cocoa produces on average 400 to 500 kg (Anonymous, 2010).

The responsibility of individual farmers and the farmers' cooperatives is significant in the search for the quality. Indeed, they must follow the cultural itinerary adapted and advised by farm advisors to appropriately perform the harvest and post-harvest practices namely fermentation and drying which are crucial steps that influence the final quality of the cocoa. However, at each step of post-harvest practices, several choices of modalities are available to both producer groups. Is there a significant difference between these two groups from the modalities of post-harvest practices? Many studies have shown the impact of the post-harvest processing methods on quality of merchantable cocoa (Cros & Jeanjean; 1995; Lainé, 2001; Mounjouenpou et al., 2011; Barel, 2013; Levai, Meriki, Adiobo, Awa-Mengi & Akoachere, 2015). However, none of them evaluated quantitatively the effects of each producer group on quality. And the quantitative influence of the modalities used by the producers is not known about the quality of cocoa. It is therefore with good reason this study was conducted.

The main purposes of this work were firstly to compare the modalities of post-harvest processing methods of these two groups, and secondly to statistically identify the modalities responsible of non-quality (under grade) of merchantable cocoa on the basis of results of Pareto chart and proportion of successes calculated for each modality.

2. Material and Methods

2.1 Material

The material was composed of 90 samples of merchantable cocoa of the Forastero variety. These samples were purchased from producers of seven production areas of the Soubré region which is also the first cocoa-producing among the regions in Côte d'Ivoire. The collection of samples was carried out during the great cocoa campaign 2013.

2.2 Methods

2.2.1 Sampling for the Comparison of Post-harvest Practices

Data collection was performed using a digital questionnaire which allowed to identify different forms of post-harvest practices of individual farmers and the farmers' cooperatives. A simple random sample of 30 farmers was selected by area (Table 1). Using the direct questionnaire during individual interviews, each producer gave information about its post-harvest practices. In total, 210 cocoa farmers were surveyed in 31 villages.

Table 1. Visited areas and number of farmers in the study

Sub-prefectures of Soubré	areas of production	Number of village	Number of farmers surveyed
Oupoyo	Petit Bondoukou	3	30
Méagui	Kragui	7	30
	Krohon	6	30
Okrouyo	Ottawa	4	30
	Kouamékro	4	30
Guéyo	Bobouo	3	30
	Djenandou	4	30
Total		31	210

2.2.2 Identification of Modalities of the Post-Harvest Practices Responsible for the Non-Quality (Under Grade)

2.2.2.1 Sampling

Taking into account the results of the survey, a stratified sampling plan based on the modalities of the fermentation equipment's (Banana Leaves; black plastic films; nylon tarps) allowed to classify the types of producer (Table 2).

Table 2. Sampling plan

Group of producers	Modalities of the fermentation	Number of samples	Total
Individual farmers	Banana leaves;	15	
	Black plastic films	15	45
	Nylon tarps	15	
Farmers in cooperatives	Banana leaves;	15	
	Black plastic films	15	45
	Nylon tarps	15	
General total			90

At each group of producers 45 samples of 1 kg were taken because of 15 samples by fermentation method. Each sample was weighed with a dial dynamometer spring from Ducatillon brand and packaged in a food bag jute of size 30cm / 40cm. For each sample, the time for breaking cocoa pods, the number of brewing during fermentation, the fermentation time, the fermentation materials, the drying time and the drying materials were noted.

2.2.2.2 Sample Classification Based on International Standard

2.2.2.2.1 Humidity

Post-harvest practices traceability of samples was carried out only for those whose average humidity level were less than 8%. The Humidity of the cocoa beans has been measured on the 90 samples with a moisture meter from Dickey John "mini GAC plus" brand.

2.2.2.2.2 Cut Test

The cocoa beans of each sample underwent a mixing. One hundred beans were counted then cut longitudinally with a bean guillotine (Barel, 2013). Examination of cocoa beans was done in the light of day when the weather was clear (International trade centre [UNCTAD/WTO], 2001). The cocoa beans with several defaults were classified in the same category of defects (moldy, moth-eaten or slate) according to the international standard for cocoa export.

2.2.2.3 Construction of Pareto Chart for Under-Grades and Determination of Proportion of Successes (Grade I and Grade II) for Each Modality

On the basis of surveys conducted, the traceability of the 90 samples (the modalities applied and quality obtained) according to the international standard was known. It was possible to construct Pareto chart for under-grades and to determine of proportion of successes (grade I and grade II) for each modalitiy.

2.2.2.3.1 Construction of Pareto Chart

The Pareto chart has allowed to better identify the modalities of post-harvest responsible for 80% of all non-quality observed. The construction of the Pareto chart was first to determine the modalities which have resulted in a non-quality samples. Then, for each post-harvest practice, the number of observations for the modalities was ranked in descending order of their size by histogram. For the equal sizes, the modality unadvised according to experts or literature was chosen first. Finally, the modalities that cumulated size were more than 80% have been identified as responsible for the poor quality.

2.2.2.3.2 Determination of Proportion of Successes

The proportion of successes (PS) was determined for each modality. It is the percentage of each modality leading to the quality samples (grade I or grade II). It was determined as following:

$$PS(\%) = \frac{Number\, of\, observations\, for\, the\, modality\, leading\, the\, quality\, samples \times 100}{Total\, number\, of\, observations\, for\, the\, modality}$$

2.3 Statistical Analyses

Data analysis, figures and Pareto charts were done using the softwares Microsoft Office Excel 2010 and Statistica 7.1. The chi square test was performed to verify the similarity of post-harvest practices between the two groups of producers at significance levels of 0.05, 0.01 and 0.001.

3. Results and Discussion

3.1 Results

3.1.1 Comparison of Post-harvest Practices

3.1.1.1 Time for Breaking Pods

The *survey conducted* among producers, shows that the time breaking pods varies between 1 and 15 days (figure 1) with distributions of both producer groups significantly similar according the chi square test (p-value > 0.05) (table 3). The modalities of the time for breaking pods change little from one type of group to another. The modalities for 2 to 7 are represented with 93% and 90% by most conditions individual farmers and farmers' cooperatives respectively. The other modalities are either absent or marginally represented. Furthermore, the area harvested seems to increase with the time breaking pods particularly between 1 and 7 day (Figure 1).

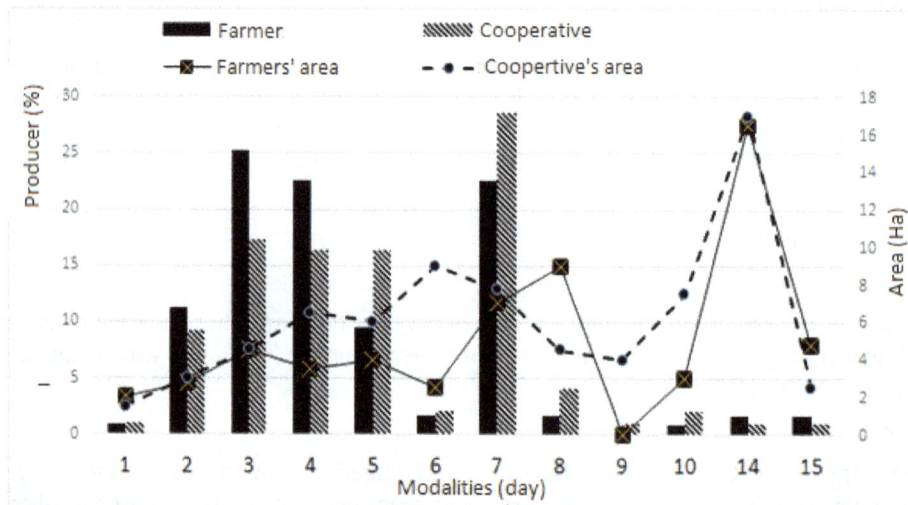

Figure 1. Comparisons of Time breaking cocoa pods and area harvested

Table 3. Chi-square tests of concordance for post-harvest practices

Post-harvest practices	Time breaking cocoa pods	Number of brewing	Fermentation time	Equipment of fermentation	Drying time	Equipment of drying	Quality (grades)
p-value	0.659	0.000*	0.000*	0.000*	0.478	0.000*	0.028*

p-value gives the probability of accepting the concordance in the post-harvest practices between the individual farmers and the farmers' cooperatives. An asterisk (*) indicates a significant difference (p<0.05), a highly significant differences (p<0.01) or a very highly significant differences (p<0.001) among the modalities of the post-harvest practices.

3.1.1.2 Number of Brewing

The analysis of figure 2 shows that the distributions of the brewing modalities between the two groups of producer are different. According the chi square test (Table 2) this difference is very highly significant (p value < 0.001). Also the number of brewing varies inversely from individual farmers to farmers' cooperative (Figure 2). The most brewing modalities applied by farmers' cooperative is 2 to 3 brewings which represents 69% and 0 brewing for individual farmers which represents 26%.

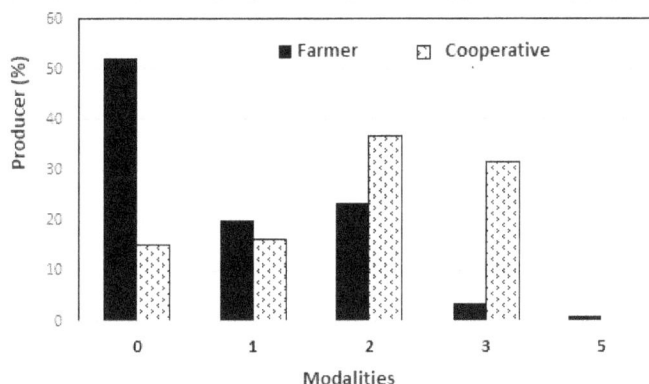

Figure 2. Comparisons of brewing during fermentation

3.1.1.3 Fermentation Time

The average time of fermentation for farmers' cooperative is higher than those for individual farmers. Indeed, 88% of farmers' cooperative conduct the fermentation in 6 or 7 days against 37% among farmers (Figure 3). These observations are confirmed statistically with a p-value less than 0.001 showing a very highly significant difference of distributions of the modalities of fermentation times between the two groups of producer (Table 2). The recommended fermentation times (4 and 7 days) are carried out at 99% in farmers in cooperatives and 85% in individual farmers.

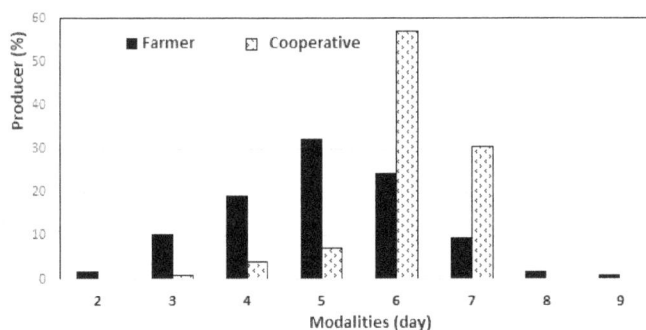

Figure 3. Comparisons of fermentation time

3.1.1.4 Fermentation Materials

The most commonly used fermentation materials are banana leaf, black plastic film and nylon tarp. They are used in 93% of cases by both groups of producers (Figure 4). However, 65% of farmers in cooperatives mainly use the banana leaves (Figure 4). Instead, individual farmers use all types of equipment with 40% for black plastic film, 27% for the banana leaf and 26% for the nylon tarp. The results of the chi square test show that there a very highly significant difference (p value < 0.001) between the two groups of producer (Table 3).

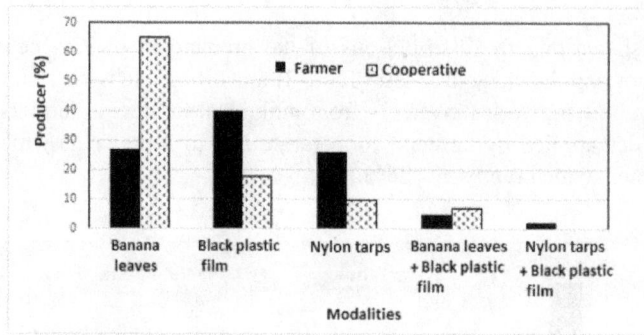

Figure4. Comparisons of fermentation materials

3.1.1.5 Drying Time

The distributions of drying time are similar for the both groups of producer (Figures 5). This observation was confirmed using the chi square test (p-value > 0.05) (Table 2). However the modalities of drying time vary from 3 to 8 days for individual farmers and up to 10 days among farmers' cooperative. The predominant modality is 7 days for both groups of producer which corresponds to 37% of individual farmers and 36% of farmers' cooperative.

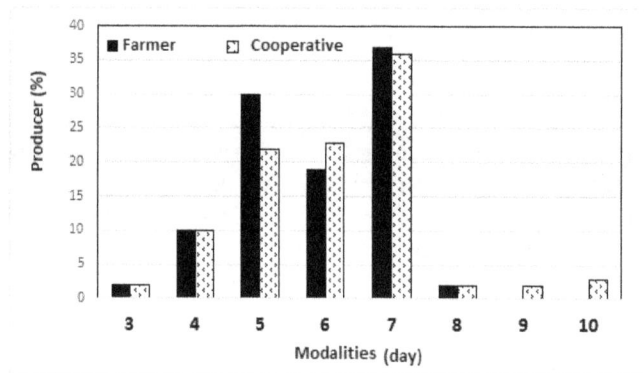

Figure 5. Comparisons of drying time

3.1.1.6 Drying Materials

The drying materials appear to be different for the both groups of producer (Figure 6). The proportion of using of the drying rack (recommended drying equipment) remains generally low among individual farmers (3%) against 32% among farmers' cooperative. The predominant modality is nylon tarp with a proportion of 50% and 39% among individual farmers and farmers' cooperative respectively. The result of chi square test was very highly significant difference (Table 3). Thus Individual farmers use different drying equipment from those used by farmers in the cooperative.

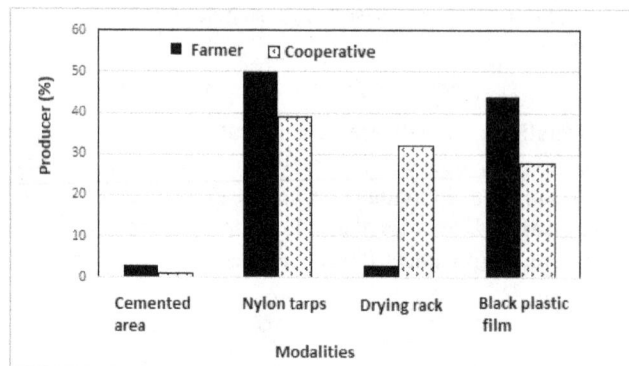

Figure 6. Comparison of drying equipment

3.1.2 Identification of Modalities of the Post-Harvest Practices Responsible for the Non-Quality (Under Grade)

3.1.2.1 Comparison of Different Quality Levels According to International Standard

The figure 7 gives the comparison of different quality levels according to international standard. It also gives the effects of each producer group on each level of merchantable cocoa quality after post-harvest processing methods.

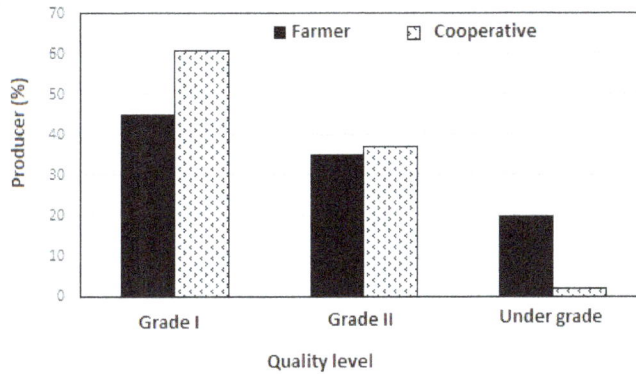

Figure7. Comparison of different quality levels according to international standard or effects of each producer group on cocoa quality

The samples from the farmers in cooperative are better than those collected from individual farmers (Figure 7). The chi square test shows a significant difference (p-value < 0.05) in the quality of cocoa beans between the two groups of producers. Only 2% of the samples coming from farmers in cooperative were under grades against 20% of the samples from individual farmers (Figure7).

3.1.2.2 Identification of the Time Breaking Cocoa Pods Causing Non-Quality

The *survey conducted* among producers, shows in order of importance that the modalities of time breaking cocoa pods of 1 day (30%); 3 days (20%); 4 days (20%) and 2 days (10%) represent 80% of samples coming from non-qualities (under grade). The modalities of time breaking cocoa pods for 5 and 7 days represent 20% of under grades. No samples from the modalities of 6; 8; 12 and 14 days were under grade. According Figure 8B, more than 80% of each modality applied have achieved the quality samples except the modality of 1 day (less than 60%). Also the comparison of these two figures (figure 8A and 8B) seem to show a decreasing of the under grade (so the increasing of the quality) with the augmentation of time breaking cocoa pods.

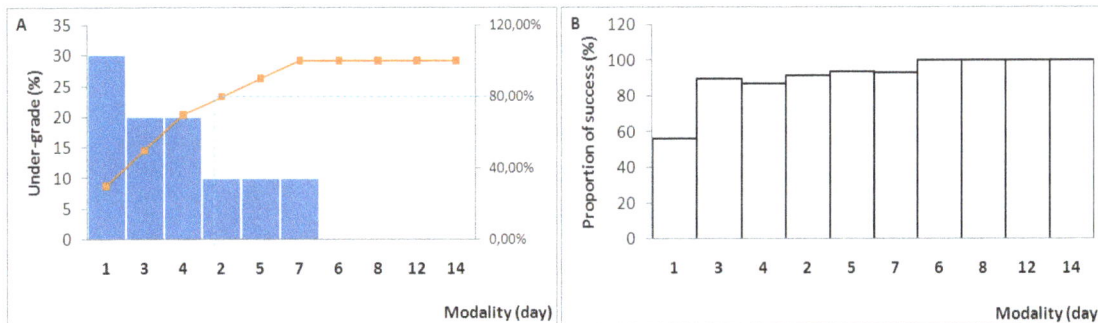

Figure 8. Time breaking cocoa pods (A: Pareto chart for modalities; B: Proportion of successes)

3.1.2.3 Identification of Number of Brewing Causing Non-quality

Eighty percent of samples in under grade are from the modalities of 0 and 2 brewings (Figure 9A). The maximum under grade (60%) was observed for beans that have undergone no brewing (Figure 9A). This corresponding to the lowest proportion of successes (Figure 9B). Therefore, no brewing can apparently having a detrimental impact on the quality of merchantable cocoa. The highest proportion of successes is more than 96% which is obtained by one brewing.

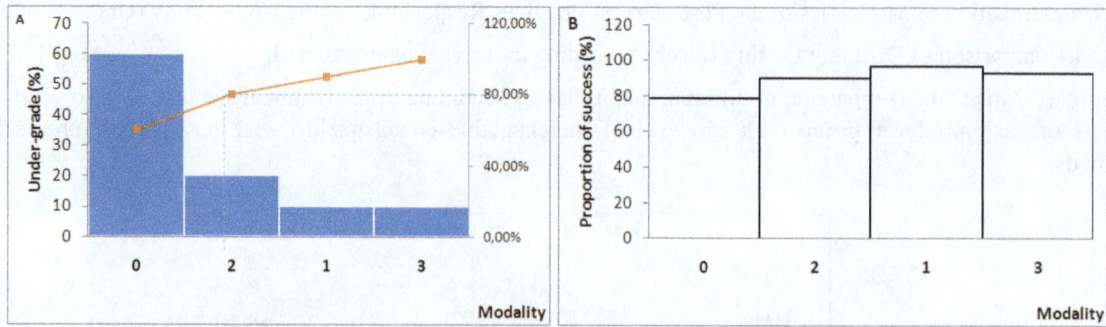

Figure 9. Number of brewing (A: Pareto chart for modalities; B: Proportion of successes)

3.1.2.4 Identification of the Fermentation Time Causing Non-quality

Fermentation times ranging from 2 to 5 days provide 80 % of the total of samples in under grade (Figure 10A). The proportions of successes for these modalities are 24%; 75%; 49% and 93% for 2; 3; 1 and 5 days respectively (Figure 10B). Among them, the fermentation times of 1, 2 and 3 days which have given the lowest percentages of success, help to lower the quality of merchantable cocoa.

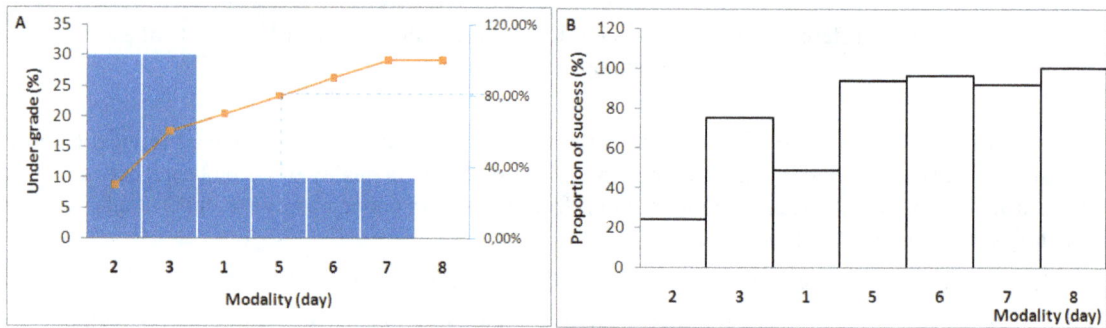

Figure 10. Fermentation time (A: Pareto chart for modalities; B: Proportion of successes)

3.1.2.5 Identification of Fermentation Materials Causing Non-quality

Figure 11A shows that 90% of the total of samples in under grade were fermented with banana leaves (50%) and black plastic films (40%). However, the proportion of successes of each of these materials remains above 80 % (Figure 11B). Unlike the latter, nylon sheeting provide 10% of the total of samples in under grade with 97% success in his use.

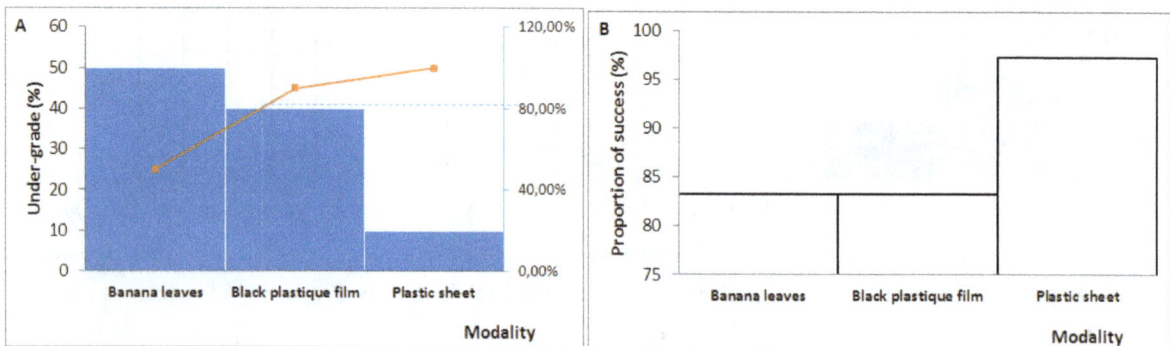

Figure 11. Fermentation mterials (A: Pareto chart for modalities; B: Proportion of successes)

3.1.2.6. Identification of the Drying Time Causing Non-quality

The modalities of drying times less than or equal to 6 days provide 80% of the total of samples in under grade (Figure 12A). The other twenty percent of the total under grades are from modalities of 7 and 8 days. No samples in under grade was observed at 10 day. Apart from the modalities of drying times of 2 (0% success) and 3 (74% success) days, all drying times applied have higher success to 80% (Figure 12B).

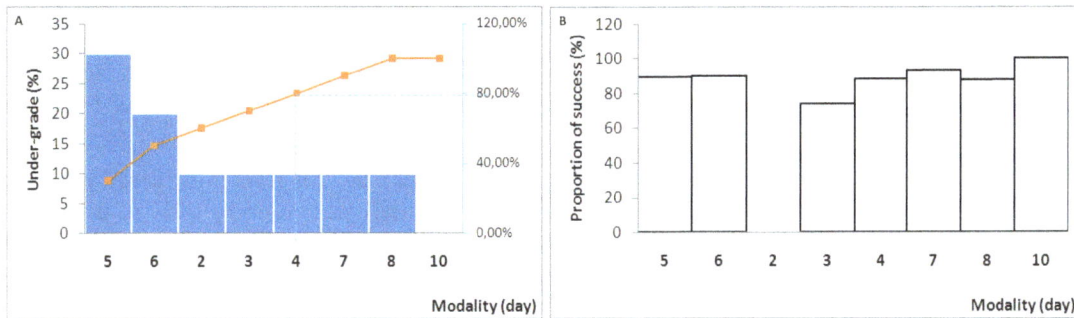

Figure 12. Drying time (A: Pareto chart for modalities; B: Proportion of successes)

3.1.2.7 Identification of Drying Materials Causing Non-quality

Drying materials such as nylon tarp and black plastic film represent 80% of the total of samples in under grade against 20% from the cemented area and drying rack (Figure 13A). The lowest proportion of successes (88%) is observed in the use of black plastic. Drying rack and cemented area gave respectively 97% and 94% of success in their use.

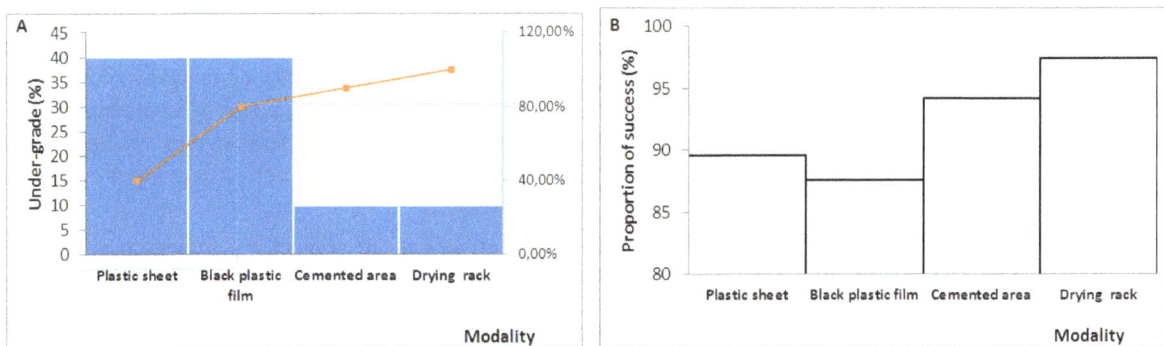

Figure 13. Drying materials (A: Pareto chart for modalities; B: Proportion of successes)

3.2 Discussion

According to the chi square tests, the comparison of post-harvest practices shows that apart from the time of pods breaking and drying time, post-harvest practices are inconsistent significantly between individual farmers and farmers in cooperative. We can assume that within the modalities of breaking pods and drying time, the producers have similar reactions in the face of the constraints of such practices. At the level of time of pods breaking, these constraints could be due to the limited need for labour that is the same in all areas of production. Because harvest and breaking pods can not be achieve only by the farmer. Because harvest and breaking pods can not be achieve only by the farmer. At the level of drying time, the constraints are related to climatic factors (sunshine, humidity) that may adversely the drying, regardless of the action of producers. In this study the times collected have varied from 3 to 10 days. In some areas, drying may last up to three weeks (Barel, 2013). The influence of the seasons of year on the drying of cocoa beans Côte d'Ivoire was mentioned by Akmel (2010).

Very high significant differences between the two groups of producers, were observed in the number of brewing; in the fermentation time; materials of fermentation and drying. Concerning the modalities of brewing, the most applied were 2 and 3 among farmers in cooperative (69%) and no brewing among individual farmers (52%). This could be due the fact that unlike farmers in cooperatives, individual farmers are not trained in good post-harvest practices and don't understand the usefulness of brewing during fermentation (Anonymous, 2006). Indeed, the brewing permit more air to get into the cocoa so that the rate of fermentation increase and temperatures will go up a few hours later (AusAid, 2010). This operation encourage the formation of a range of acetic bacteria which is crucial to ensure a good fermentation.(Cros, & Jeanjean, 1995). The lack of training is also observed on the recommended time of fermentation (4 and 7 days) (Cros, & Jeanjean, 1995). Individual farmers take less time in the fermentation that farmers in cooperatives. According Barel (2013) and others (Anonymous, 2000), to make a good cocoa, fermentation must stop between 5 and 7 days for the Forastero variety. This recommendation is followed by 95% of the farmers in cooperative against 66% of the individual

Food Processing Technologies: Quality and Safety

farmers. There may be a risk of insufficient fermentation in the case of individual farmers. Regarding fermentation materials, 65% of farmers in cooperative have used of the banana leaves against 27% among individual farmers. They prefer to use other materials (black plastic film and nylon tarp) due to lack of banana leaves that seem to be the most recommended material for fermentation (Cruz, Troude, Griffon & Hébert, 1988; Anonymous, 2000).

Regarding the drying materials, individual farmers have used different materials from those used by the farmers in cooperative. Indeed the drying racks (recommended drying material) were used by 32% of the farmers in cooperative and have been increasingly abandoned among individual farmers (3%). However, the user rate of drying racks remains low (17.5%) of producers as a whole. This is due to the fact that according to the majority of farmers, it is very difficult to find the raffia bamboo (plant family Arecaceae) to manufacture the drying racks. For this reason, other materials were preferred over the drying racks.

Regarding the modalities of the post-harvest practice responsible of the under-grades and their proportion of successes, the results showed that the samples of farmers in cooperatives had fewer defects than those of individual farmers. This is evidence that farmers in cooperatives are better trained and coordinate better the modalities of post-harvest practices for quality. In addition, the results showed that most of the modalities can lead to under-grade beans and they have different proportions of success. This is explained by the fact that quality can not depend on a single modality but it depends on a good combination of the modalities of the post-harvest practices.

At the level of time breaking pods, the modality of one day must not to be encouraged because it gave poor results, unlike other methods, which the number of success remained above 80%. The tendency to decrease in under-grade with increasing of time breaking pods is consistent with the literature. Indeed, delay the time for the breaking of pods is always advantageous. Water loss during storage of pods reduces about 50% the quantity of juice and promotes better aeration during fermentation, which accelerates the temperature rise and thus fermentation (Barel, 1987). However, some authors recommend not to exceed 5 or 6 days to prevent the germination of beans, their decay or the appearance of foreign tastes (Mossu 1990; Barel, 2013).

At the level of number of brewing, the maximum under-grade (60%) was observed for the cocoa beans which have not undergone any brewing. This result may be due to the lack of individual farmers awareness concerning the brewing. The modality with a single brewing is encouraging (96% success) especially for the heap fermentations. Indeed, for the fermentation most of ivorians farmers carry out the heap fermentations. This type of fermentation is simplest of all and does not require a farmer to have to construct fermentation boxes (AusAID, 2010). According Barel (2013) under these conditions, the number of brewing can be reduced to one or two. But for box fermentations, two to three brewings are needed for proper fermentation.

At the level of fermentation times, durations of less than five days must be not encouraged because they represented more than 70% of all the total of under- grades. These durations may be responsible for an insufficient fermentation. For the modalities longer than or equal 5 days there were over 90% of success on each modality. However for good fermentation 5 to 7 days are required depending on the size of the fermenting cocoa mass (Mossu, 1990; Anonymous, 2000; Barel 2013). At the level of fermentation materials, the fact that the banana leaves have the proportion of under-grade relatively high could be due to unavailability or a bad combination of this material with other modalities non-compliant. For indeed banana leaves are the most recommended (AusAID, 2010). Nylon tarps are very successful in their use it may due to good control of these farmers who find them more available.

At the level of fermentation materials, the fact that the banana leaves have the proportion of under-grade relatively high could be due to unavailability or a bad combination of this material with other modalities non-compliant. For indeed banana leaves are the most used (Lainé, 2001; AusAID, 2010). Nylon tarps are becoming popular, the reason is the good control of this material by the farmers who find them more available.

At the level of drying of materials, even if all the materials are very successful in their use, the drying rack and the cemented area is largely characterized by a relatively small proportion of under-grade. Among both materials, the drying rack has an advantage because air can flow horizontally and vertically through the cocoa beans that are spread on the rack out in the sun. There is at this level more convective heat exchange between the air and the product. Moreover, in the early hours of drying, the too wet beans (50-60% wb) can sometimes drip off through the racks without the retaining of the water on its surface, which does is not the case for non-perforated materials such as cemented area (Akmel, 2010; Barel, 2013).

4. Conclusion

The comparison of post-harvest practices shows that there is no correlation between individual farmers and farmers in cooperative. Highly significant differences (p-value < 0.001) between the two groups of producers have been observed in the number of brewing; the time of fermentation; the materials of fermentation and drying. These differences have confirmed that post-harvest practices have an impact on the quality of merchantable cocoa. The samples collected from farmers in cooperative have under well defects than those of individual farmers. Regarding the modalities of post-harvest practice responsible of under-gade, the results show that most of conditions can lead to the under-grades. And these modalities have different proportions of success. Thus, in order to achieve a production of high quality cocoa, farmers must do a good combination of each recommended modality. However certain modality should be avoided. These are: the time breaking pods of one day; the absence of brewing during the fermentation; the time of fermentation less than or equal to three days and the time of drying less than or equal to two days.

Acknowledjements

This study was possible thanks to the support of leaders of ICRAF (International Centre for Research in Agroforestry) Côte d'Ivoire.

References

Akmel, D. C. (2010). Séchage solaire des fèves de cacao : Etude expérimentale et modélisation de la cinétique de séchage. Thèse unique de l'Université d'Abobo Adjamé, Abidjan, Côte d'Ivoire.

Anonymous. (2000). De l'école lenôtre: Chocolat et confiserie, Tome 1. Ed. Jerôme villette, De Agostini, Milan, pp 13-16. Retrieved from http://c3d.94.free.fr/ebooks/Desserts/Chocolats%20et%20Confisserie%20-Tome%201%20-%20Ecole%20Lenotre.pdf

Anonymous. (2006). Elaboration d'un plan national café cacao : étude diagnostic. Bourse du café et du cacao (BCC). Abidjan, 60.

Anonymous. (2010). Chocolat : 3 entreprises (Cemoi, Blommer et Petra Food) familiales et internationales s'inussent pour développer une filière éthique et de qualite en Côte d'Ivoire. ADOCOM, Dossier de presse, Paris, 7 pages.

Anonymous. (2015). Bilan de la réforme de la filière café-cacao. Journées nationales du cacao et du chocolat, Octobre 2015, Yamoussoukro-Côte d'ivoire, pp 1-20

Anonymous. (2016). Côte d'ivoire: inquiétudes redoublées autour de la qualité des fèves de cacao. Jeune Afrique. Retrieved from http://www.jeuneafrique.com/332225/economie/cote-divoire-inquietudes-autour-de-qualite-feves-de-cacao/

AusAid. (2010). Cocoa processing methods for the production of high quality cocoa in Vietnam. Retrieved from www.canacacao.org/uploads/smartsection/19_Cocoa_fermentation_manual_Vietnam.pdf

Banzio, D., (2003). Comprendre et operer dans les filières café cacao en dix modules. Ed. Eburnie; Abidjan, Côte d'Ivoire, 10-23.

Barel, M. (1987). Délai d'écabossage: Influence sur les rendements et la qualité du cacao marchand et du cacao torréfié. 31.

Barel, M. (2013). Qualité du cacao : impact du traitement post-récolte. Ed. Quae, Savoir faire, 104 p.

Guehi, T. S., Dingkuhn M., Clément-Vidal, A., Cros, E., Fourny, G., Ratomahenina, R., & Moulin, G. (2005). Détermination de l'origine de l'activité enzymatique impliquée dans la formation des acides libres du cacao. Cirad-Agritrop, 14th International Cocoa Research Conference. Lagos: Cocoa Producers' Alliance, 869-873. Retrieved from http://catalogue-bibliotheques.cirad.fr/cgi-bin/koha/opac-detail.pl?biblionumber=193229

International trade centre [UNCTAD/WTO], (2001). Cocoa: Commercial Practices Guide. Ed : Genève: *UNCTAD/WTO, cop. Vol. XI*, 190.

Cros, E., & Jeanjean, N., (1995). Cocoa quality: effect of fermentation and drying. CIRAD-CP, Montpellier, France, 24. Retrieved from www. agritrop.cirad.fr/387560/1/document_387560.pdf

Cruz, J. F., Troude, F., Griffon, D., & Hébert J. P. (1988). Conservation des grains en régions chaudes-2. ed.- «Techniques rurales en Afrique». Paris, France, *Ministère de la Coopération et du Développement, 542.* Retrieved from http://www.fao.org/Wairdocs/x5164F/X5164f00.htm

Janny, G. M., Ritchie, B. J., & Flood, J. (2003). A la découverte du cacao. Un Guide pour la formation des facilitateurs, 1-50.

Koko, L. (2014). Teractiv Cacao as a New Fertilizer based Reactive Phosphate Rock for Cocoa Productivity in Côte d'Ivoire: A Participatory Approach to Update Fertilization Recommendation. *Procedia Engineering, 83*, 348-353. http://dx.doi.org/10.1016/j.proeng.2014.09.14.

Lainé, K. (2001). Enquête sur les pratiques culturales dans les cacaoyères en Côte d'Ivoire 29 Octobre-10 Novembre 2001 Projet PACCC / ICCO / Industrie sur l' amélioration de la qualité du cacao en Côte d' Ivoire. Rapport. (pp. 1-27).

Levai, L. D., Meriki, H. D., Adiobo, A., Awa-Mengi, S., & Akoachere, J. F. (2015). Postharvest practices and farmers' perception of cocoa bean quality in Cameroon. *Agriculture & Food Security, 4*, 28. http://dx.doi.org/10.1186/s40066-015-0047-z

Mossu, G. (1990). Le cacaoyer. Editions Maisonneuve et Larousse. 159. Retrieved from *www.ecofog.gf/giec/doc_num.php?explnum_id=774*

Mounjouen, P., Gueule, D., Ntoupka, M., Durand, N., Fontana-Tachon, A. Guyot, B., & Guirand, J. (2011). Influence of post-harvest processing on ochratoxin A content in cocoa and on consumer exposure in Cameroon. *World Mycotoxin Journal, 4*(2), 141-146. http://dx.doi.org/10.3920/WMJ2010.1255

16

Phenotypic and Genotypic Characterization of *Cronobacter* isolated from Powdered Infant Formula Retailed in Nigeria

Abimbola Rashidat Ezeh[1,2], Olusimbo Olugbo Aboaba[1], Barbara E Murray[2], Ben Davies Tall[3] & Stella Ifeanyi Smith[4]

[1] University of Lagos, Nigeria

[2] University of Texas Medical School, Houston, Texas, USA

[3] Center for Food Safety and Applied Nutrition, US Food and Drug Administration, USA

[4] Nigerian Institute of Medical Research, Nigeria

Correspondence: Abimbola Rashidat Ezeh, University of Lagos, Nigeria; University of Texas Medical School, Houston, Texas, USA. E-mail: bimsal@yahoo.com

Abstract

Cronobacter is a genus with emerging pathogens that has been associated with life threatening diseases in neonates, infants and immunocompromised adults. Three *Cronobacter* species were isolated from powdered infant formula retailed in Nigeria. Different methods of phenotypic and genotypic characterization were carried out. All the isolates were identified biochemically by Microscan identification analysis as *Enterobacter sakazakii* (98.87%). The Vitek MALDI-TOF system identified the isolates as *Cronobacter sakazakii*. 16S rRNA sequencing identified the isolates as *C. sakazakii*. In contrast the use of species-specific PCR assays targeting *rpo*B, and *cgc*A, helped to identify two of the three strains as *C. sakazakii* and the last strain was identified as *C. malonaticus*. Multi locus sequence typing (MLST) analysis was used to identify each strain's sequence type and the results identified three new sequence types: 303, 304 and 296. *C. sakazakii* BAA 894 served as a positive control for all the experiments. Biochemical methods and commercial identification systems are not sensitive enough to identify *Cronobacter* strains to the species level. Molecular methods are needed to confirm the species identity of strains.

Keywords: *Cronobacter*, genotyping, phenotyping, powdered infant formula milk

1. Introduction

Cronobacter, (former *Enterobacter sakazakii)* is an emerging, opportunistic pathogen that causes infections such as sepsis, meningitis and necrotizing enterocolitis in neonates and infants, and can sometimes lead to death. It is ubiquitous in various food products such as dairy based products, adult and infant cereals, and spices. It has been associated with the ingestion of contaminated reconstituted powdered infant formula (PIF), and can be found in various environments, in particular PIF production facilities. *Cronobacter* spp have shown high resistance to osmotic stress and this contributes to its persistence in PIF factories, dried products and environments (Osaili & Forsythe, 2009). The frequency of disease caused by *Cronobacter* is very low but the mortality rate has been reported to be as high as 80% with surviving patients often suffering severe neurological sequelae (Alsonosi et al., 2015), hydrocephalus and permanent mental damage. The World Health Organization (WHO) has recognized all *Cronobacter* species as microorganisms pathogenic for human beings of all ages (FAO/WHO, 2008) although this organism is also part of the normal human flora (Holy et al., 2014). Current international microbiological standards require the absence of all *Cronobacter* species in PIF (test volume 10 g) (Jackson et al., 2014). Of all *Cronobacter* species only *C. sakazakii, C. malonaticus* and *C. turicensis* have been linked to infantile infections (Alsonosi et al., 2015).

Cronobacter spp are Gram-negative motile rods of the family Enterobacteriaceae. They were formerly known as the yellow pigmented *Enterobacter sakazakii*. Reclassification based on the results of independent molecular methods and biochemical markers resulted in a new genus with seven species being described: *Cronobacter sakazakii, C. malonaticus, C. muytjensii, C. dublinensis, C. turicensis, C. condimenti* and *C. universalis* (Joseph et al., 2012). Target genes for PCR probe-based methods for *Cronobacter* identification include *cgcA, gyrB, ompA, rpoB, gluA, dnaG, zpx,* iron acquisition genes, the macromolecular synthesis operon, the 16S rDNA gene, and the 16S-23S intergenic transcribed spacer region (Carter et al., 2013, Grim et al., 2013, Lehner et al., 2012, Stoop et al., 2009). Molecular based techniques such as Multi locus sequence typing (MLST), random amplification of polymorphic DNA, pulsed-field gel electrophoresis (PFGE) have also been successfully applied to the characterization of *Cronobacter* spp (Fields et al., 2011). Pulsed field gel electrophoresis (PFGE) is considered the 'gold standard' method for subtyping foodborne bacteria and the most discriminatory technique for genotyping, and in 2012, a PulseNet protocol was validated for subtyping *Cronobacter* spp. (Brengi et al., 2012).

The aim of this study was to identify *Cronobacter* spp isolated from powdered infant formula products retailed in Nigeria using various phenotypic and genotypic methodologies.

2. Materials and Methods

2.1 Samples Collection

A total of 154 different samples of PIF were purchased from local markets and super markets in different geopolitical zones of Nigeria.

2.2 Bacterial Strains and Cultivation

Cronobacter strains used in this study were isolated from the PIF using the FDA method in combination with the method described by Iversen et al. (2004). Briefly, 25g of powdered infant formula was added to 225ml of buffered peptone water (BPW; pH 7.2+ 0.2). The suspended powdered infant formula was then incubated at 37^0C for 24h without shaking. Four aliquots of 40ml each were removed from the solution and placed into 50ml centrifuge tubes. The tubes were subjected to centrifugation at 3000 x g for 10 min. The supernatants of each tube were discarded and the resultant pellets were suspended in 200µL of phosphate-buffered saline (PBS; pH 7.2+ 0.2). The aliquots were cultured on Druggan-Forsythe-Iversen (DFI) agar. The control strain, *C. sakazakii* ATCC BAA 894 was a gift from the Division of Virulence Assessment, Office of Applied Research and Safety Assessment, Center for Food Safety and Applied Nutrition, U. S. Food and Drug Administration (MD, USA).

2.3 Phenotyping

Cronobacter isolates were phenotyped using the MicroScan WalkAway identification panel (Beckman Coulter Inc., CA, USA) according to the manufacturer's instructions. A sterile applicator stick was used to touch the surface of 4-5 morphologically similar, well isolated colonies from an 18-24 h Brain Heart Infusion (BHI) (Difco, New Jersey) agar plate. This was emulsified in 3 mL of sterile deionized water and the suspension was vortexed for 2-3 s to achieve a final turbidity similar to the 0.5 McFarland Turbidity Standard scale. One-hundred microliters (100 µL) of the standardized suspension was pipetted into 25 mL of inoculum buffer, capped tightly and inverted 8-10 times to mix. An oxidase test was performed using tetramethyl-p-phenylenediamine – dihydrochloride (Sigma-Aldrich, USA) prior to inoculating the panels. The panel was rehydrated and inoculated using the RENOK system with Inoculators-D (Siemens, Frimley, Camberley, UK). A dropper bottle was used to overlay the GLU, URE, H2S, LYS, ARG, ORN and DCB (Glucose, Urea, Hydrogen sulphide, Lysine, Arginine, Ornithine, and Decarboxylase Base) with 3 drops of sterile mineral oil. The panel was incubated in the Walkaway System for 16-20 h after which the results were read.

2.4 Species Identification Using the VITEK®MS MALDI-TOF

Identification of the *Cronobacter* isolates was carried out using VITEK®MS MALDI-TOF (Matrix Assisted Laser Desorption Ionization Time-of-Flight) technology (bioMerieux, France), according to the manufacturer's instructions. Two milliliters of bacterial broth was added to 1.0 ml of lysis buffer (0.6% polyoxyethylene 10 oleoyl ether (Brij 97) in 0.4 M (3-(cyclohexylamino)-1-propane sulfonic acid) (CAPS) filtered through a 0.2-µm-pore-size filter, pH 11.7), vortexed for 5 s, and allowed to incubate for 2 to 4 min at room temperature. The resulting lysate was filtered through a 25-mm 0.45-µm-pore-size filter (catalog no. HPWP02500; Millipore Express PLUS, Billerica, USA), MA and the microbial cells remaining on the filter were removed and washed three times with wash buffer (20 mM Na3PO4, 0.05% Brij 97, and 0.45% NaCl), using a 0.2-µm-pore-size filter, pH 7.2); washed three times with deionized water; and removed from the surface by scraping the filter with a micro-swab (Texwipe CleanTips swabs; catalog no. TX754B; Kernersville, NC). Sample processing time was approximately 10–15 min for up to three samples. Microorganisms recovered from the filter were directly applied to VITEK MS target plates and covered with 1 µL of CHCA (α-cyna-4-hydroxycinnamic acid) matrix. If the VITEK MS was unable to identify an isolate on the first attempt, the sample was repeated using double the volume of culture broth and corresponding buffers. A sample was considered to have a valid VITEK MS ID if at least one spot on the target slide gave a confidence level of ≥75% without conflicting identifications from replicate spots of the same sample. Samples that did not generate an ID on the first attempt were repeated only once.

2.5 DNA Template and PCR Analysis

Polymerase chain reaction (PCR) was performed using 18-24 h culture. DNA templates were prepared by using a sterile inoculating wire to remove a colony and the cells were suspended into 20 µL of sterile distilled water (dH$_2$O) dispensed into Eppendorf tubes. The reaction mixture (25 µL) was prepared as followed: dH$_2$O (15.75 µL), Phusion Buffer (5.0 µL; New England BioLabs, MA), Forward primer (1.25 µL; Sigma-Aldrich, USA), Reverse primer (1.25 µL; Sigma-Aldrich, USA), dNTPS (0.5 µL), Phusion DNA polymerase (0.25 µL; New England BioLabs, MA), and Template (1.0 µL).

2.6 PCR Probe Assays of Cronobacter spp

PCR was carried out using the 16S rDNA gene (Sigma-Aldrich, USA), β-subunit of RNA Polymerase gene (rpoB) (Sigma-Aldrich, USA), and Diguanylate Cyclase-Encoding gene (cgcA) (Sigma-Aldrich, USA). The PCR conditions and primers used are stated on Table 1. PCR amplicons were subjected to agarose gel electrophoresis using 0.8% Tris-borate-EDTA buffer (TBE; Sigma-Aldrich, USA) in a Bio-Rad sub cell GT (Bio-Rad, Belgium) horizontal electrophoresis unit and were photographed with transilluminated UV light using an Alpha Imager system (Alpha Innotech Corp, San Leandro, CA, USA). The PCR products were purified using Wizard SV Gel and PCR clean up system (Promega, USA) and sequenced by GENEWIZ (New Jersey, USA) in a Sanger DNA Sequencer System. The obtained nucleotide sequences were compared with the corresponding sequences of C. sakazakii strain ATCC BAA 894 (NCBI accession no. 290339) and with other Enterobacteriaceae using DNASTAR-MegAlign.

Table 1. Primers and PCR conditions used for allelic profiling of Cronobacter spp

Target gene	Primer Sequence	PCR conditions	Amplicon size (bp)	Species identification	References
16S rDNA	16SUNI-L 8F- AGAGTTTGATCATGGCTCAG 1492R- GGT TAC CTT GTT ACG ACT T 515F - GTG CCA GCA GCC GCG GTA 1100R - GTT GCG CTC GTT G	C. sakazakii: Initial denaturation step at 94^{0}C for 3 min, 30 cycles of 94^{0}C for 60 s, 67^{0}C for 30 s, 72^{0}C for 1 min, with a final extension step at 72^{0}C for 10 min. C. malonaticus: Initial denaturation step at 94^{0}C for 3 min, 25 cycles of 94^{0}C for 60 s, 60^{0}C for 30 s, 72^{0}C for 30 s, with a final extension step at 72^{0}C for 10 min	1500	Cronobacter spp	Kuhnert et al., 1996
rpoB	C. sakazakii– Csakf 5'-ACG CCA AGC CTA TCT CCG CG-3' Csakr 5'-ACG GTT GGC GTC ATC GTG-3' C. malonaticus - Cmalf 5'-CGT CGT ATC TCT GCT CTC-3' Cmalr 5'-AGG TTG GTG TTC GCC TGA-3'	C. sakazakii: Initial denaturation step at 94^{0}C for 3 min, 30 cycles of 94^{0}C for 60 s, 67^{0}C for 30 s, 72^{0}C for 1 min, with a final extension step at 72^{0}C for 5 min. C. malonaticus: Initial denaturation step at 94^{0}C for 3 min, 30 cycles of 94^{0}C for 60 s, 60^{0}C for 30 s, 72^{0}C for 1 min, with a final extension step at 72^{0}C for 10 min	514 251	C. sakazakii: C. malonaticus	Stoop et al., 2009
cgcA	Cmstu-825F- GGTGGCSGGGTATGACAAAGAC Csak-1317R- GGCGGACGAAGCCTCAGAGAGT Cmal1410R- GGTGACCACACCTTCAGGCAGA	Initial denaturation step at 98^{0}C for 2 min, 30 cycles of 98^{0}C for 60 s, 50^{0}C for 30 s, 72^{0}C for 1 min, with a final extension step at 72^{0}C for 10 min.	492 585	C. sakazakii: C. malonaticus	Carter et al., 2013

2.7 Pulsed Field Gel Electrophoresis (PFGE)

PFGE analysis of *Cronobacter* isolates was performed as previously described by Ribot et al. (2006) with some modification. The restriction enzyme *XbaI* was used for DNA digestion. Bands were separated using a CHEF-DR III System (BIO-RAD, Belgium) at 14 – 17° C. Electrophoretic conditions of initial switch time 1.8 s to a final switch time of 25 s, at 6 volts/cm for a run time of 17- 18 h were used. Gels were stained for 30 min in 500 mL dH2O containing 25 µl ethidium bromide (10 mg/mL) and visualized under UV light using an Alpha Imager system (Alpha Innotech Corp., San Leandro, CA, USA).

2.8 Multi Locus Sequence Typing

MLST was performed as previously described by Baldwin et al. (2009) and the *Cronobacter* PubMLST open access database (http://pubmlst.org/Cronobacter/). The seven housekeeping genes amplified were ATP synhase beta chain (*atpD*), elongation factor G (*fusA*), glutaminyl-tRNA synthetase (*glnS*), glutamate synthase large subunit (*gltB*), DNA gyrase subunit B (*gyrB*), translation initiation factor IF-2 (*infB*) and phosphoenol-pyruvate synthase (*ppsA*) (Sigma-Aldrich, USA).

Table 2. Oligonucleotide nested primer sequences for the amplification and sequencing of genes from *C. sakazakii* and *C. malonaticus*, with gene number and location corresponds to *C. sakazakii* strain ATCC BAA-894 genome (Baldwin et al., 2009)

Gene (Gene label)	Putative Gene Product	Chromosome location (bp)	Gene Size (bp)	Locus Primers (5'→3')	
				Amplification	Sequencing
atpD	ATP synthase β chain	3,689,177 - 3,690,559	1,382	CGACATGAAAGGCGACAT TTAAAGCCACGGATGGTG	CGAAATGACCGACTCCAA GGATGGCGATGATGTCTT
fusA	Elongation factor	3,275,843 - 3,277,957	2,114	GAAACCGTATGGCGTCAG AGAACCGAAGTGCAGACG	GCTGGATGCGGTAATTGA CCCATACCAGCGATGATG
glnS	Glutaminyl-tRNA	660,368 - 662,035	1,667	GCATCTACCCGATGTACG TTGGCACGCTGAACAGAC	GGGTGCTGGATAACATCA CTTGTTGGCTTCTTCACG
gltB	Glutamate synthase	3,538,713 - 3,542,921	4,208	CATCTCGACCATCGCTTC CAGCACTTCCACCAGCTC	GCGAATACCACGCCTACA GCGTATTTCACGGAGGAG
gyrB	DNA gyrase B	3,719,848 - 3,722,262	2,414	TGCACCACATGGTATTCG CACCGGTCACAAACTCGT	CTCGCGGGTCACTGTAAA ACGCCGATACCGTCTTTT
infB	Translation initiation	4,139,051 - 4,141,762	2,711	GAAGAAGCGGTAATGAGC CGATACCACATTCCATGC	TGACCACGGTAAAACCTC GGACCACGACCTTTATCC
ppsA	Phosphoenol-pyruvate	1,218,599 - 1,220,977	2,378	GTCCAACAATGGCTCGTC CAGACTCAGCCAGGTTTG	ACCCTGACGAATTCTACG CAGATCCGGCATGGTATC

2.9 Molecular Characterization of Cronobacter Lipopolysaccharide O-Antigen Gene Clusters Using Serotype Specific PCR Primers

Primers designed based on the *wzx* and *wzy* gene (O antigen) sequences (Sun et al., 2012) were used. Multiplex PCR was performed by mixing all primers in a final volume of 50 µL containing the following components: 1× Taq Reaction buffer; 2.5 mM MgCl$_2$; 400 µM (each) of dATP, dCTP, dGTP, and dTTP; 0.06 to 0.10 µM primer sets listed in Table 3, 2.5 U of *Taq* DNA polymerase and 50 to 100 ng of template DNA. The following PCR conditions were used for amplification: an initial denaturation step at 95°C for 5 min, followed by 30 amplification cycles at 94°C for 30 s, 53°C for 30 s, and 72°C for 1 min, with a final extension step at 72°C for 5 min. Samples (5 µl) of the PCR products were subjected to agarose gel electrophoresis for examination.

Table 3. Primers used for serotyping (Sun et al., 2012)

C. sakazakii serotype	Target gene	Primer	Sequence (5'- 3')	Final conc (uM)	Amplicon size bp
O1	wzy	wl-35646	CCCGCTTGTATGGATGTT	0.10	364
		wl-35647	CTTTGGGAGCGTTAGGTT	0.10	
O2	wzy	wl-37256	ATTGTTTGCGATGGTGAG	0.06	152
		wl-37257	AAAACAATCCAGCAGCAA	0.06	
O3	wzy	wl-37258	CTCTGTTACTCTCCATAGTGTTC	0.10	704
		wl-37259	GATTAGACCACCATAGCCA	0.10	
O4	wzy	wl-39105	ACTATGGTTTGGCTATACTCCT	0.06	890
		wl-39106	ATTCATATCCTGCGTGGC	0.06	
O5	wzy	wl-39873	GATGATTTTGTAAGCGGTCT	0.10	235
		wl-39874	ACCTACTGGCATAGAGGATAA	0.10	
O6	wzy	wl-40041	ATGGTGAAGGGAACGACT	0.06	424
		wl-40042	ATCCCCGTGCTATGAGAC	0.06	
O7	wzx	wl-40039	CATTTCCAGATTATTACCTTTC	0.06	615
		wl-40040	ACACTGGCGATTCTACCC	0.06	

3. Results

Out of the 154 different samples of powdered infant formula analyzed, only 3 (1.95%) were positive for *Cronobacter spp.*

3.1 Biochemical Identification of the Cronobacter Strains Using Phenotypic Assays

MicroScan WalkAway analysis identified all the isolates as *Enterobacter sakazakii* with 98.87% probability. Supplemental Table 1.

3.2 Cronobacter spp. Identification Using the VITEK®MS MALDI-TOF

VITEK MS MALDI-TOF (bioMerieux) analysis identified all the isolates as *C. sakazakii*. Supplemental Figures 1-4.

3.3 Identification of the Cronobacter Strains Using 16S rDNA Gene Sequencing

Amplification of the 16S rDNA gene from the *Cronobacter* isolates gave PCR products of about 1,500 bp. The PCR products were sequenced and identified using BLAST analysis (http://www.ncbi.nlm.nih.gov/). All amplification products were identified as the 16S rDNA gene of *Cronobacter* and had a 97-98% identity with *Cronobacter* spp.

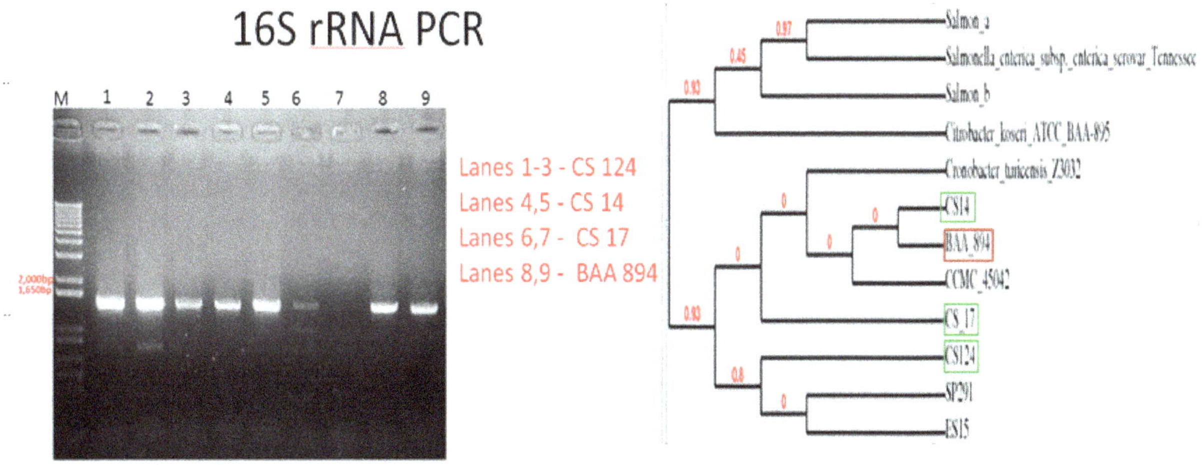

Figure 1. Gel image of the PCR products obtained from the amplification of the 16S rDNA gene from *Cronobacter* samples (CS) isolated from PIF. Lane M, DNA ladder (Invitrogen); lanes 1-3, CS 124; lanes 4 and 5, CS 14; lanes 6 and 7, CS 17; lanes 8 and 9, BAA 894 (positive control). The PCR products were about 1500 bp. Phylogenetic tree derived from the 16S rDNA sequence showing the relationship of CS 14, CS 17 and CS 124 to other *Cronobacter* spp.

3.4 β-subunit of RNA Polymerase Gene (rpoB) Based PCR Identification

Amplification of the *rpoB* gene with *C. sakazakii* primers yielded PCR products of about 514 bp (Figure 2). The four isolates including BAA 894 produced same sized bands suggesting that all the isolates were *C. sakazakii*. *Escherichia coli* which served as a negative control produced no band. When the *C. malonaticus rpo*B primers were used for amplification only CS14 and CS 124 produced bands of 251 bp suggesting that CS 14 and CS 124 are *C. malonaticus*. All other isolates including *E. coli* which served as a negative control, produced no band.

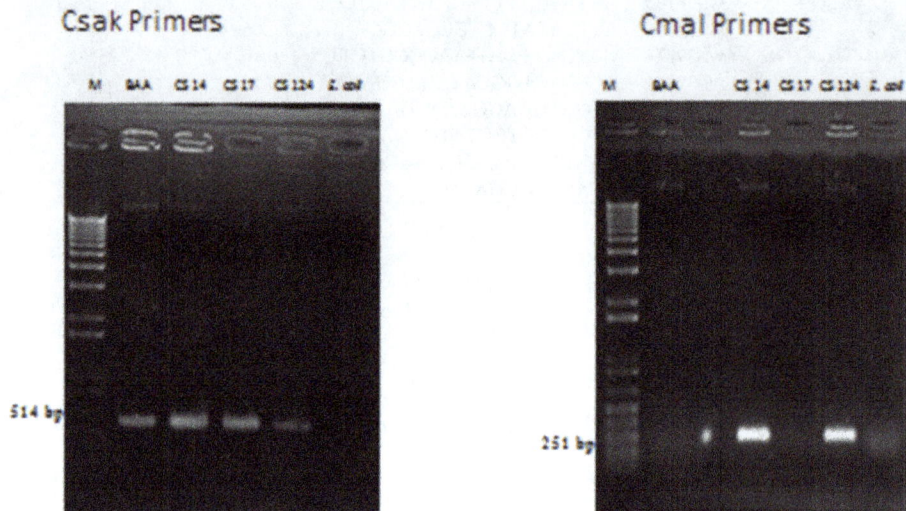

Figure 2. Gel image of the PCR products obtained from the amplification of the *rpoB* gene using *C. sakazakii* and *C. malonaticus rpoB* primers. C.sak: Lane M, DNA ladder (Invitrogen); lane 1, BAA 894; lane 2, CS 14; lane 3, CS 17; lane 4, CS 124 and lane 5, *E. coli*. The PCR products were about 514 bp. C.mal: Lane M, DNA ladder (Invitrogen); lane 1, BAA 894; lane 3, CS 14; lane 4, CS 17; lane 5, CS 124 and lane 6, *E. coli*. The PCR products were about 251 bp. C.mal primer identified CS 124 as *C. malonaticus* instead of *C. sakazakii*

3.5 Diguanylate Cyclase-Encoding Gene (cgcA) Based PCR Identification

Amplification of *cgc*A gene produced two bands that were characteristic of *C. sakazakii* (492 bp) and *C. malonaticus* (585 bp). BAA 894, CS 17 and CS 124 showed a band of approximately 490 bp characteristic of *C. sakazakii* while CS14 showed a fragment around 500-600 bp characteristic of *C. malonaticus* (Figure 3).

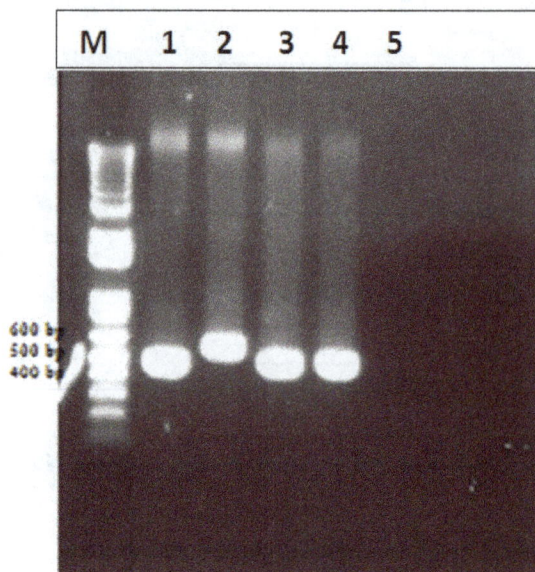

Figure 3. Gel image of the PCR products obtained from the amplification of the *cgc*A gene. Lane 1, BAA 894; lane 2, CS 14; lane 3, CS 17; lane 4, CS 124 and lane 5, sterile distilled water (negative control). *C. sakazakii* produced a band of about 492 bp while *C. malonaticus* produced a band of 585 bp. These results suggest that sample CS 14 (lane 2) corresponds to *C. malonaticus* and samples CS 17 (lane 3) and CS 124 (lane 4) correspond to *C. sakazakii*

3.6 Pulsed Field Gel Electrophoresis (PFGE)

Genomic DNA from *Cronobacter* isolates were analyzed by PFGE using *Xba*I as restriction enzyme. Eleven (11) to 16 fragments ranging from 40 to 1000 kbp were captured. All the pulsotypes obtained from each isolate were different from one another (Figure 4).

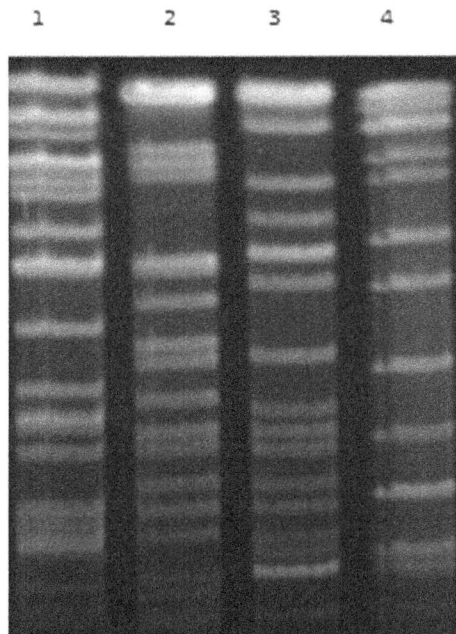

Figure 4. *Xba*I-pulsed-field gel electrophoresis (PFGE) patterns obtained from *Cronobacter* isolates. Lane 1, CS 14; lane 2, CS 17; lane 3, BAA 894, and lane 4, CS 124. Distinct PFGE profiles correspond to epidemiologically unrelated sources

3.7 Multi Locus Sequence Typing

Sequence Types (STs) and allelic profiles numbers were assigned to the isolates in accordance to the scheme initially established for *C. sakazakii* and *C. malonaticus* by Baldwin et al. (2009). The seven housekeeping genes were successfully sequenced from each isolate. MLST analysis showed that all three strains possessed new sequence type allelic variants and were different from those available on the *Cronobacter* PubMLST database (Table 4). CS17 and CS124 shared the allelic profile of *atp*D while all the three isolates shared the allelic profile of *pps*A.

Table 4. Multi locus sequence typing for *Cronobacter* isolates using the seven housekeeping genes

	atpD	*fusA*	*glnS*	*gltB*	*gyrB*	*inf*B	*ppsA*	Sequence Type
CS14 C. mal	10	13	64	75	72	14	1	303
CS 17 C. sak	3	17	13	57	58	63	1	304
CS 124 C. sak	3	15	28	22	5	38	1	296

3.8 Molecular Characterization of Cronobacter Lipopolysaccharide O-Antigen Gene Clusters Using Serotype Specific PCR Primers

Serotyping assays based on PCR specific to O-antigen genes have become acceptable methods for typing many Gram-negative bacteria. At the present work, this method was successfully applied to identify *Cronobacter* serotypes. PCR analysis using the primers described by Sun *et al.* (2012) identified sample CS 14 as *C. sakazakii* serotype O:6, which is also the same as *C. malonaticus* serotype O:2 according to Yan et al. (2015).

PCR-products obtained from the analysis of samples CS 17 and CS 124 correspond to *C. sakazakii* serotype O:2 and O:4, respectively. As expected, the O-antigen serotyping scheme obtained for the control strain of *C. sakazakii* ATCC BAA 894 agrees with that reported for the O1 serotype (Figure 5).

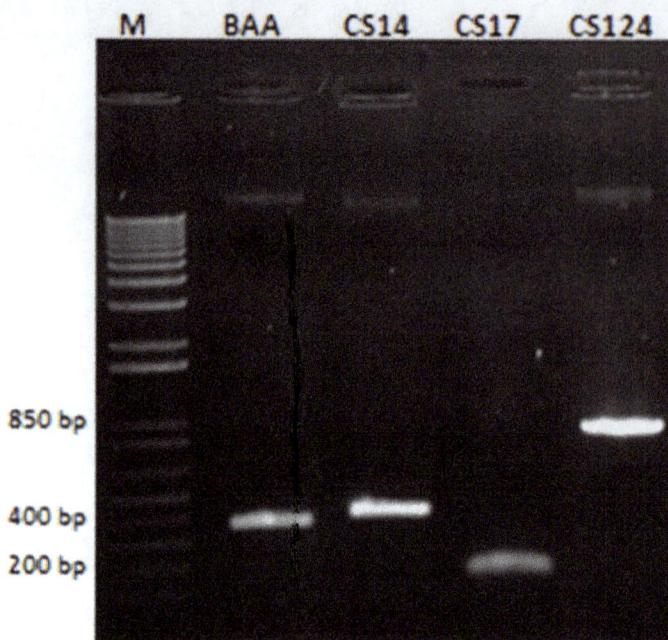

Figure 5. Gel electrophoresis showing the O-antigen binding patterns for all C. sakazakii isolates. Bands of different sizes define dstinct serotypes. Lane 1, BAA 894; lane 2, CS 14; lane 3, CS 17; lane 4, CS 124. *C. sakazakii* O:1 (364 bp); *C. sakazakii* O:2 (152 bp); *C. sakazakii* O:4 (840 bp); *C. malonaticus* O:2 (424 bp)

4. Discussion

The risk of *Cronobacter* infection to neonates and immunocompromised individuals is very high. The use of correct methods for identification of this bacterium will provide accurate results on the contamination of *Cronobacter* in food products and help in understanding the epidemiology of infections.

MicroScan WalkAway analysis (Beckman Coulter) is designed for use in determining antimicrobial agent susceptibility and/or identification of an organism to the species level. All the isolates were identified by the MicroScan as *Enterobacter sakazakii* (98.87% probability). *Cronobacter sakazakii* was formerly referred as *E. sakazakii,* and most likely the MicroScan WalkAway system database used in this study needs to be updated with the correct *Cronobacter* taxonomy.

Matrix Assisted Laser Desorption Ionization Time-of-Flight analysis (VITEK®MS MALDI-TOF) (bioMerieux) is a rapid and cost-effective system that is replacing conventional phenotypic methods for routine identification of bacteria. Microbial identification is based on the comparison of a protein spectrum generated from intact whole bacterial cells to a database of species-specific reference protein profiles using a particular algorithm (Dubois et al., 2012). All three of the isolates were identified by the MALDI-TOF system as *C. sakazakii* but CS14 was later identified as *C. malonaticus* using the *cgcA* and *rpoB* species-specific PCR assays. *C. sakazakii* and *C. malonaticus* are very closely related with 99% gene sequence similarity (Li et al., 2012). This could be the reason why CS14 was identified as *C. sakazakii* instead of *C. malonaticus* by the MALDI-TOF system. Jamal et al. (2014) reported 99.9% (n= 806) correct identification by VITEK®MS MALDI-TOF to the genus level and 99.0% to the species level of 507 Gram negative bacilli, 16 Gram negative cocci, 267 Gram positive cocci and 16 Gram positive bacilli, made up of 39 genera and 70 species.

The 16S rDNA gene is widely used as a phylogenetic target as it is a highly conserved gene, ubiquitous in all organisms and contains variable and hypervariable sequence regions (Kuhnert et al., 1996). A BLAST comparison of the 16S rDNA nucleotide sequences of the isolates showed a sequence similarity of about 97% to 98% with *Cronobacter* species. The phylogenetic tree generated with the 16S rDNA sequences showed all of the isolates grouping within *Cronobacter* species clusters. The 16S rDNA gene sequence analysis however, has limitations for discrimination between very closely related organisms such as *C. sakazakii* and *C. malonaticus* because of minimal sequence diversity or the presence of multiple copies of 16S rDNA gene loci (Carter et al., 2013) CS 14 which was previously characterized as *C. malonaticus* using the *cgcA* and *rpoB* species-specific PCR assays was closely clustered to *C. sakazakii* ATCC BAA 894. This is highly misleading, thus showing the limitation of the 16s rDNA gene.

The *rpoB* gene sequence analysis has been proposed as a method for inferring relationships among very closely related species (Adekambi et al., 2009). Li et al. (2012) reported that phylogenetic analysis based on partial *rpoB* gene sequence analysis cannot distinguish between *C. sakazakii* and *C. malonaticus* even though it can differentiate these two species from other *Cronobacter* species. All isolates including BAA 894 showed similar band patterns using the *C. sakazakii rpo*B primers while CS 14 and CS 124 showed a similar pattern using the *C. malonaticus rpo*B primers. This result confirms the reported by Li et al. (2012) because CS 124 was identified as *C. sakazakii* using the multiplex *cgc*A primers.

Cyclic diguanylate (c-di-GMP) is a bacterial second messenger signal transduction molecule recognized for its involvement in the regulation of a number of complex physiological processes, including bacterial virulence, biofilm formation, and persistence (long-term survival) (Sondermann et al., 2012). Carter *et al.*, (2013) reported the use of *Cronobacter* multiplex *cgc*A PCR assay to identify *Cronobacter* strains in a single reaction. This PCR assay was found to be 100% specific (n=305) and 100% sensitive (n=20). The multiplex *cgc*A primers were used to identify sample CS14 as *C. malonaticus,* and CS17 and CS124 as *C. sakazakii.* The control strain BAA 894 was correctly identified as *C. sakazakii.*

Applying discriminatory molecular subtyping methods to characterize foodborne pathogens facilitates the detection of outbreaks, sources of infection, and transmission pathways (Fields et al., 2011). Epidemiologically related isolates share the same profile (Kuhnert et al., 1996). All *Cronobacter* isolates and the control strain BAA 894 demonstrated distinct PFGE profiles, indicating different sources.

The set of seven housekeeping genes in MLST has greater sequence diversity than the 16S rRNA gene and has been applied to identify many bacteria (Urwin & Maiden, 2003). New sequence types were created for each of the isolates based on the allelic combination of the seven loci. None of the *fus*A profiles were shared between any of the isolates, thus proving that all the isolates are different. This locus has been recommended for use with two PCR primer sets to define species of *Cronobacter* without the ambiguity of 16S rRNA gene sequence analysis. MLST has proven to be an effective and robust typing scheme for the *Cronobacter* genus and has exhibited a high level of discrimination between the isolates (Joseph et al., 2012). The new sequence types (303, 304 and 296) are not clustered with any other known *Cronobacter* pathovars but further work will be needed to analyze the genes that are present in these strains.

The O-antigen is a highly variable component of the lipo-polysaccharide of Gram-negative bacteria and is used for the development of both serological and molecular typing methods (Jarvis et al., 2013). *C. sakazakii* O-antigen gene clusters of all seven serotypes are located on the chromosome between the housekeeping genes *galF* and *gnd*. All the isolates in this study proved to be of different serogroups. Sun et al. (2012) reported the high-level identity (99.3%) of *C. sakazakii* O6 and *C. malonaticus* O2, implying the recent lateral transfer of the respective O-antigen gene cluster between these two species. However, Yan *et al.* (2015) reported that the strains originally used to design the Sun-based serotype primers may have been misidentified as *C. sakazakii.*, which subsequently led to an incorrect identification of the corresponding serotypes. So, *C. sakazakii* O6 serotype should be corrected to *C. malonaticus* O2. Interestingly, *C. sakazakii* O2, to which CS 17 belongs, was isolated between 2010 and 2011 from infant clinical cases of *Cronobacter* infections in the USA (FDA/CFSAN, 2011).

5. Conclusion

The emerging pathogenic genus *Cronobacter* was identified as a contaminant in powdered infant formula retailed in Nigeria. The use of PCR assays, biochemical identification tests, and gene sequenced-based methods gave a reliable identification and profiling of all the *Cronobacter* isolates thus limiting the misidentification of false positive and negative results. Our results suggest that more severe measures must to be taken in order to improve the quality control on powdered infant formula production in order to protect neonates and infants from diseases caused by *Cronobacter* spp., which sometimes can be fatal.

Acknowledgements

We thank Kavindra Singh, Karen Jacques-Palaz, Jung Roh, Isabel and other members of the BEM Lab, Dr. Caesar Arias and other members of the ARIAS Lab, University of Texas Medical School for the assistance rendered in the course of this research. We thank Dr. Randall J. Olsen, M.D. for his assistance with the MALDI-TOF analysis. We also thank Dr. Audrey Wanger for her assistance with the Microscan Walkaway.

References

Adekambi, T. Drancourt, M., & Raoult, D. (2009). The *rpo*B gene as a tool for clinical microbiologists. *Trends in Microbiology, 17*, 37-45.

Alsonosi, A., Hariri, S., Kajsik, M., Orieskova, M., Hanulik, V., Roderova, M., ... Holy, O. (2015) The speciation and genotyping of *Cronobacter* isolates from hospitalized patients. *European Journal of Clinical Microbiololgy of Infectious Diseases, 34*(10), 1979-88.

Baldwin, A., Loughlin, M., Caubilla-Barron, J., Kucerova, E., Manning, G., Dowson, C., & Forsythe, S. (2009) Multilocus sequence typing of *Cronobacter sakazakii* and *Cronobacter malonaticus* reveals stable clonal structures with clinical significance which do not correlate with biotypes. *BMC Microbiololgy, 9*, 223-231.

Brengi, S. P., O'Brien, S. B., Pichel, M., Iversen, C., Arduino, M., Binsztein, N., ... Fanning, S. (2012) Development and validation of a PulseNet standardized protocol for subtyping isolates of *Cronobacter* species. *Foodborne Pathogens and Diseases, 9861-867.*

Carter, L., Lindsey, L. A., Grim, C. J., Sathyamoorthy, V., Jarvis, K. G., Gopinath, G., ... Hu, L. (2013) Multiplex PCR assay targeting a diguanylate cyclase-encoding gene, cgcA, to differentiate species within the genus *Cronobacter*. *Applied and Environmental Microbiology, 79*, 734-737.

Dubois, D., Grare, M., Prere, M., Segonds, C., Marty, N., & Oswald, E. (2012). Performances of Vitek MS Matric-Assisted laser desorption ionization-time of flight mass spectrometry system for rapid identification of bacteria in routine clinical microbiology. *Journal of Clinical Microbiology, 50*(8), 2568-2576.

FAO/WHO. (2008). *Enterobacter sakazakii (Cronobacter* spp.) in powdered follow-up formulae. Microbiological risk assessment series No. 15. Food and Agriculture Organization of the United Nations, World Health Organisation. Retrieved September 3, 2016, from http://www.who.int/foodsafety/publications/mra_followup/en/

Fields, P. I., Fitzgerald, C., & McQuiston, J. R. (2011). Fast and high-throughput moleular typing methods. In J. Hoorfar (ed.), *Rapid Detection, Identification and Quantification of Foodborne pathogens*. Washington, DC: ASM Press.

Grim, C. J., Gopinath, G. R., Mammel, M. K., Sathyamoorthy, V., Trach, L. H., Chase, H. R., ... Stephan, R. (2013) Genome sequences of an *Enterobacter helveticus* strain, 1159/04 (LMG 23733), isolated from fruit powder. *Genome Announcements, 1*, e01038–13.

Holy, O., Petrzelova, J., Hanulik, V., Chroma, M., Matouskova, I., & Forsythe, S. J. (2014). Epidemiology of *Cronobacter* spp. isolates from patients admitted to the Olomuoc University Hospital (Czech Republic). *Epidemiology, Microbiology, Immunology, 63*, 69-72.

Iversen, C., Druggan, P., & Forsythe, S. J. (2004). A selective differential medium for *Enterobacter sakazakii*. *International Journal of Food Microbiology, 96*,133-139.

Jackson, E. E., Sonbol, H., Masood, N., & Forsythe, S. J. (2014). Genotypic and phenotypic characteristics of *Cronobacter* species, with particular attention to the newly reclassified species *Cronobacter helveticus*, *Cronobacter pulveris*, and *Cronobacter zurichensis*. *Food Microbiology, 44*, 226-235.

Jamal, W., Albert, M. J., & Rotimi, V. (2014). Real-time comparative evaluation of bioMerieux VITEK MS versus Brucker Microflex MS, two matrix assisted laser desorption-ionization time-of-flight mass spectrometry systems, for identification of clinically significant bacteria. *BMC Microbiology, 14*, 289-297.

Jarvis, K. G., Yan, Q. Q., Grim, C. J., Power, K. A., Franco, A. A., Hu, L., ... Tall, B. D. (2013). Identification and Characterization of five new molecular serogroups of *Cronobacter* spp. *Foodborne Pathogens and Disease, 10*(4), 343-52.

Joseph, S., Sonbol, H., Hariri, S., Desai, P., McClelland, & Forsythe, S. J. (2012). Diversity of the *Cronobacter* Genus as Revealed by Multilocus Sequence Typing. *Journal of Clinical Microbiology, 50*(9), 3031-3039.

Kuhnert, P., Capaul, S. E., Nicolet, J., & Frey, J. (1996). Phylogenetic positions of *Clostridium chauvoei* and *Clostridium septicum* based on 16S rRNA gene sequences. *International Journal of Systemic Bacteriology, 46*(4), 1174-1176.

Lehner, A., Fricker-Feer, C., & Stephan, R. (2012). Identification of the recently described *Cronobacter condimenti* by an rpoB-gene-based PCR system. *Journal of Medical Microbiology, 61,* 1034-1035.

Li, Y., Cao, l., Zhao, J. Cheng, Q., Lu, F., Bie, X., & Lu, Z. (2012) Use of *rpo*B gene sequence analysis for phylogenetic identification of *Cronobacter* species. *Journal of Microbiological Methods, 88,* 316-318.

Osaili, T., & Forsythe, S. (2009). Desiccation resistance and persistence of *Cronobacter* species in infant formula. *International Journal of Food Microbiology, 136*(2), 214-20. http://dx.doi.org/10.1016/j.ijfoodmicro.2009.08.006

Ribot, E. M., Fair, M. A., Gautom, R., Cameron, D. N., Hunter, S. B., Swaminathan, B., & Barret, T. J. (2006). Standardization of Pulsed-Field Gel Electrophoresis Protocols for the subtyping of *Escherichia coli* O157:H7, *Salmonella,* and *Shigella* for PulseNet. *Foodborne Pathogens and Disease, 3*(1), 59-67.

Sondermann, H., Shikuma, N. J., Yildiz, F. H. (2012). You've come a long way: c-di-GMP signaling. *Current Opinion in Microbiology, 15*(2), 140-146.

Stoop, B., Lehner, A., Iversen, C., Fanning, S., & Stephan, R. (2009). Development and evaluation of rpoB based PCR systems to differentiate the six proposed species within the genus *Cronobacter. International Journal of Food Microbiology, 136,* 165-168.

Sun, Y., Wang, M., Wang, Q., Cao, B., He, X., Li, K., ... Wang, L. (2012). Genetic analysis of the *Cronobacter sakazakii* O4 to O7 O-antigen gene clusters and development of a PCR assay for identification of all *C. sakazakii* serotypes. *Applied and Environmental Microbiology, 78*(11), 3966-3974.

Urwin, R., & Maiden, M. C. (2003). Multi-locus sequence typing: a tool for global epidemiology. *Trends in Microbiology, 11,* 479-487.

Yan, Q, Jarvis, G. K., Chase, R. H., Hebert, K., Trach, L. H., Lee, C., ... Tall, D. B. (2015). A proposed harmonized LPS molecular-subtyping scheme for *Cronobacter* species. *Food Microbiology, 50,* 38-43.

Appendix A

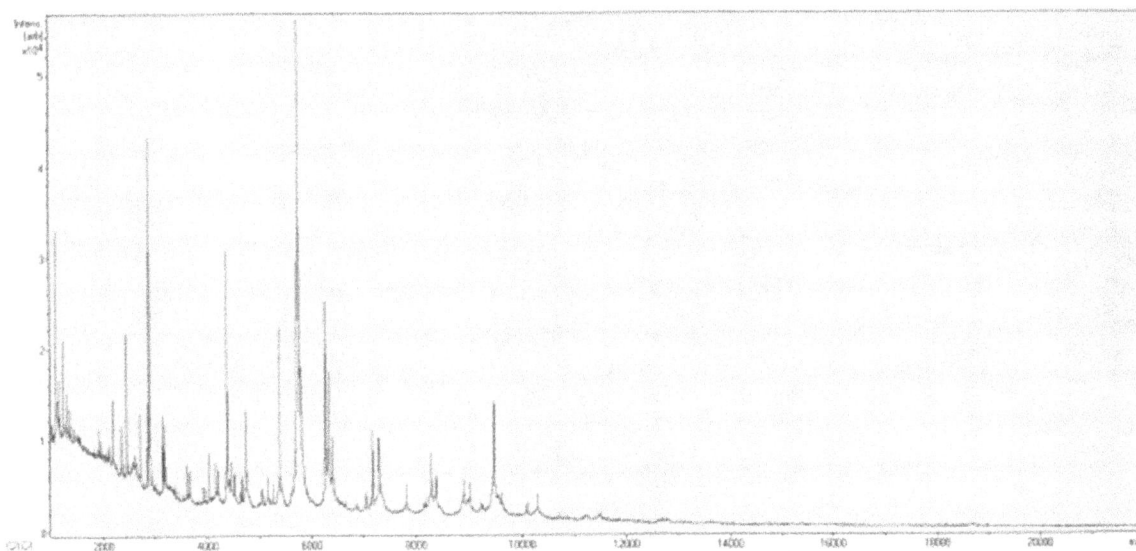

Figure A1. Mass Spectrometry of VITEK MALDI-TOF for CS 17

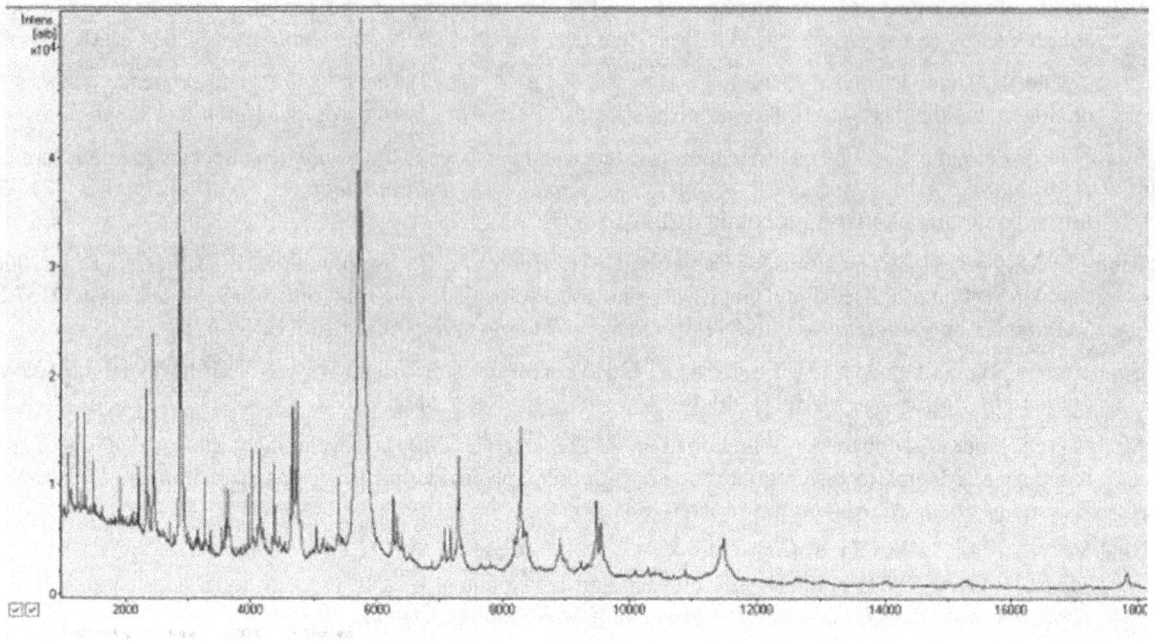

Figure A2. Mass Spectrometry of VITEK MALDI-TOF for CS 14

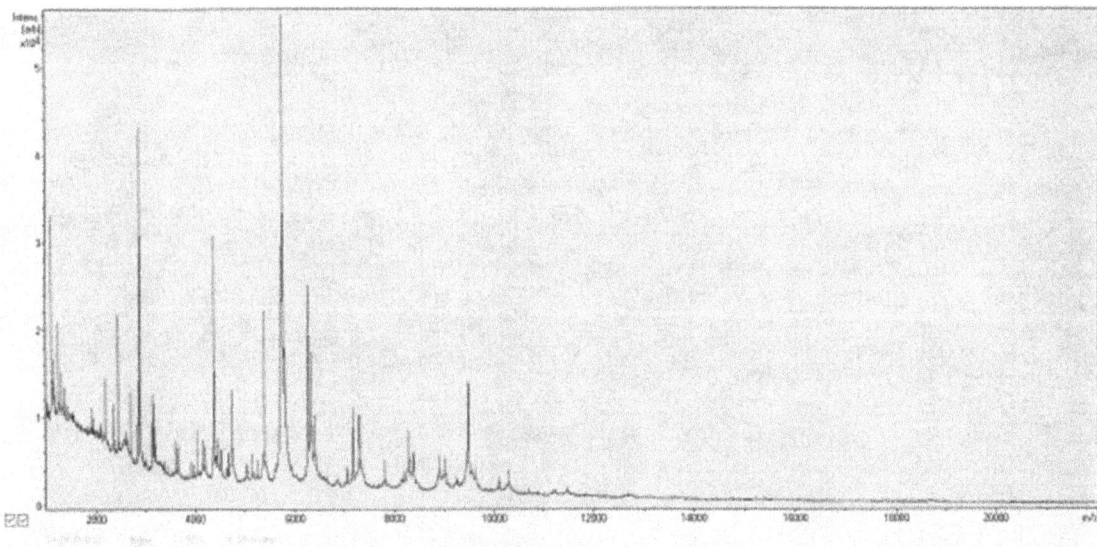

Figure A3. Mass Spectrometry of VITEK MALDI-TOF for CS 124

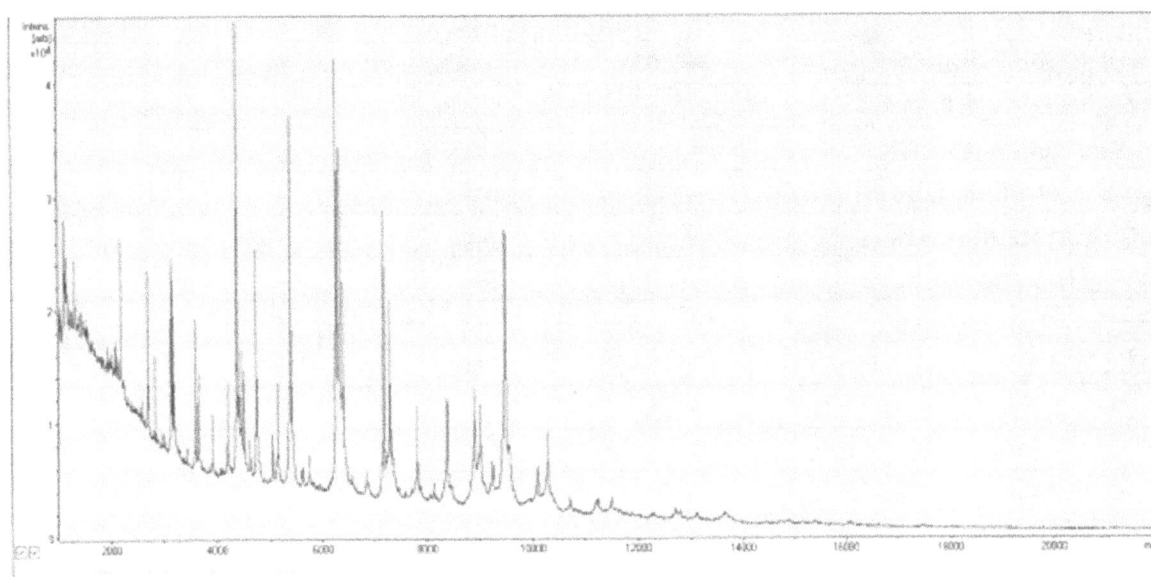

Fig A4. Mass Spectrometry of VITEK MALDI-TOF for BAA 894

Table A1. Biochemical characterization of isolates based on Microscan walkaway analysis.

Biochemicals	BAA 894	CS 14	CS 17	CS 124
Gram Reaction	-	-	-	-
Glucose-GLU	+	+	+	+
Sucrose-SUC	+	+	+	+
Sorbitol-SOR	-	-	-	-
Raffinose-RAF	+	+	+	+
Rhamnose-RHA	-	-	-	-
Arabinose-ARA	+	+	+	+
Inositol-INO	+	+	+	+
Adonitol-ADO	-	-	-	-
Mellibiose-MEL	+	+	+	+
Urea-URE	-	-	-	-
H_2S	-	-	-	-
Indole-IND	-	-	-	-
Lysine-LYS	-	-	-	-
ArginineARG	+	+	+	+
Ornithine-ORN	+	+	+	+
Tryptophan Deaminase-TDA	-	-	-	-
Esculin Hydrolysis-ESC	+	+	+	+
Vogues-Proskauer-VP	+	+	+	+
Citrate-CIT	+	+	+	+
Malonate-MAL	-	+	-	-
Galactosidase-ONPG	+	+	+	+

Colistin-Cl$_4$	-	-	-	-
Cephalothin-CF$_8$	+	+	+	+
Oxidase-OXI	-	-	-	-
Acetamide-ACE	-	-	-	-
Cetrimide-CET	-	-	-	-
Nitrofurantoin-Fd$_{64}$	-	-	-	-
Kanamycin-K$_4$	-	-	-	-
Nitrate-NIT	+	+	+	+
Oxidation-Fermentation-OF/G	+	+	+	+
Penicillin-P$_4$	+	+	+	+
Tartrate-TAR	-	-	-	-
Tobramycin-TO$_4$	-	-	-	-
CAT	+	+	+	+
Identification	*E. sakazakii*	*E. sakazakii*	*E. sakazakii*	*E. sakazakii*

17

Teff-Based Complementary Foods Fortified with Soybean and Orange-Fleshed Sweet Potato

Mesfin W. Tenagashaw[1, 2], Glaston M. Kenji[1], Eneyew T. Melaku[3], Susanne Huyskens-Keil[4] & John N. Kinyuru[1]

[1]Department of Food Science and Technology, Jomo Kenyatta University of Agriculture and Technology, Nairobi, Kenya

[2]Faculty of Chemical and Food Engineering, Bahir Dar Institute of Technology, Bahir Dar University, Bahir Dar, Ethiopia

[3]Department of Food Science and Applied Nutrition, Addis Ababa Science and Technology University, Addis Ababa, Ethiopia

[4]Division Urban Plant Ecophysiology, Research Group Quality Dynamics/Postharvest Physiology, Faculty of Life Sciences, Humboldt-Universitätzu Berlin, Berlin, Germany

Correspondence: Mesfin W. Tenagashaw, Faculty of Chemical and Food Engineering, Bahir Dar Institute of Technology, Bahir Dar University, P.O. Box: 26, Bahir Dar, Ethiopia.
E-mail: mesfinwogayehu@gmail.com

Abstract

The macronutrient composition of teff-based complementary foods (ComFs) prepared through extrusion cooking and a combination of household-level strategies were evaluated. In extrusion, teff, soybean and orange-fleshed sweet potato were separately processed into their respective flours and composited in a percentage ratio of 70:20:10, respectively. It was then extruded into a complementary food (ComF1). In the case of household-level methods, portions of teff grains separately germinated for 24 and 48 h were dried and ground to fine flours. Similarly, small portions of soybean grains were separately blanched and roasted; then each were ground to fine flour. Four ComFs (ComF2, ComF3, ComF4, ComF5) were developed by blending flours of ungerminated teff, germinated teff, blanched or roasted soybean and sweet potato using the 70:20:10 ratio. The extrusion cooked ComF had significantly ($p < 0.05$) high protein (17.92 g/100 g) while the household-level ComFs had lower protein contents. Energy content of the ComFs ranged from 391.63 to 400.60 kcal/100 g. All ComFs met the requirements of protein and energy for 6 to 8 month-old infants. There was no significant difference in the dietary fiber contents of the developed complementary foods despite increased values of insoluble dietary fiber due to germination of teff and blanching or roasting of soybean. The study revealed the potential of developing complementary foods from teff-soybean-sweet potato blends with improved protein and energy contents.

Keywords: teff, soybean, sweet potato, extrusion cooking, germination, complementary food

1. Introduction

Protein-energy malnutrition (PEM), the silent emergency (Anuonye, Onuh, Egwim, & Adeyemo, 2010), coupled with other problems such as lack of proper hygiene, improper food preparation and storage practices, infectious diseases and dietary taboos, have continued tempting mankind, especially in the Sub-Saharan Africa, the world's most food-insecure region (Makeri, Bala, & Kassum, 2011). The problem is even worth among infants and children. According to a recent report, globally 50 million infants were wasted in 2014 (WHO, UNICEF, & WORLD BANK GROUP, 2015). The problem of PEM begins when a child is introduced to additional foods known as complementary foods. Infants are usually introduced to such transitional foods around the age of six month. According to the FAO/WHO category, those infants from 6 to 8 months of age are very vulnerable to malnutrition (Dewey & Brown, 2003; WHO/UNICEF, 1998) as the vital nutrients are insufficient to maintain the fast growth and development (Konyole, Kinyuru, Owuor, Kenji, & Onyango, 2012). At the same time, their biological systems didn't grow fully yet to digest and utilize the foods they receive. Therefore, a special attention is required to the nutritional and functional properties of the complementary foods to be fed to this group of infants.

Owing to economic and to some extent dietary taboos, the majority of the poor community in developing countries cannot afford commercially processed and/or fortified complementary foods or animal origin foods to feed their infants. Thus, they usually depend on the locally-available crops (usually cereals) that are not supplemented with legumes and/ or tubers and thus lack the required level of macro- and micro-nutrients. Cereals are a good source of the essential amino acids, methionine and cysteine, and also B-complex vitamins but limiting in lysine. On the other hand, most legumes are rich in lysine but low in sulfur-containing amino acids. Thus, compositing of cereals and legumes results in a good complementarity of a number of nutrients (Mensa-Wilmot, Phillips, Lee, & Eitenmiller, 2003). Moreover, starchy foods usually produce gruels/porridges of high bulk and not of the desirable (easy-to-swallow) semi-liquid consistency for infants (Nout, unpublished document). Thus, the porridge is usually diluted with more water for easy feeding of infants which in turn results in reduced nutrient and energy density, a problem referred to as nutrient thinning, a cause for poor infant growth and development (Tumwebaze, Gichuhi, Rangari, Tcherbi-Narieh, & Bovell-Benjamin, 2015). Thus, cereals and legumes cannot effectively be used in complementary food preparation unless modified to some extent. In fact, the use of locally available crops for processing complementary foods for low-income communities through the application of household-level methods has been promoted for decades (Castell-Perez, Griffith, Castell-Perez, & Griffith, 1998; Elias, 1974; Molina, Braham, & Bressani, 1983).Therefore, infant foods need proper formulation and processing to improve their nutritional values and consistency as well.

Keeping these facts into account, complementary foods were formulated and processed from a composite of teff, soybean and sweet potato (orange-fleshed) through an industrial-level approach called extrusion cooking, the most common and novel food processing technique (Ramachandra & Thejaswini, 2015), and a household-level approach referred to as combined strategies, a combination of traditional food processing practices (Hotz & Gibson, 2007), such as soaking and germination [teff], blanching or roasting and dehulling [soybean], peeling, slicing, blanching [sweet potato], followed by drying and milling of each ingredient and then blending them into a composite flour. The aim of the study was, therefore, to develop complementary foods with improved macronutrient composition and energy content as well.

2. Materials and Methods

2.1 Raw Materials Used for Processing the Complementary Foods

Red teff [*Eragrostis tef (Zucc.)* Trotter], soybean [*Glycine max (L.) Merr.*] and orange-fleshed sweet potato [*Ipomoea batatas, (L.) Lam*], which will simply be referred to as sweet potato hereafter, were purchased and collected from a local farmers' market, Pawe Agricultural Research Centre and Awassa Agricultural Research Centre, respectively, all from Ethiopia.

2.2 Processing Approaches of the Complementary Foods

Two processing approaches were employed to prepare the complementary foods: Industrial-level approach (extrusion cooking) and household-level approach (combined household strategies).

2.2.1 Industrial-level Approach: Extrusion Cooking

Before the actual extrusion process was carried out, the raw materials were separately prepared. Teff grain samples were cleaned for some extraneous materials and then washed with tap water until all soil and other undesirable components were removed completely. This was followed by drying the grains in the sun. Similarly, soybean grain samples were manually cleaned, sun dried and dehulled using a laboratory-scale dehulling mill (Alvan Blanch, England). The hulls and undehulled grains were removed by manual sieving. The cleaned teff and soybean grains were ground to fine flour using a local stone mill. Sweet potato tubers were processed according to the method described by Haile, Admassu, & Fisseha (2015). They were manually washed with tap water, peeled with kitchen knives and sliced into 2 mm thick slices with a vegetable slicer (model CL30, Robot coupe, Vincennes, France). This was followed by blanching with water at 60 °C for 5 min in a water bath. Finally, the slices were spread on racks and dried in a solar drier (Alvan Blanch, Wiltshire, England). The dried potato slices were ground using a laboratory-scale mill (Zhejiang Top Instrument Co., Ltd., China). All the flour samples were packaged in polyethylene bags, labelled and stored in a cold room until the extrusion process was carried out.

The composite flour (described in Section 2.3), the mixture of the three raw materials - teff, soybean and sweet potato - was subjected to extrusion cooking using a pilot-scale twin-screw extruder (model Clextral, BC-21 No. 124, Clextral, Firminy, France). The barrel of the extruder has three 100 mm long temperature zones fitted with 25 mm diameter screws. The temperatures of the last two zones were regulated by electrical heating (thermocouples) and a water cooling system. The extrusion temperature is controlled by a Eurotherm controller (Eurotherm Ltd., Worthing, U.K.). The composite flour was manually transferred to the feed hopper and then automatically conveyed into the extruder inlet by a twin-screw volumetric feeder (type K-MV-KT20). Tap water was injected into the extruder close to the feed section of the extruder barrel via an inlet port by a metering pump (DKM-Clextral) where the volume of the water is read in units of stroke. A die plate with four circular openings each with a diameter of 3 mm was used. Both the feeder and the pump were calibrated prior to running the extrusion to determine the set points required for desired flow rates of flour and water, respectively. The extrusion operating conditions were moisture content (22%), barrel temperature (120 °C) and screw speed (200 rpm) selected based on recommended values in the literature and a preliminary experiment. The feed rate was maintained constant at 64 g/min. Three extrusion runs were performed. Finally, the extrudates were ground using a laboratory-scale mill (Type ZM 100, F. Kurt Retsch GmbH & Co.KG, Germany) fitted with a 0.5 mm sieve. The product was a pre-cooked complementary food (ComF1). It was stored in a cold room until laboratory analysis was carried out.

2.2.2 Household-level Approach: Combined Strategies

Teff grains were germinated following the method described by Badau, JideanI, & Nkama (2006) for finger millets with some modifications. The grains were cleaned, washed and rinsed first in distilled water and then in 5% (w/v) sodium chloride solution to disinfect them. The grains were then soaked in distilled water for 12 h using a ratio of 1:3 (grain:water; w/v) in a small plastic bucket. The soaking water was changed in four hours interval. After soaking, the water was drained off and the grains separated into two portions, spread on and covered with a moistened cotton cloth and allowed to germinate at room temperature (25 ± 2 °C) for two durations, 24 h and 48 h. At the end of the germination period, the grains were dried in a drying oven (model DHG-9140) at 70 °C for 8 h. The vegetative parts were discarded by gentle abrasion between hands followed by manual sieving. Soybean grains were processed using two different methods: blanching and roasting. Blanching was conducted according to the method described by Iombor, Umoh, & Olakumi (2009). The grains were cleaned and then blanched in a boiling pot (model RE300B, Bibby Sterlin Ltd, UK) at 100 °C for 10 min (1:5 ratio of seeds to boiling water). The blanched grains were drained, cooled to room temperature, manually dehulled and rinsed to remove the seed coat. The rinsed seeds were then oven dried at 80 °C for 5 h. Roasting was carried out in a mini baking oven according to the WFP procedure (WFP, 2004), at 170 °C for 15 min, used in the production of Super Cereal (corn soya blend). The roasted grains were then dehulled using a lab-scale decorticator (Alvan Blanch, England) and the hulls manually removed. The clean teff and soybean grains were finally ground to fine flour using a laboratory-scale mill (Type ZM 100, F. Kurt Retsch GmbH & Co.KG, Germany). Sweet potato tubers were processed the same way as done for extrusion cooking (Section 2.2.1). Some part of the flour samples were blended to get the final complementary foods (Section 2.3) while a small portion of each of them (germinated teff, blanched soybean, roasted soybean, processed sweet potato) were left for analysis of their composition. All samples were packaged in ziplock polyethylene bags, labelled and stored in a cold room (10 °C) until laboratory analysis.

2.3 Formulation of the Complementary Foods

The formulation of the complementary foods was performed using previously published nutrient compositions of the respective raw materials (teff, soybean, sweet potato) so as to approximately meet recommended levels of some vital nutrients (protein, fat, vitamin A) and energy that are suggested in some guidelines for complementary foods for infants and children (Codex Alimentarius Commission, 1991; WFP, 2014). A computer programme known as NutriSurvey for Windows (ProNut-HIV, 2005), was used to approximate the above-mentioned nutrients. The recommendation to use a blend formulation of 75% cereal and 25% legume in the formulation of infant complementary foods (Gopaldas, 1991; Plahar, Okezie, & Gyato, 2003) was taken into account at the beginning of the formulation. The addition of the sweet potato (10%) was meant to increase the vitamin A level of the complementary foods as it is a rich source of β-carotene (precursor of vitamin A) (Amagloh et al., 2012). Accordingly, the percentage proportion of 70:20:10 for teff, soybean and sweet potato, respectively, was found to closely meet the recommendations and thus employed in the formulation of the complementary foods (refer to **Table 1**).

Table 1. Ingredients used in the formulations of the complementary foods with their proportions

ComF Formulation	Processing method	Ingredients	Proportion (%)
ComF1	Extrusion cooking	Ungerminated teff	70
		Unprocessed soybean	20
		Processed sweet potato	10
ComF2	Combined strategies	Ungerminated teff	60
		Germinated teff - 24 h	10
		Blanched soybean	20
		Processed sweet potato	10
ComF3	Combined strategies	Ungerminated teff	60
		Germinated teff - 24 h	10
		Roasted soybean	20
		Processed sweet potato	10
ComF4	Combined strategies	Ungerminated teff	60
		Germinated teff - 48 h	10
		Blanched soybean	20
		Processed sweet potato	10
ComF5	Combined strategies	Ungerminated teff	60
		Germinated teff - 48 h	10
		Roasted soybean	20
		Processed sweet potato	10

ComF— Complementary Food

In case of the industrial-level approach (extrusion cooking), the flours from teff, soybean and sweet potato (Section 2.2.1) were mixed to obtain a composite flour using the above-mentioned proportion, 70:20:10, respectively. This composite flour was finally extruded into ComF1. Some part of each flour was kept for analysis of the ingredients themselves.

Meanwhile, the flour samples prepared using the combined household-level methods (germinated teff, blanched soybean, roasted soybean, processed sweet potato) and also ungerminated teff were blended to produce complementary foods using the same proportion (70:20:10) employed in extrusion cooking. However, in this approach, 10% of the germinated teff flour was used thus reducing the ungerminated teff to 60%. This was based on the fact that addition of a small quantity of germinated flour (amylase-rich flour) reduces the dietary bulk (gruel viscosity) and increases energy density of a complementary food (Gopaldas, Mehta, Patil, & Gandhi, 1986; Hossain, Wahed, & Ahmed, 2015). Moreover, the fact that germination leads to a significant loss of dry matter due to oxidation and leaching, was taken into account (Egli, 2001). In this case, four household-level complementary foods (ComF2, ComF3, ComF4 and ComF5) were developed through blending of the ingredients mentioned above. The mixing was carried out first manually with the help of a ladle in a plastic bowel and then using a laboratory-scale mixer (model B20-B, H.L Universal Mixing Machine, China) for 15 min (at position 2) to ensure homogeneity of the blends.

2.5 Laboratory Analysis

The macronutrient composition of the flour samples of each ingredient and those of the developed complementary foods were analyzed according to AOAC International standard methods (AOAC International, 2000). Moisture content was determined by the air-oven method using a hot-air circulating oven (method #925.09). Ash content was determined through incineration (550 °C) of a known weight of sample in a muffle furnace (method #923.03). Crude protein was determined by the micro-Kjeldahl method (method #979.09). The crude protein was obtained by multiplying the corresponding total nitrogen content by a factor of 6.25 (FAO, 2003). Crude fat was determined by extracting a known weight of sample in petroleum ether (boiling point, 40 to 60 °C) in a Soxhlet extractor (method #930.09). Crude fiber content was determined following method #962.09. Available carbohydrate was obtained by difference while energy was calculated using the Atwater's calorie conversion factors: 4 kcal/g for crude protein, 9 kcal/g for crude fat and 4 kcal/g for available carbohydrate (FAO, 2003). The dietary fiber contents (soluble, insoluble and total) were measured using the enzymatic-gravimetric method as described in the AOAC method #991.43 and AACC method #32-05-01 as described in the Megazyme Assay Procedure: K-TDFR 06/14 (Megazyme International, 2013). The water-soluble and water-insoluble fractions were determined through digestion with enzymes. A 0.5 g of the sample was weighed followed by addition of 50 ml of phosphate buffer. Then enzymatic hydrolysis of starch and protein was carried out through a sequential addition of three enzymes: α-amylase (50 μl, 95 °C, 30 min, pH 6.0), protease (100 μl, 60 °C, 30 min, pH 7.5) and amyloglucosidase (200 μl, 60 °C, 30 min, pH 4.5). Insoluble dietary fiber was recovered by filtration using celite as the filter aid. Soluble dietary fiber was then precipitated from the filtrate with 15 ml each of 78% ethanol, 96% ethanol and acetone. The fiber values were corrected for indigestible protein (N x 6.25) through determination by the Kjeldahl method and ash which was determined by ignition at 525 °C for 8 h. The total dietary fiber was calculated as the sum of the soluble and insoluble fractions.

2.6 Statistical Analysis

One-way analysis of variance (ANOVA) was applied to the data by the use of IBM SPSS for Windows (version 21). The significant difference was analyzed by Tukey's HSD (honest significant difference) test. The level of significance was set at 5% probability level.

3. Results and Discussion

The results for macronutrient composition of the ingredients used for formulating the complementary foods are shown in **Table 2**. The moisture and ash contents were in the range of 2.51 to 7.36 and 2.12 to 5.03 g/100 g, respectively. The ash content in a given food sample indicates the level of minerals present (Kavitha & Parimalavalli, 2014). It was affected by the processes employed. Germination of teff, both for 24 and 48 h, had significantly ($p < 0.05$) decreased the ash content. As it was found out in an earlier research, this decrease might be due to leaching of minerals during washing and soaking of the grains (Inyang & Zakari, 2008). However, ash content of soybean significantly increased due to blanching (Ugwuona, Awogbenja, & Ogara, 2012) and roasting (Kavitha & Parimalavalli, 2014) which could be due to volatilization of the organic matter. As expected, significantly high crude protein ($p < 0.05$) was found for the unprocessed soybean (34.57 g/100 g), very close to that reported by Anuonye et al. (2010). The blanched or roasted soybean also gave slightly higher protein contents of 35.59 and 36.22 g/100 g, respectively. The lowest value of protein was that of the sweet potato (5.70 g/100 g) which is nearly similar to what was reported by Omodamiro et al. (2013) for some varieties of sweet potato. A similar trend was also observed for crude fat. The highest crude fat (22.99 g/100 g) was found in unprocessed soybean which agrees with the available literature (Anuonye et al., 2010) whereas the lowest was recorded for sweet potato (0.84 g/100 g). Germination of teff, both for 24 and 48 h, had significantly ($p < 0.05$) decreased the crude fat contents. Similar results were found by Megat Rusydi, Noraliza, Azrina, & Zulkhairi (2011) for some legumes and rice varieties. The decrease in fat content might be due to the increased activities of the lipolytic enzymes which hydrolyze fats to fatty acids and glycerol during germination (Kavitha & Parimalavalli, 2014).The fatty acids are further oxidized to carbon dioxide and water to generate energy for the germination process (Hahm, Park, & Martin Lo, 2009). Blanching or roasting of soybean had also significantly ($p < 0.05$) reduced fat contents which could be due to loss of some fat during the heating process; especially, the reduction caused by roasting was very high.

Table 2. Macronutrient composition of the ingredients used in the formulations of complementary foods

ComF Ingredients		Macronutrient Composition (g/100g)					
		Moisture	Ash	Crude Protein	Crude Fat	Crude Fiber	Available CHO
Teff	Ungerminated	6.35 ± 0.12^d	3.28 ± 0.03^{bc}	7.80 ± 0.05^{ab}	5.71 ± 0.84^c	3.87 ± 0.27^a	79.34 ± 0.85^c
	Germinated-24 h	4.87 ± 0.69^c	2.12 ± 0.25^a	10.40 ± 0.76^b	2.72 ± 0.28^{ab}	4.98 ± 0.49^b	79.78 ± 0.65^c
	Germinated-48 h	4.53 ± 0.05^c	2.24 ± 0.19^a	8.01 ± 0.37^{ab}	2.50 ± 0.13^a	5.19 ± 0.06^b	82.05 ± 0.07^c
Soybean	Unprocessed	3.16 ± 0.32^{ab}	3.08 ± 0.02^b	34.57 ± 0.83^c	22.99 ± 1.11^e	5.34 ± 0.70^b	34.02 ± 1.81^a
	Blanched	3.48 ± 0.19^b	3.61 ± 0.09^c	35.59 ± 2.04^c	10.22 ± 1.05^d	5.35 ± 0.29^b	45.22 ± 2.76^b
	Roasted	2.51 ± 0.43^a	5.03 ± 0.11^d	36.22 ± 0.76^c	4.57 ± 0.31^{bc}	5.76 ± 0.33^b	48.43 ± 1.00^b
Sweet potato	Processed	7.36 ± 0.03^e	3.61 ± 0.04^c	5.70 ± 0.31^a	0.84 ± 0.04^a	3.76 ± 0.17^a	86.09 ± 0.18^d
P-value		< 0.001	< 0.001	< 0.001	< 0.001	< 0.001	< 0.001

Values are mean ± standard deviation of three independent determinations on dry matter basis. The different superscripts in the same column with different letters are significantly different ($p < 0.05$).

CHO - Carbohydrate

Crude fiber contents ranged from 3.76 for sweet potato to 5.76 g/100 g for roasted soybean. There was a significant increase due to germination of teff which could be, as reported by Megat et al. (2016), because of synthesis of new polysaccharides during the process. Only a slight change was observed from blanching or roasting of soybean. The sweet potato gave the highest value of carbohydrate (86.09 g/100 g), which is in good agreement with a previous report (Oke & Workneh, 2013) followed by those of 48 h and 24 h germinated teff with 82.05 and 79.78 kcal/100 g, respectively. The significant increase in the carbohydrate content of soybean after blanching or roasting could be due to starch hydrolysis to simple sugars. This increase in carbohydrate content is advantageous in infant feeding from a number of perspectives. On one hand, the molecules are more soluble in water, and on the other, the porridges from the resulting final blends would be of lower viscosity and thus need less water for dilution thereby preventing nutrient and energy thinning (Amagloh et al., 2013).

Similarly, the macronutrient composition of the complementary foods are presented in **Table 3**. No significant difference ($p > 0.05$) was found in the moisture and crude fiber contents of the complementary foods, but they are lying within the range prescribed for complementary foods (Codex Alimentarius Commission, 1991; WFP, 2014). The industrial-level complementary food (ComF1) had the lowest ash content (3.20 g/100 g) compared to those of the household-level complementary foods (ComF2 to ComF5). Minerals are heat stable and thus unlikely to be lost by the extrusion process (Singh, Gamlath, & Wakeling, 2007). Thus the reduction in the mineral content of ComF1 could probably be because of physical loss of some minerals during the extrusion process as suggested by Reddy & Love (1999) who reported that milling and extrusion can cause physical removal of minerals during processing. The highest ash contents were obtained for ComF3 (3.55 g/100 g) and ComF5 (3.56 g/100 g) which could be due to the increased ash contents in the roasted soybean (**Table 2**).

The industrial-level complementary food (ComF1) had significantly higher ($p < 0.05$) protein content (17.92 g/100 g). Obatolu et al. (2000) reported that such an increase in protein content could be because of denaturation of protein molecules as a result of the high extrusion temperature and that this makes the protein molecules more susceptible to proteolysis thereby improving protein digestibility. There is no significant difference ($p > 0.05$) in the protein content among the household-level complementary foods, the values ranging from 13.30 to 13.53 g/100 g. However, all the complementary foods can closely meet the recommended protein content (15 g/100 g) (Codex Alimentarius Commission, 1991). The fat contents of all complementary foods were within the range prescribed for complementary foods (6 g/100 g) as described in the WFP technical specifications for the manufacture of Super Cereal (Corn-Soya-Blend) (WFP, 2014). It can be seen that the fat contents are safer both from nutrition and shelf stability point of view. The fat content of ComF1 (6.22 g/100) was significantly higher ($p < 0.05$) compared to those of ComF3 (4.69 g/100) and ComF5 (4.68 g/100 g) while it had no difference from those of ComF2 (5.83 g/100) and ComF4 (5.80 g/100).

Table 3. Macronutrient composition (g/100 g) and energy content (kcal/100 g) of the developed complementary foods

ComF Formulation	Macronutrient Composition and Energy Content						
	Moisture	Ash	Crude Protein	Crude Fat	Crude Fiber	Available CHO	Gross Energy
ComF1	6.53 ± 0.93[a]	3.20 ± 0.03[a]	17.92 ± 0.41[b]	6.22 ± 0.59[b]	4.43 ± 0.34[a]	68.23 ± 0.81[a]	400.60 ± 3.49[b]
ComF2	5.77 ± 0.05[a]	3.27 ± 0.01[b]	13.41 ± 0.49[a]	5.83 ± 0.47[ab]	4.27 ± 0.11[a]	73.23 ± 0.35[b]	398.99 ± 2.23[ab]
ComF3	5.58 ± 0.17[a]	3.55 ± 0.02[c]	13.53 ± 0.09a	4.69 ± 0.53a	4.35 ± 0.07[a]	73.87 ± 0.59[b]	391.88 ± 2.51[a]
ComF4	5.70 ± 0.04[a]	3.27 ± 0.01[b]	13.17 ± 0.43[a]	5.80 ± 0.51[ab]	4.29 ± 0.14[a]	73.46 ± 0.32[b]	398.74 ± 2.55[ab]
ComF5	5.50 ± 0.01[a]	3.56 ± 0.01[c]	13.30 ± 0.16[a]	4.68 ± 0.57a	4.37 ± 0.08[a]	74.10 ± 0.66[b]	391.63 ± 2.82[a]
P-value	0.078	< 0.001	< 0.001	0.015	0.823	< 0.001	0.005
Reference value	10[a]	4.1[a]	15[δ]	6[a]	5[δ]	60-75[μ]	400[δ]

Values are mean ± standard deviation of three independent determinations on dry matter basis. The different superscripts in the same column with different letters are significantly different ($p < 0.05$).

[a](WFP, 2014); [δ](Codex Alimentarius Commission, 1991); [μ](Amagloh, 2012)

A high fat content in a complementary food provides more energy to the infant. However, if it exceeds the desirable level, it would be disadvantageous for stability of the product as the unsaturated fatty acids are vulnerable to oxidative rancidity (Lohia & Udipi, 2015) that would shorten its shelf life. On the other hand, a lower fat content in complementary foods results in a poor energy density which is an issue of a big concern in complementary feeding (Suri, Tano-Debrah, & Ghosh, 2014). For 6 to 8 month-old infants in low-income countries, the required percentage of energy from fat in complementary foods is 0-34% depending on the level of breast milk intake and the fat content of the breast milk. Fat is important in the diets of infants and young children because it provides essential fatty acids, facilitates absorption of fat soluble vitamins, and enhances dietary energy density and sensory qualities (Pan American Health Organization (PAHO) and World Health Organization (WHO), 2001). The carbohydrate contents of the complementary foods are in the range of 68.23

g/100 g to 74.10 g/100 within the range recommended (60-75 g/100) for infants and children (Amagloh, 2012). The gross energy contents of all complementary foods are very close to that recommended in the Codex standard (400 kcal/100 g) (Codex Alimentarius Commission, 1991). The lower energy contents of comF3 and ComF5 are results of the corresponding lower values of crude fat.

The dietary fiber contents of the flour samples used in the development of the complementary foods are presented in **Table 4**. Dietary fibers are parts of plant materials that are not digested by the endogenous secretions of the human digestive tracts (Azizah & Zainon, 1997). They include cellulose, non-cellulosic polysaccharides such as hemicellulose, pectic substances, gums, mucilages and a non-carbohydrate component lignin and are subdivided into water-soluble and water-insoluble fractions (Dhingra, Michael, & Rajput, 2012; Ötles & Ozgoz, 2014). Soluble fibers dissolve in water and form viscous gels. They bypass the digestion of the small intestine and are easily fermented by the microflora of the large intestine. However, insoluble fibers do not dissolve in water and fermentation is limited (Ötles & Ozgoz, 2014).

No significant difference ($p > 0.05$) was found in the soluble dietary fiber (SDF) contents of the ingredients used for developing the complementary foods. Very low SDF were obtained ranging from 0.55 to 1.35 g/100 g. That is, both germination of teff and blanching or roasting of soybean didn't significantly affect the SDF values. The highest SDF (3.57 g/100 g) was obtained for sweet potato which is an indication that orange-fleshed sweet potato is a rich source of SDF. This finding agrees with the work of Astawan & Widowati (2011) who reported dietary fibers contents for different varieties of sweet potato. The values for insoluble dietary fiber (IDF) varied to some extent though there were no detectable trends. Germination of teff for 48 h and blanching of soybean had significantly ($p < 0.05$) increased the IDF, 10.26 and 16.74 g/100 g, respectively. A similar trend of increment in dietary fiber values were obtained for some germinated legumes in a previous study (Benitez et al., 2013). The increase in dietary fiber after germination is, according to a previous report, due to synthesis of new polysaccharides (Megat et al., 2016). There was no significant difference in the IDF contents of the unprocessed soybean (11.50 g/100 g) and roasted soybean (12.22 g/100 g) whereas the blanched soybean had significantly higher IDF content (17.30 g/100 g). A similar trend was observed by Azizah & Zainon (1997) for boiled soybean. They reported that boiling significantly increased the IDF content of soybean and this may be attributed to the production of Maillard reaction products due to the high protein content of soybean. The lowest IDF value was that of sweet potato (5.59 g/100 g). A similar trend, as for the IDF, was found for the total dietary fiber (TDF) of the flour samples. The lowest TDF was obtained for the ungerminated teff (8.63 g/100 g) whereas the highest was that of the blanched soybean (17.30 g/100 g).

Table 4. Soluble, insoluble and total dietary fiber contents of the ingredients

ComF Ingredient		Dietary Fiber (g/100 g)		
		Soluble	Insoluble	Total
Teff	Ungerminated	0.74 ± 0.25^a	7.89 ± 0.50^b	8.63 ± 0.77^a
	Germinated-24 h	0.56 ± 0.04^a	8.29 ± 0.15^{bc}	8.85 ± 0.19^{ab}
	Germinated-48 h	0.55 ± 0.09^a	10.26 ± 0.99^{cd}	10.81 ± 0.90^{bc}
Soybean	Unprocessed	0.93 ± 0.05^a	11.50 ± 0.63^d	12.43 ± 0.70^{cd}
	Blanched	0.57 ± 0.12^a	16.74 ± 0.14^e	17.30 ± 0.03^e
	Roasted	1.35 ± 0.07^a	12.22 ± 0.15^d	13.57 ± 0.07^d
Sweet potato	Processed	3.57 ± 0.47^b	5.59 ± 0.34^a	9.15 ± 0.13^{ab}
P-value		< 0.001	< 0.001	< 0.001

Values are mean ± standard deviation of duplicate determinations on dry matter basis. The different superscripts in the same column with different letters are significantly different ($p < 0.05$).

SDF = Soluble Dietary Fiber, IDF = Insoluble Dietary Fiber, TDF = Total Dietary Fiber

The dietary fiber results for the developed complementary foods are presented in **Table 5**. All soluble, insoluble and total dietary fiber contents didn't show significant variation ($p > 0.05$) among all the complementary foods. The SDF values were in the range of 0.97 to 1.13 g/100 g and are very low compared to the 1:4 or 1:3 (soluble to insoluble dietary fiber) recommendation (Williams, 1995) whereas the IDF values ranged from 8.23 to 9.66 g/100 g which is largely a contribution from the soybean ingredients and also to some extent from teff components. The TDF values are close to the recommended value of about 10 g/100 g (Egli, 2001; Williams, 1995) which represents a safe and tolerable level for most children. This higher level of TDF could be attributed to the higher IDF values in soybean and teff grains (**Table 4**).

Table 5. Soluble, insoluble and total dietary fiber contents of the complementary foods

ComF Formulation	Dietary Fiber (g/100 g)		
	Soluble	Insoluble	Total
ComF1	1.30 ± 0.20^a	8.23 ± 1.25^a	9.53 ± 1.05^a
ComF2	0.97 ± 0.12^a	9.47 ± 0.28^a	10.44 ± 0.41^a
ComF3	1.13 ± 0.09^a	8.57 ± 0.36^a	9.69 ± 0.44^a
ComF4	0.97 ± 0.14^a	9.66 ± 0.21^a	10.63 ± 0.35^a
ComF5	1.13 ± 0.10^a	8.76 ± 0.26^a	9.89 ± 0.37^a
Reference value	---	---	10^β
P-value	0.255	0.245	0.378

Values are mean ± standard deviation of duplicate determinations on dry matter basis. The different superscripts in the same column with different letters are significantly different ($p < 0.05$).

$^\beta$(Egli, 2001)

According to earlier reports (Agostoni, Riva, & Giovannini, 1995; Brooks, Mongeau, Deeks, & Lampi, 2006), dietary fiber has a number of health benefits during early and future lives of a child. Although there are no specific recommendations as such, both soluble and insoluble fibers are strongly associated with improving health issues and are recommended to be taken in generous amounts (Williams, 1995). Soluble fibers serve as substrate for health-promoting bacteria (lactobacilli and bifidobacteria) (Anderson et al., 2009; Gustafson & Anderson, 1994). They also increase fecal mass and their fermentation byproducts have a laxative effect (Williams, 1995). Insoluble fibers soften and enlarge the stool in the colon by absorbing water, increasing bacterial proliferation and gas reduction all of which resulting in decreasing stool transit time and increasing frequency of bowel movements. Generally, consumption of foods with good dietary fiber content are associated with a higher satiety, lower incidences of obesity and improved micronutrient intake in children (Brooks et al., 2006). Thus, it is important that infants be fed with weaning foods that contain enough dietary fibers so that tastes and eating patterns become established as early as possible (Brooks et al., 2006). However, despite all these benefits of dietary fiber, presence of excess dietary fiber in weaning foods may have undesirable effects such as lower caloric density and irritation of the gut mucosa (Asma, El Fadil, & El Tinay, 2006). Dietary fiber may also prevent bioavailability of minerals as high-fiber foods may contain phytates and oxalates that can form insoluble compounds with minerals, thereby preventing normal absorption and metabolism (Williams, 1995). However, decreased bioavailability of minerals is likely to occur only when mineral intake is inadequate (Asma et al., 2006). Nonetheless, the desirable or optimal level of dietary fiber content that should be present in the foods of infants and children is not yet clearly defined as there are varying reports in the literature showing the need for further investigations on the subject (Egli, 2001).

4. Conclusion

In low-income countries, the complementary feeding period has been observed to be a very vulnerable period in one's life. It is the period when the PEM disorder begins and affects the normal growth and development of an infant. In addition to the lack of protein- and energy-rich foods, there are a number of factors contributing to the problem and the consequences are multidimensional ultimately affecting the socio-economic aspects of a given community and the nation at large.

The problem of malnutrition among infants and children in resource-poor communities can be tackled through appropriately feeding the infant using locally available materials and practices. Complementary foods prepared from blends of teff, soybean and orange-fleshed sweet potato, either through extrusion cooking or a combination of household practices (soaking, germination, blanching, roasting, dehulling, milling, blending) can meet recommended levels of protein, energy and other nutrients to a 6 to 8 month-old infant. Both the compositing of the raw materials and processes employed (industrial level and household level) resulted in improved macronutrient compositions of the developed complementary foods. Rural communities in the Sub-Saharan Africa, especially Ethiopia, can acquire a multitude of benefits by using the composite flours to prepare complementary foods for their infants and children as the raw materials are locally available and the processing approaches are easily applicable at home- and/or community-level.

Acknowledgement

We are thankful to DAAD/RUFORUM for the scholarship awarded to Mesfin Tenagashaw under which this research was carried out. We also wish to thank Bahir Dar Institute of Technology (Bahir Dar University) and Jomo Kenyatta University of Agriculture and Technology where the complementary foods were processed and analyzed, respectively. The kind assistance received from Dr. Susanne Huyskens-Keil, Humboldt-Universität zu Berlin, and also the generous cooperation of iASP (Institute for Agricultural and Urban Ecological Projects), Berlin, Germany, to conduct the dietary fiber analysis, are all highly acknowledged.

References

References

Agostoni, C., Riva, E., & Giovannini, M. (1995). Dietary fiber in weaning foods of young children. *Pediatrics*, *96*(5 Pt 2), 1002-1005.

Amagloh, F. K. (2012). *Sweetpotato-based complementary food for infants in Ghana (Unpublished doctoral dissertation).* Massey University, Palmerston North, New Zealand.

Amagloh, F. K., Hardacre, A., Utukumira, A. N., Weber, J. L., Brough, L., & Coad, J. (2012). A household-level sweet potato-based infant food to complement vitamin A supplementation initiatives. *Maternal and Child Nutrition*, *8*(4), 512-521. https://doi.org/10.1111/j.1740-8709.2011.00343.x

Amagloh, F. K., Mutukumira, A. N., Brough, L., Weber, J. L., Hardacre, A., & Coad, J. (2013). Carbohydrate composition, viscosity, solubility, and sensory acceptance of sweetpotato- and maize-based complementary foods. *Food and Nutrition Research*, *57*, 1-9.

Anderson, J. W., Baird, P., Jr, R. H. D., Ferreri, S., Knudtson, M., Koraym, A., … Williams, C. L. (2009). Health benefits of dietary fiber, Nutrition Reviews, *67*(4), 188-205. https://doi.org/10.1111/j.1753-4887.2009.00189.x

Anderson, J. W., Smith B. M., & Gustafson, N. J. (1994). Health benefits and practical aspects of high-fiber diets. *The American Journal of Clinical Nutrition*, *59*(Suppl), 1242S-7S.

Gopaldas, T. (1991). Technologies to improve weaning foods in developing countries. *Indian Pediatrics*, *28*(March), 217-221.

Anuonye, J. C., Onuh, J. O., Egwim, E., & Adeyemo, S. O. (2010). Nutrient and antinutrient composition of extruded acha/soybean blends. *Journal of Food Processing and Preservation*, *34*(2010), 680-691. https://doi.org/10.1111/j.1745-4549.2009.00425.x

AOAC International. (2000). *Official Methods of Analysis* (18th ed.). Washington, DC: Association of Official Analytical Chemists.

Asma, M. A., El Fadil, E. B., & El Tinay, A. H. (2006). Development of weaning food from sorghum supplemented with legumes and oil seeds. *Food and Nutrition Bulletin*, *27*(1), 26-34.

Astawan, M., & Widowati, S. (2011). Evaluation of nutrition and glycemic index of sweet potatoes and its appropriate processing to hypoglycemic foods. *Indonesian Journal of Agricultural Science 12*(1), 40-46.

Azizah, A., & Zainon, H. (1997). Effect of processing on dietary fiber content of cereals and pulses. *Malaysian Journal of Nutrition*, *3*, 131-136.

Badau, M. H., JideanI, I. A., & Nkama, I. (2006). Rheological behaviour of weaning food formulations as affected by addition of malt. *International Journal of Food Science and Technology*, *41*(10), 1222-1228. https://doi.org/10.1111/j.1365-2621.2006.01189.x

Benitez, V., Cantera, S., Aguilera, Y., Molla, E., Esteban, R. M., Diaz, M. F., & Martin-Cabrejas, M. A. (2013). Impact of germination on starch, dietary fiber and physicochemical properties in non-conventional legumes. *Food Research International*, *50*(1), 64-69. https://doi.org/10.1016/j.foodres.2012.09.044

Bressani, R., Murillo, B. & Elias, L. G. (1974). Whole soybeans as a means of increasing and calories in maize-based diet and protein. *Journal of Food Science*, *39*, 577-580.

Brooks, S. P. J., Mongeau, R., Deeks, J. R., & Lampi, B. J. (2006). Dietary fibre in baby foods of major brands sold in Canada. *Journal of Food Composition and Analysis*, *19*, 59-66. https://doi.org/10.1016/j.jfca.2005.02.002

Castell-Perez, M. E., Griffith, L. D., Castell-Perez, M. E., & Griffith, M. E. (1998). Effects of blend and processing method on the nutritional quality of weaning foods made from select cereals and legumes. *Cereal Chemistry*, *75*(1), 105-112. https://doi.org/10.1094/CCHEM.1998.75.1.105

Codex Alimentarius Commission. (1991). *Guidelines on formulated supplementary foods for older infants and young children: CAC/GL 08-1991*. Rome, Italy.

Dewey, K. G., & Brown, K. H. (2003). Update on technical issues concerning complementary feeding of young children in developing countries and implications for intervention programs. *Food and Nutrition Bulletin*,

24(1), 5-28. Retrieved from
http://www.who.int/entity/mip/2003/other_documents/en/FNB_24-1_WHO.pdf#page=5

Dhingra, D., Michael, M., & Rajput, H. (2012). Dietary fibre in foods: a review. *Journal of Food Science and Technology*, *49*(3), 255-266. https://doi.org/10.1007/s13197-011-0365-5

Egli, I. M. (2001). *Traditional food processing methods to increase bioavailability from cereal and legume based weaning foods (Unpublished doctoral dissertation)*. Swiss Federal Institute of Technology, Zurich, Switzerland.

FAO. (2003). *Food energy - Methods of analysis and conversion factors: Report of a technical workshop, Rome, 2002. FAO Food and Nutrition Paper No. 77*. Rome. Retrieved from http://www.fao.org/docrep/006/y5022e/y5022e00.htm#Contents\nftp://ftp.fao.org/docrep/fao/006/y5022e/y5022e00.pdf

Gopaldas, T., Mehta, P., Patil, A., & Gandhi, H. (1986). Studies on reduction in viscosity of thick rice gruels with small quantities of an amylase-rich cereal malt. *Food and Nutrition Bulletin*, *8*(4).

Hahm, T.-S., Park, S.-J., & Martin Lo, Y. (2009). Effects of germination on chemical composition and functional properties of sesame (*Sesamum indicum L.*) seeds. *Bioresource Technology*, *100*(4), 1643-7. https://doi.org/10.1016/j.biortech.2008.09.034

Haile, F., Admassu, S., & Fisseha, A. (2015). Effects of pre-treatments and drying methods on chemical composition, microbial and sensory qualities of orange-fleshed sweet potato flour and porridge. *American Journal of Food Science and Technology*, *3*(3), 82-88. https://doi.org/10.12691/ajfst-3-3-5

Hossain, M. I., Wahed, M. A., & Ahmed, S. (2015). Increased food intake after the addition of amylase-rich flour to supplementary food for malnourished children in rural communities of Bangladesh. *Food and Nutrition Bulletin*, *26*(4), 323-329.

Hotz, C., & Gibson, R. S. (2007). Traditional food-processing and preparation practices to enhance the bioavailability of micronutrients in plant-based diets. *The Journal of Nutrition*, *137*(4), 1097-1100.

Inyang, C. U., & Zakari, U. M. (2008). Effect of germination and fermentation of pearl millet on proximate chemical and sensory properties of instant "Fura" - A Nigerian cereal food. *Pakistan Journal of Nutrition*, *7*(1), 9-12. https://doi.org/10.3923/pjn.2008.9.12

Iombor, T. T. T., Umoh, E. J. E., & Olakumi, E. (2009). Proximate composition and organoleptic properties of complementary food formulated from millet (*Pennisetum psychostachynum*), soybeans (*Glycine max*) and crayfish (*Euastacus spp*). *Pakistan Journal of Nutrition*, *8*(10), 1676-1679. https://doi.org/10.3923/pjn.2009.1676.1679

Kavitha, S., & Parimalavalli, R. (2014). Development and evaluation of extruded weaning foods. *European Academic Research*, *II*(4), 5197-5210. https://doi.org/10.1007/978-1-59745-530-5_6

Konyole, S. O., Kinyuru, J. N., Owuor, B. O., Kenji, G. M., & Onyango, C. A. (2012). Acceptability of amaranth grain-based nutritious complementary foods with dagaa fish (*Rastrineobola argentea*) and edible termites (*Macrotermes subhylanus*) compared to corn soy blend plus among young children / mothers dyads in Western Kenya. *Journal of Food Research*, *1*(3), 111-120. https://doi.org/10.5539/jfr.v1n3p111

Lohia, N., & Udipi, S. A. (2015). Use of fermentation and malting for development of ready-to-use complementary food mixes. *International Journal of Food and Nutritional Sciences*, *4*(1), 1-4.

Makeri, M. U., Bala, S. M., & Kassum, A. S. (2011). The effects of roasting temperatures on the rate of extraction and quality of locally-processed oil from two Nigerian peanut (*Arachis hypogea L* .) cultivars. *African Journal of Food Science*, *5*(4), 194-199.

Megat, R. M. R., Azrina, A., & Norhaizan, M. E. (2016). Effect of germination on total dietary fibre and total sugar in selected legumes. *International Food Research Journal*, *23*(1), 257-261. Retrieved from http://www.ifrj.upm.edu.my/23 (01) 2016/(38).pdf

Megat Rusydi, M. R., Noraliza, C. W., Azrina, A., & Zulkhairi, A. (2011). Nutritional changes in germinated legumes and rice varieties. *International Food Research Journal*, *18*(2), 705-713.

Megazyme International. (2013). Total Dietary Fibre: Assay Procedure K-TDFR 06/14. Ireland.

Mensa-Wilmot, Y., Phillips, R. D., Lee, J., & Eitenmiller, R. R. (2003). Formulation and evaluation of cereal/legume-based weaning food supplements. *Plant Foods for Human Nutrition*, *58*, 1-14.

Molina, M. R., Braham, J. E., & Bressani, R. (1983). Some characteristics of whole corn: whole soybean (70:30) and rice: whole soybean (70:30) mixtures processed by simple extrusion cooking. *Journal of Food Science*,

48(2), 434-437. https://doi.org/10.1111/j.1365-2621.1983.tb10759.x

Obatolu, V. A., Cole, A. H., & Maziya-Dixon, B. B. (2000). Nutritional quality of complementary food prepared from unmalted and malted maize fortified with cowpea using extrusion cooking. *Journal of the Science of Food and Agriculture*, *80*(6), 646-650. https://doi.org/10.1002/(SICI)1097-0010(20000501)80:6<646::AID-JSFA509>3.0.CO;2-L

Oke, M. O., & Workneh, T. S. (2013). A review on sweet potato postharvest processing and preservation technology. *African Journal of Agricultural Research*, *8*(40), 4990-5003. https://doi.org/10.5897/AJAR2013.6841

Omodamiro, R. M., Afuape, S. O., Njoku, C. J., Nwankwo, I. I. M., & Echendu, T. N. C. (2013). Acceptability and proximate composition of some sweet potato genotypes: Implication of breeding for food security and industrial quality. *International Journal of Biotechnology and Food Science*, *1*(5), 97-101.

Ötles, S., & Ozgoz, S. (2014). Health effects of dietary fiber. *Acta Scientiarum Polonorum Technologia Alimentaria*, *13*(2), 191-202.

Pan American Health Organization (PAHO) and World Health Organization (WHO). (2001). *Guiding principles for complementary feeding of the breastfed child. Global consultation on complementary feeding.* Washington, D.C.

Plahar, W. A. A., Okezie, B. O., & Gyato, C. K. K. (2003). Development of a high protein weaning food by extrusion cooking using peanuts, maize and soybeans. *Plant Foods for Human Nutrition*, *58*(3), 1-12. https://doi.org/10.1023/B:QUAL.0000041157.35549.b3

ProNut-HIV. (2005). Linear programming module of NutriSurvey. Retrieved from http://www.pronutrition.org/archive/200509/msg00027.php

Ramachandra, H. G., & Thejaswini, M. L. (2015). Extrusion technology: A novel method of food processing. *International Journal of Innovative Science, Engineering & Technology*, *2*(4), 358-369.

Reddy, M. B., & Love, M. (1999). The impact of food processing on the nutritional quality of vitamins and minerals. In L. S. Lackson, M. G. Knize, & L. N. Morgan (Eds.), *Impact of processing on food safety.* (pp. 99-106). New York: Springer Science+Business Media.

Singh, S., Gamlath, S., & Wakeling, L. (2007). Nutritional aspects of food extrusion: a review. *International Journal of Food Science & Technology*, *42*(8), 916-929. https://doi.org/10.1111/j.1365-2621.2006.01309.x

Suri, D. J., Tano-Debrah, K., & Ghosh, S. A. (2014). Optimization of the nutrient content and protein quality of cereal-legume blends for use as complementary foods in Ghana. *Food and Nutrition Bulletin*, *35*(3), 372-381.

Tumwebaze, J., Gichuhi, P., Rangari, V., Tcherbi-Narieh, A., & Bovell-Benjamin, A. (2015). Rheological properties and sugar profile of a maize-based complementary food for Ugandan children 12 to 23 months of age. *International Journal of Nutrition and Food Sciences*, *4*(6), 631-638. https://doi.org/10.11648/j.ijnfs.20150406.15

Ugwuona, F. U., Awogbenja, M. D., & Ogara, J. I. (2012). Quality evaluation of soy-acha mixes for infant feeding. *Idian Journal of Scientific Research*, *3*(1), 43-50.

WFP. (2004). *Fortified blended food: Good manufacturing practice and HACCP principles. A handbook for processors in partnership with the U.N. World Food Programme.*

WFP. (2014). Technical specifications for the manufacture of Super Cereal (corn soya blend), Version 14.1.

WHO/UNICEF. (1998). *Complementary feeding of young children in developing countries: A review of the current scientific knowledge. WHO/NUT/98.1.* Geneva: World Health Organization. https://doi.org/10.1017/CBO9781107415324.004

WHO, UNICEF, & WORLD BANK GROUP. (2015). *Levels and trends in child malnutrition: UNICEF-WHO-World Bank Group joint child malnutrition estimates.* https://doi.org/10.1016/S0266-6138(96)90067-4

Williams, C. L. (1995). Importance of dietary fibre in childhood. *Journal of the American Dietetic Assosciation*, *95*(10), 1140-1149.

18

Food Safety and Aflatoxin Control

Philippe Villers[1]

[1]President, GrainPro, Inc., Concord MA, USA

Correspondence: Philippe Villers, GrainPro, Inc., 200 Baker Ave, Suite 309, Concord MA, USA.
E-mail: pvillers@grainpro.com

Abstract

This paper examines the prevention of the exponential growth of aflatoxin occurring in multi-month, postharvest storage in tropical countries, with examples from field experience and scientific data. Four approaches to modern, safe, postharvest storage methods are described, the most successful being the use of flexible, UltraHermetic™ airtight structures that create an unbreatheable atmosphere (low oxygen, high carbon dioxide) through insect and microorganism respiration alone, without use of chemicals, fumigants, vacuum, or refrigeration.

The increase in aflatoxin levels during multi-month, postharvest storage is a serious health hazard affecting several major crops. During postharvest crop storage by conventional methods in tropical conditions, molds existing within crops can produce aflatoxin levels many times greater than at harvest, often vastly exceeding the international safety standards of 20 ppb (parts per billion). For example, field data from Mali documents that during just two months of conventional peanut storage, average aflatoxin levels rose 200%. In Uganda, aflatoxin levels in conventionally stored maize rose 300% in three months. By contrast, laboratory and field data from Mali and Uganda show that the organically modified atmosphere created using various forms of sufficiently hermetic (airtight) storage containers (ranging from 25kg to 1000-tonne capacity) prevents the exponential growth of aflatoxin-producing molds in various grains, peanuts, and seeds.

Keywords: aflatoxin, Cocoon™, GrainSafe™, grain storage, hermetic, pesticide-free, postharvest, SuperGrainbag™, UltraHermetic

1. Consequences of High Aflatoxin Levels

Dr. Williams of the University of Georgia, USA, describes aflatoxins as follows: "Aflatoxin (AF) is a toxin produced by fungi acting on staple crops (like maize, rice, cassava and peanuts) that constitute a large part of the diet of people living in developing countries. For these people, there is little management of food quality and they are at risk of uncontrolled exposure to AF. In the US and many other countries, human foods must have less than 20 ppb (the threshold for cumulative genetic toxicity) but the threshold for diagnosable symptoms of acute aflatoxicosis (jaundice, vomiting, abdominal pain, hemorrhage, pulmonary edema and death) is much higher. Veterinary toxicology has shown chronic, moderate exposure results in suppressed immunity, nutrition and increased infectious diseases. These results from animal studies are being found relevant to humans living in developing countrie (Williams et al., 2004).

Since the health consequences of high levels of aflatoxins (*Aspergillus flavus* and *Aspergillus parasiticus*) are widely recognized as a major problem, the international community and many individual countries have set strict limits on acceptable levels of aflatoxin – most commonly 10-20 parts per billion (ppb). In practice these limits are often greatly exceeded with serious health effects.

In human beings, high aflatoxin levels depress the immune system, thereby contributing to many health problems ranging from cancer and susceptibility to HIV, to stunted growth among children. In African countries, Dr. Williams cites a sampling survey of several local markets showed that in 40% of the commodities found that levels of aflatoxins in foods exceeded the international standard of 10 to 20 ppb, putting an estimated 4.5 billion people in developing African countries at risk. A cross-sectional study conducted in Ghana and also cited by Dr. Williams showed that the immune systems of recently HIV-infected people had above-median levels of aflatoxins and that "people with a high aflatoxin biomarker status in the Gambia and Ghana were more likely to have active malaria." Small holder farmers were particularly affected: "A major area of neglect and opportunity is foods stored by small farmers for their own consumption. A very common consequence of quality control in markets is for farmers to retain, for their own use, grains that would reduce the price offered in the market place. Studies of peanuts in local storage facilities show a steady increase in contamination levels and these differences are observed in the cyclical variation in the biomarkers of rural African people." (Williams, 2011).

At an October 2016 meeting of the Partnership for Aflatoxin Control in Africa (PACA), Ugandan President, H.E. Yoweri Museveni called for a collective effort among African countries to address aflatoxin challenges as they are a major health risk and impede the agricultural and trade sectors of the continent. President Museveni pointed to diseases such as cancer, pneumonia and Hepatitis B, among other illnesses that could be caused by consuming foods contaminated by aflatoxins (Museveni, 2016).

In the case of domestic animals, various studies in Africa have attributed severe health impact to high aflatoxin levels. Excessive aflatoxin levels also cause failure to thrive (or even death) in farm animals such as chickens, turkeys and cattle. Per Dr. Oladele Dotun, a Veterinarian at the Animal Care Laboratory in Nigeria, research has shown that aflatoxins cause infertility, abortions, and delayed onset of egg production in birds as well as sudden losses in egg production. Furthermore, loss of appetite, skin discoloration or even yellowish pigmentation on skin can be observed in fish (Oladele, 2014).

In 2014, the Global Forum for Innovations in Agriculture (GFIA) convened a high-level meeting in Abu Dhabi, UAE. There, Frank Rijsberman, the CEO of the Global Agricultural Research Partnership (CGIAR Consortium), concluded on the basis of a Benin study (Gong et al., 2004) that post-weaning exposure to aflatoxin has impaired growth in children and is costing African farmers over $450 million USD per year in lost exports (Rijsberman, 2014).

In 2010, 10% of the Kenyan maize crop was condemned because of excessive aflatoxin levels. In the same year, a Kenyan laboratory tested 130 maize samples, out of which only 47 samples had aflatoxin levels less than 10 ppb. The highest level of aflatoxin recorded in that year was 830 ppb (FAO, 2011).

In addition to health problems, quantitative losses from insects, rodents and molds in grains such as maize conventionally stored for many months can and often do exceed 25%. A World Bank report shows postharvest losses for maize in Eastern Africa (Figure 1).

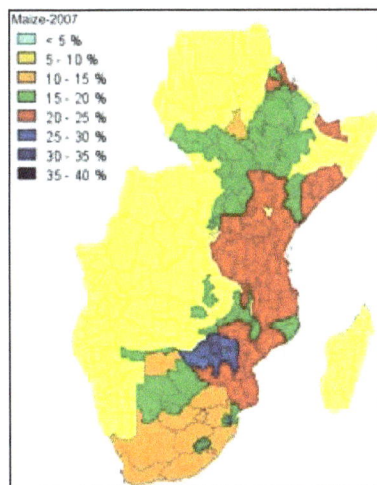

Figure 1. Estimated percentage of cumulative postharvest weight loss from production of maize in East and Southern Africa for 2007 (Zorya et. al., 2011)

2. Preventing Growth of Aflatoxins

Aflatoxins produced by strains of the Aspergillus fungus (*Aspergillus flavus* and *Aspergillus parasiticus)* are now a well-documented threat to public health, especially in tropical climates. Much work has been done in the field to control aflatoxin levels, with some useful results. To date, partial solutions such as selective breeding of crops for greater resistance, modified growing conditions, and biological controls have shown positive effects.

Biological competition, through use of molds that are similar to aflatoxin-generating molds but do not generate aflatoxins, has been shown to be particularly important. In the field, these molds can be used to outcompete the aflatoxin producing molds. A product called AflaSafe™, which uses this competitive approach, is now used in eleven African countries (Bandyopadhyay et al., 2016).

However, postharvest storage is the most overlooked stage for effectively preventing aflatoxin growth. In hot, humid climates, long term conventional storage can produce exponential growth of aflatoxins as shown in Table 1. It shows that restricting the increase in aflatoxin levels during both drying and long-term storage is a major challenge, particularly in hot and humid conditions. Table 2, following, shows field data from Mali where the aflatoxin in peanuts increased on average 200% in just two months of conventional storage (Gou, 2013). In addition, Figure 2 (De Bruin, et al., 2014) shows the exponential growth of fungus (mold) density at relative humidity above 65%. Such levels are common in the hot humid climates found in Africa, Asia and Latin America.

Table 1. Effect of interventions on aflatoxin levels (ppb) in peanuts for Drobonso Village, Ashanti Region, Ghana, 2014/2015 major season (Appaw, 2016).

	Field (Harvesting Stage)	Drying Stage * (Ground vs Tarpaulin)		Storage Stage [+] (Poly sac vs Hermetic Bag)	
Practice	Aflatoxin level	Aflatoxin level	% Reduction	Aflatoxin level	% Reduction
Farmer (conventional)	Not detected	4.68 – 51.90 (38.24 average)	50 – 97 (85% average)	6.61 – 438.79 (133.22 ave.)	86 – 99 (95% average)
Improved (hermetic)	Not detected	1.49 – 21.21 (5.94 average)		0.88 – 31.36 (10.89 average)	

Aflatoxin analysis done using HPLC based on AOAC official methods.

*Solar dried to reach average moisture content of 6.25%.

[+]Storage approximately 9 months at ambient temperature.

Table 2. Increased aflatoxin levels in peanuts during conventional storage in farmers' fields in Mali.

Village	Aflatoxin content (ppb)		
	At harvest	1 month in storage	2 months in storage
Bamba (5)	101.3	168.9	275.5
Gouak (5)	61.4	118.0	174.7
Kolokani (5)	119.2	352.6	400.0
Sido (5)	53.7	93.6	166.2

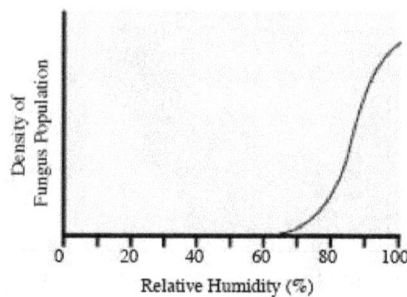

Figure 2. Typical fungus density versus humidity

3. Key Susceptible Crops

The crops that most commonly exhibit excessive levels of aflatoxin in hot, humid climates include maize, wheat, peanuts, rice, millet and cassava. As stated earlier, in Kenya in 2010, maize suffered catastrophic contamination due to excessive aflatoxin levels; 10% of the Kenyan maize crop was condemned for aflatoxin levels exceeding international standards (20ppb). In some cases, the product was reported to cause deaths (FAO, 2011).

4. Forms of Aflatoxin Control in Storage

4.1 Non-Chemical Methods

What can be done to protect susceptible commodities during storage? One very costly approach is refrigeration or air conditioning, which can inhibit significant growth of aflatoxins by keeping down both humidity and temperature levels. It has been used successfully in some countries, including in Japan for storing rice. However, this method cannot be successfully applied everywhere due to both the cost and uncertainty of available electrical power. A new, and still experimental, method has been introduced by the Ozonextrade Kft. company in Hungary (Figure 3) using a small ozone generator to destroy microbes, insects and molds in large metal silos (Lippai, 2016). Ozone, however, is not compatible with storage in many plastics, including PVC.

Figure 3. Ozone Generator, courtesy of Ozonextrade, Hungary

4.2 Chemical Methods

A chemical treatment was described by Lunven of FAO, as follows: "Chakrabarti showed that aflatoxin levels could be reduced to less than 20 ppb using separate treatments with 3 percent hydrogen peroxide, 75 percent methanol, 5 percent dimethylamine hydrochloride or 3 percent perchloric acid. These treatments, however, induced losses in weight and also in protein and lipids. Other methods include the use of carbon dioxide plus potassium sorbate and the use of sulphur oxide." Lunven also noted another process that had received some attention, namely the use of calcium hydroxide, a chemical used for lime cooking of maize: "Studies have shown a significant reduction in aflatoxin levels, although the extent of reduction is related to the initial levels. Feeding tests with moldy maize treated with calcium hydroxide have shown a partial restoration of its nutritional value" (Lunven, 1992).

However, as stated in the FAO report, "To date, no chemical or additive method has gained general acceptance. Concerns about using chemicals or additives is still growing, encouraging the expanded use of methods that have no contamination potential, and methods that cannot harm consumers, producers, or warehouse personnel." (Lunven, 1992). One example is flexible, UltraHermetic storage with or without the use of carbon dioxide as an accelerant to create controlled humidity, low oxygen, with a high carbon dioxide, modified atmosphere.

4.3 Biological Method Using GMO Technology

Biological methods include using GMO technology such as providing insect resistance using *Bacillus thuringiensis* (Monda, 2016). The focus on resistance to insects is due to the high correlation existing between insect damage and aflatoxin contamination. GMO technological approaches to aflatoxin management mainly have focused on expression of recombinant insecticidal proteins from *Bacillus thuringiensis*, expression of antifungal peptides and proteins and the use of Host-Induced Gene Silencing technology. The use of *Bacillus thuringiensis* toxin technology against the European corn borer has enabled a reduction of aflatoxin levels in maize (Monda, 2016).

4.4 UltraHermetic Storage Method

An increasingly popular and inexpensive alternative method for controlling aflatoxin growth during multi-month storage is through use of flexible, UltraHermetic storage containers. This storage technology relies on creating a condition such that insect plus microflora respiration, and sometimes respiration of the commodity itself, is greater than residual intake of oxygen through ultra-low permeability container material. In some cases initial CO_2 injection can be added in order to speed up the process of reducing oxygen levels to an unbreatheable level, typically 3% O_2. On a commercial basis, UltraHermetic storage is an environmentally friendly solution now being used in more than 110 countries. It is lightweight, flexible, and available in a growing variety of forms to store 25kg to 1000-tonne capacity Cocoons™ and SuperGrainbags™.

Growth of aflatoxins in this type of storage is largely inhibited, by both oxygen deprivation and preventing the ingress of humidity. This denies aflatoxin producing molds what they need to grow, namely oxygen and high relative humidity. The molds themselves survive but grow very slowly.

In most other commodities stored hermetically, respiration alone remains sufficient to deplete oxygen levels. In the special case of ground nuts (peanuts), however, it has been found that the level of insect infestation is often too low to prevent significant growth of aflatoxins through microflora and insect respiration alone, before oxygen levels approach an unbreathable 3%. To combat this, two supplemental methods have proved successful. One method is to introduce CO_2 immediately after closing the container to drive out the air and fill it largely with CO_2 – typically up to 90% CO_2 or more. To accomplish the same goal in farm settings, the other method is to insert into the storage container the readily available small commercial oxygen absorbers "sachets" such as seen in Figure 4. These absorb available oxygen, typically in less than a day. In the case of ground nuts (peanuts), this method also creates the desired low-oxygen atmosphere much faster than can be done using insect and microflora respiration alone.

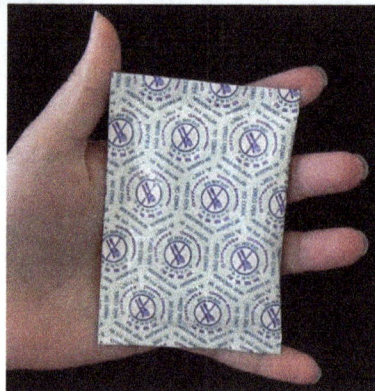

Figure 4. 1,000cc-capacity oxygen absorber

5. Field Data on the Control of Aflatoxin

In recent years, several controlled experiments have shown the advantage of hermetic storage in preventing growth of aflatoxins during storage. The most recent example is the recent USAID study previously cited in Section 1 and shown in Table 1.

In addition, a Millenium Village study in Ruhira, Uganda (Figure 5) shows the difference in growth of aflatoxin levels between conventional and UltraHermetic storage (De Bruin, T., Villers, P., Navarro, S., 2014). This is similar to the data on peanuts previously cited in Section 2 and Table 1.

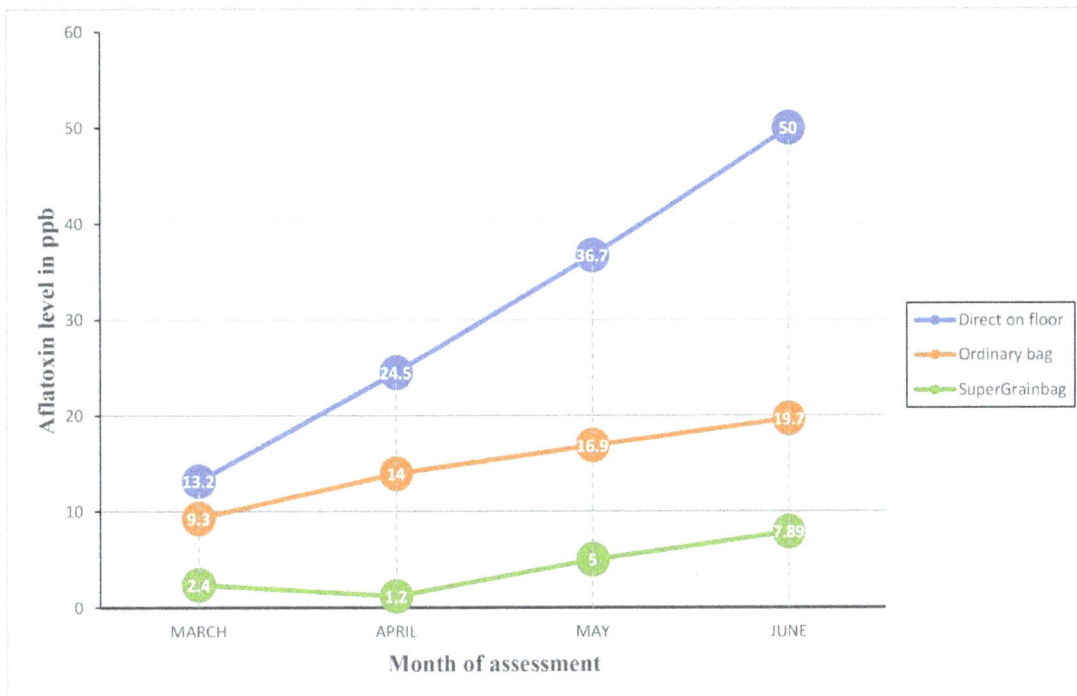

Figure 5. Monthly aflatoxin concentration in maize stored in different storage conditions in Ruhira, Uganda, Millennium Villages study, 2013

6. Proper Drying of Grains

For successful use of UltraHermetic storage as well as for other storage methods, crops must be adequately dried, typically to a point below their "critical moisture level" (in equilibrium with 65% relative humidity). However, the drying period is often a source of growth in aflatoxins. Long drying cycles or re-wetting because of rain or dew encourage growth of aflatoxins both directly and indirectly—directly because of long drying exposure, and indirectly because extended drying and rewetting creates more broken kernels (i.e., "brokens") which make the spread of aflatoxins far more rapid. This can be prevented in several ways, including by thermal dryers using fossil fuels such as wood or petroleum products; these allow the commodity to dry more rapidly than it does when solar energy is used. However, for high value crops such as coffee or seeds, care must be taken not to exceed the maximum safe temperature to avoid product deterioration: 40°C in the case of parchment coffee, and 45°C for coffee cherries.

Figure 6. First generation Solar Dryer (courtesy of GrainPro, Inc.)

Figure 7. Solar Bubble Dryer with optional, solar panels (courtesy of GrainPro, Inc.)

The product being dried can be easily contaminated during sun drying, which is often done on patios or on paved or unpaved roads. Further, drying is slowed on patios or roads, particularly when there is dew in the morning or rain, because the porous surface on which the product is being dried must be dried along with the commodity. Recently, use of portable solar dryers such as the Collapsible Dryer Case (CDC™) (Figure 6) and the Solar Bubble Dryer (SBD™) (Figure 7) have resolved this problem by separating the underlying surface such as a patio or road from the grains being dried with an impermeable, coated, woven polyethylene surface, which provides both a moisture barrier and a thermally insulated surface. When there is rain or heavy dew, the simpler Collapsible Dryer Case™ (CDC) dryer is folded back on itself manually with the commodities still inside, and is closed, awaiting the end of the rain.

The more advanced Solar Bubble Dryer™ (SBD), using a greenhouse approach, prevents rain from rewetting the crops by enclosing it in a transparent "bubble" which also speeds drying from solar gain or the so-called "greenhouse effect." These relatively simple solar dryers are becoming increasingly popular because they are relatively low cost to purchase and operate. In addition, they are both environmentally friendly and easily moved to new locations.

7. Seed Storage

Seeds of all types also benefit from hermetic storage. Maintaining seed germination capability for multiple months is difficult in tropical climates unless one uses costly, energy intensive, air conditioning or refrigeration. Without refrigeration, germination rates can drop dramatically, as shown in Figure 8.

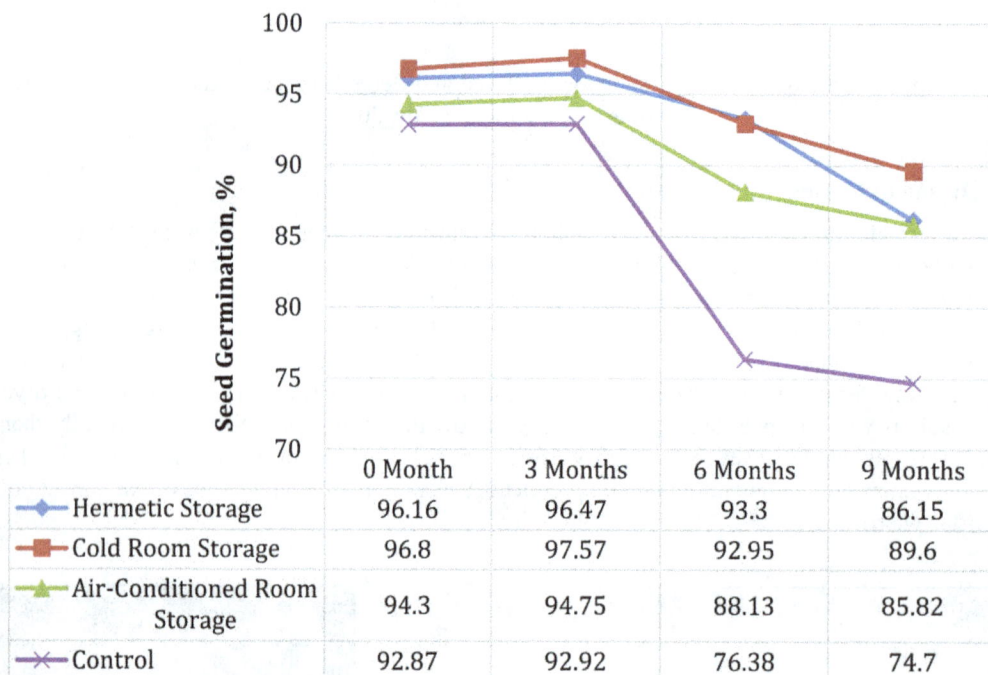

	0 Month	3 Months	6 Months	9 Months
Hermetic Storage	96.16	96.47	93.3	86.15
Cold Room Storage	96.8	97.57	92.95	89.6
Air-Conditioned Room Storage	94.3	94.75	88.13	85.82
Control	92.87	92.92	76.38	74.7

Figure 8. Preservation of rice paddy seed using hermetic and low temperature storage technologies. (Sabio, Dator et al., 2006)

Seeds in UltraHermetic storage maintain high levels of germination capacity, typically for at least a year (85% is required for certified seed). For this reason, hermetic storage is now used increasingly for seed storage, particularly in the developing world, as an equivalent alternative to the older refrigeration techniques with their higher energy requirements.

Figure 9 shows cotton seed germination when stored in a cold room, as contrasted with hermetic storage in Cocoons and conventional (unprotected) storage in jute bags over a 12-month period.

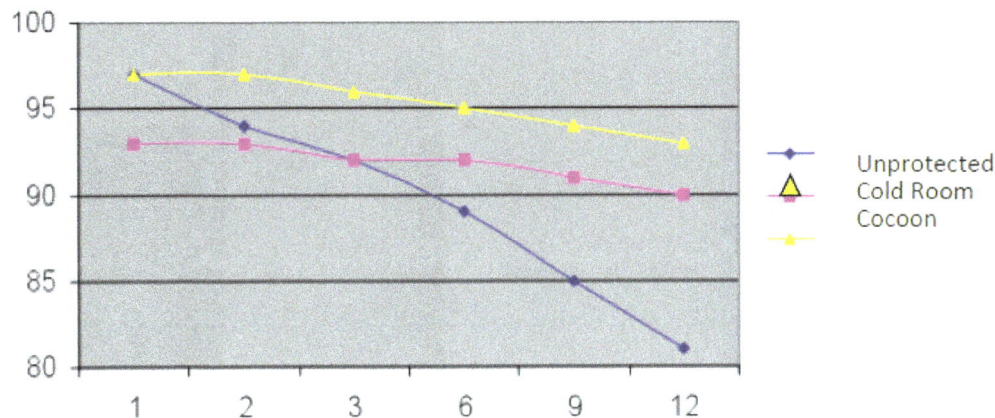

Figure 9. Cotton seed percent germination for 12-month storage, Peru, 2012 (courtesy Bayer Crop Science)

8. Other Uses of Hermetic Storage & Available Forms

Ultrahermetic storage is important for aflatoxin reduction in multi-month storage, but it also fills other needs. By design it is sufficiently airtight so that insect respiration rate from both the insects and microflora that infect all grains is higher than residual leakage of air into the UltraHermetic container. Typical values achieved for permeability to oxygen in smaller UltraHermetic containers are 2cc to 4 $cc/m^2/day$ and for water vapor 2gm/m^2/day. Not only do all insect life forms—from eggs to adult—die through asphyxiation, but the commodity is protected from humidity entering. Commodities such as milled rice and bran that otherwise are very difficult to store can now be safely held for months without developing rancidity due to oxidation.

The advantage of UltraHermetic bags in man-portable sizes is their greater airtightness with ultra-low leakage rates for oxygen typically 2 to 4 cc/m^2/day. The bags are superior to ordinary plastic bags of the same thickness (0.078 mm), which have typically a permeability to oxygen of 2,000 cc/m^2/day because they have a special high resistance barrier layer sandwiched between two polyethylene layers. Nonetheless, despite the higher permeability, in some applications by using two plain high-density polyethylene 0.078 mm bags, one inside the other, commercially useful results have been obtained for a variety of crops. The most widely known example of this is the PICS bag developed by Purdue University initially for cowpea storage, but now used for several commercially important crops. (Sudini, et al., 2014, Williams, Baributsa, & Woloshuk, 2014)

Both UltraHermetic and less hermetic bags have excellent low permeability to water vapor, typically 1 to 2 gm/m^2/day or less. Other users have used plain polyethylene bags plus a modest vacuum level, but the vacuum levels reached do not approach lethal oxygen levels (normally 3% oxygen) and have not been found to perform better than UltraHermetically sealed bags. (Borem, 2013)

Because of wide variation in storage needs, the size and form of UltraHermetic storage in current use covers a broad range. Depending on the application, UltraHermetic storage forms can range from man-portable bags with capacities from 25 to 100 kilograms (Figure 10) to 1.4 tonne SuperGrainbags or Cocoons™ with large scale capacities from 5- to 1,000-tonnes of bagged grains. In cases where the product needs to be very portable, the SuperGrainbag® (SGB) from 25kg to 100kg capacity is in wide commercial use; over 2 million were delivered last year.

Figure 10 shows maize as sold in Nigeria raised and sold by a contract farming project known as Babban Gona in Kanu State, where an aggregation of 8,000 small farmers store maize in 50kg SuperGrainbags. Affixed with the label "Low Aflatoxin Maize." Many of the bags at Babban Gona are later used for commercial baby food.

Figure 10. 50-kg SuperGrainbag liner inside a protective outer bag (courtesy Babban Gona, Nigeria)

Figure 11. 1.4 tonne SuperGrainbag Ocean (courtesy GrainPro, Inc.)

The 1.4-tonne capacity SGB Ocean™ (Figure 11) and the GrainSafe 2™ (Figure 12) provide capability for continuous loading and unloading in bulk via input and output funnels without destroying the container's hermeticity. The GrainSafe™ is used for storage of bulk commodities and allows continuous in and out. Figure 12 also shows the range of some other commonly used UltraHermetic storage units.

 SuperGrainbag 25 to 100 kg	 GrainSafe Mini 250kg bulk	 GrainSafe 1 ton bulk	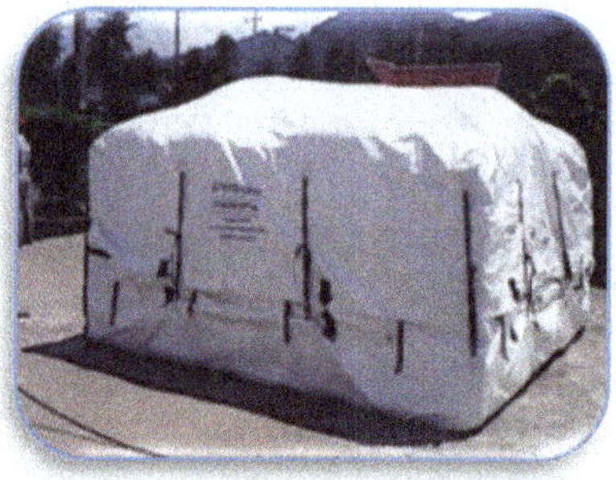 Cocoon 5 to 1,000 tons (bagged)

Figure 12. Hermetic storage capacities ranging from 25 kilos to 1000 tonnes. (GrainPro, Inc.)

There is already considerable literature on other benefits obtained from UltraHermetic storage (De Bruin, T., Villers, P., Navarro, S., 2014). These include the preservation of quality as well as prevention of quantitative losses caused by infesting insects, or additional weight loss due to change in moisture content. In the case of coffee, for example, the quality of green coffee beans is maintained for up to a year as measured by cupping tests. In the special case of coffee, insect infestation alone is small, and does not greatly reduce O_2 levels. The addition of CO_2 has only minor benefit, showing that in some specific cases hermetic storage is sufficient to maintain quality without requiring low O_2 levels. Table 3 shows the small difference between normal UltraHermetic storage of coffee and addition of CO_2.

Table 3. Mean values of the overall cupping score (indicates quality) of the coffee beans after 12-months of storage (Borem et al., 2013)

Big-bag (1-tonne hermetic)	Position	Score
With CO_2	Upper	80.00a
With CO_2	Middle	80.80a
Without CO_2	Upper	78.09a
Without CO_2	Middle	78.06a
GrainPro (SuperGrainbag, 69kg, no CO_2)	Hermetic only	78.98a
Jute sack alone	Jute sack	73.03b

In other commodities, such as cocoa, milled rice or various nuts, hermetic storage also prevents the significant growth of free fatty acids (and therefore, rancidity) for periods of up to a year. Grains and other dry commodities must be preserved with little loss in quality and low aflatoxin levels to be able to export to such major markets as the EU.

A series of controlled experiments run in Brazil compared the effect of different green coffee storage methods, measuring "sensory analysis" after one year's storage (Figure 13). (Borem, 2016) UltraHermetic storage (referred to as "paper + high barrier") with or without anti-fogging agent and CO_2 did best, with no statistically significant difference. The "a" label on some columns in Figure 13 indicates there is no statistical difference among them. Vacuum packaging and injection of CO_2 resulted in a slightly lower (but again, not statistically significant) sensory analysis score, which includes cupping tests. Plain jute or paper storage without high barrier performed much worse. Use of paper vs. jute as an outer protective bag made no difference.

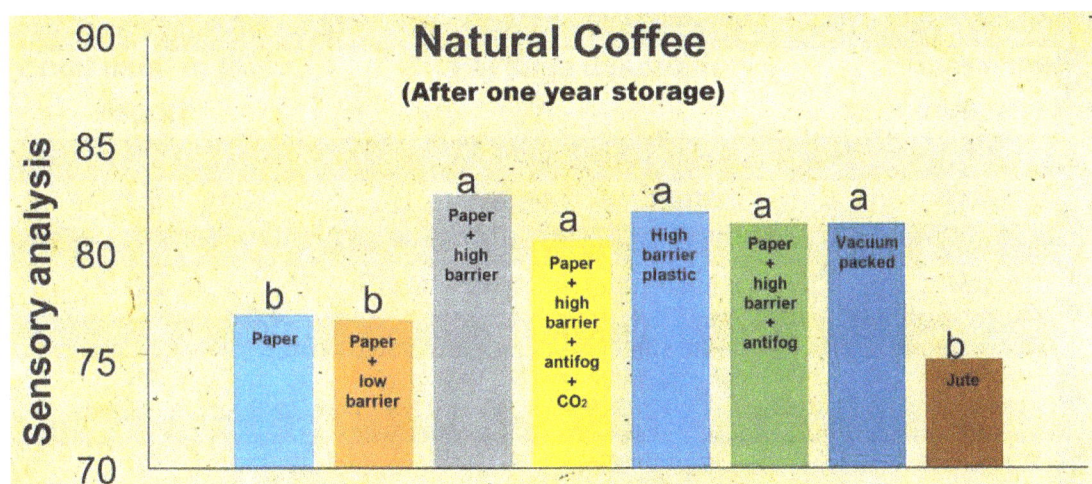

Figure 13. Sensory analysis for different coffee storage methods, measured after 12-months of storage (courtesy Prof. Flavio Meira Borem, Universidade Federal de Lavras, Lavras - Minas Gerais, Brazil)

9. Cost of UltraHermetic Storage Systems

The cost of portable UltraHermetic storage liners for insertion into conventional jute or woven polypropylene outer bags for mechanical protection vary widely based on volume, distribution chain, import costs and custom duties, if any. The least expense form with 69kg capacity and intermediate permeability to oxygen of 50 $cc/m^2/day$ is for farm use only for many commodities; it is called the SuperGrainbag Farm[TM]. In quantities of 10,000 (FOB factory) they cost $0.99 each. For premium product storage and ultra-low permeability of 2 $cc/m^2/day$, the factory quantity price starts at $1.10 each. Retail prices in smaller quantities vary widely and are generally 2 to 3 times higher.

For multi-ton storage Cocoon prices (FOB factory) range from $225/ton to $25/ton for capacities between 5 tons and 1,000 tons, respectively, and have a 10 to 15 year life. The growing acceptance of UltraHermetic storage in 110 countries suggest that the available prices have been found cost-effective by a large range of users.

An Excel-based, user-input, cost-benefit analysis program to compare hermetic storage with alternative storage solutions is available without cost from GrainPro, Inc. (GrainPro, 2016)

10. Conclusion

The problem of aflatoxin presence in food increasingly is being recognized as a major public health issue. In Africa, the African Union's PACA (Partnership for Aflatoxin Control in Africa) has led a Pan-African effort, with support from AGRA (Alliance for a Green Revolution in Africa) and from major foundations including the Rockefeller Foundation, to sensitize governments on the hazards of aflatoxins and the serious health problems they cause to their citizens, especially in the tropics.

Efforts to reduce aflatoxins through controlling their growth prior to harvest have produced encouraging improvements such as more resistant plant varieties, biological control such as AflaSafe, and breeding for increased plant resistance. However, farmers, city dwellers, and grain reserve agencies established to protect against large climatic swings suffer from enormous losses of stored grains contaminated by aflatoxins, especially if they are in hot, humid countries. This is why many look to UltraHermetic storage and solar dryers as better alternatives to conventional multi-month storage and traditional drying.

In June 2016 in Nairobi, Kenya, the Africa Strategic Grain Reserve Conference adopted the following resolution:

1   "Improvements in safe storage are possible – and necessary. Solutions exist that would make Africa self-sufficient in grains.
2   The adoption rate of modern storage technology is slower than the need."

The dominant food contamination problem in hot, humid climates continues to be exponential aflatoxin growth of grains during multi-month storage. UltraHermetic post-harvest storage is already in widespread use and recognized for its ability to dramatically reduce quantitative losses and quality deterioration, especially in tropical climates where annual storage losses can often be reduced to less than 1% from more than 25%. This same storage technology also is available commercially to prevent dangerous exponential growth of aflatoxin contamination in grains and other food products grown and stored in these same environments.

References

Appaw (2016, January). An overview of mycotoxins in the food production chain and the impact on nutrition and health. *Paper presented at the Multi-Sectoral Nutrition Strategy Global Learning & Evidence Exchange*, Accra, Ghana.

Bandyopadhyay, R., Ortega-Beltran, A., Akande, A., Mutegi, C., Atehnkeng, J., Kaptoge, L., Senghor, A., Adhikari, B., & Cotty, P., (2016). Biological control of aflatoxins in Africa: current status and potential challenges in the face of climate change. *World Mycotoxin Journal, 9*(5), 771-789. http://dx.doi.org/10.3920/WMJ2016.2130

Bayer Crop Sciences, 2012, unpublished (GrainPro Document #: PPT4051CG0314).

Borem, F., (2013). Handbook of Coffee Post-Harvest Technology. ISBN 978-0-9915721-0-6.

Borem, F., (2016). Postharvest technology and new packaging materials. PowerPoint presentation, SCAJ2016, Tokyo, Japan, 28-30 September, 2016.

Borem, F., Ribeiro, F., Figueiredo, L., Giomo, G., Fortunato, V., & Isquierdo, E., (2013) Evaluation of the sensory and color quality of coffee beans stored in hermetic packaging. *Journal of Stored Products Research, 52,* 1-6. https://doi.org/10.1016/j.jspr.2012.08.004

De Bruin, T., Villers, P., & Navarro, S., (2014). Worldwide development in Ultra Hermetic storage and solar drying techniques. *11th International Working Conference on Stored Product Protection (IWCSPP)*, Chiang Mai, Thailand, 22-28 November, 2014

De Bruin, T., Villers, P., Wagh, A., & Navarro, S., (2012). Worldwide Use of Hermetic Storage for the Preservation of Agricultural Products. *Proceedings of the 9th International Controlled Atmosphere & Fumigation Conference.* Antalya, Turkey.

FAO (2011). Situation analysis: improving food safety in the maize value chain in Kenya. *Report prepared for Food and Agriculture Organization of the United Nations [FAO] by Prof. Erastus Kang'ethe.* College of Agriculture and Veterinary Science, University of Nairobi.

Gong, Y., Hounsa, A., Egal, S., Turner, P., Sutcliffee, A., Hall, A., Cardwell, K., & Wild, C., (2004). Postweaning exposure to aflatoxin results in impaired child growth: A longitudinal study in Benin, West Africa. *Environmental Heal Perspectives, 112,* 1334-1338. https://doi.org/10.1289/ehp.6954

Gou, B. (2013). Increase in aflatoxin concentration during storages of ground nuts in the farmers' fields. (*Private communication*). USDA-ARS, Tifton, GA.

GrainPro (2016). Cost Benefit Analysis for Hermetic versus Non-hermetic Storage. *GrainPro #LT2263PV1111-3.* Available on request thru sales@grainpro.com.

Lippai, A. (2016, Oct). (*Private communication*). OzonExtrade Kft., Budapest, Hungary.

Lunven, P. (1992). Maize in Human Nutrition. *FAO Food and Nutrition Series* 25, Rome, Italy.

Monda E., & Alakonya A., (2016). A review of agricultural aflatoxin management strategies and emerging innovations in Sub-Saharan Africa. *African Journal of Food, Agriculture, Nutrition and Development, 16*(3). https://doi.org/10.18697/ajfand.75.ILRI11

Museveni, H. & Yoweri, E. (2016). President Museveni opens 2nd Partnership for Aflatoxin Control in Africa (PACA) Partnership Platform. *Press Release No. 355/2016.*

Navarro, H., Navarro, S., Finkelman, S. (2012). Hermetic and modified atmosphere storage of shelled peanuts to prevent free fatty acid and aflatoxin formation. *Proceedings of the Conf. Int. Org. Biol. Integrated Control of Noxious Animals and Plants (IOBC). Work Group on Integrated Prot. Stored Prod. Bull.* Volos, Greece.

Okoth, S., (2016). Improving the Evidence Base on Aflatoxin Contamination and Exposure in Africa: Strengthening the Agriculture-Nutrition Nexus. *CTA Working Paper 16*(13), 1-112.

Oladele, D., (2014). The effects of aflatoxins on animals. *Partnership for Aflatoxin Control in Africa (PACA), (Meridian Institute, Washington, DC), Aflatoxin Partnership Newsletter, 2,* 1-4.

Sabio, G.C. et al., (2006). Preservation of Mestizo 1 (PSB Rc72H) seeds using hermetic and low temperature storage technologies. *Proc. 9th Int. Working Conf. Stored Prod. Prot. Campinas, ABRAPOS*, Sao Paulo, Brazil, 946-955.

Sudini, H., Rao, G., Gowda, C., Chandrika, R., Margam, V., Rahore, A., & Murdock, L., (2014). Purdue Improved Crop Storage (PICS) bags for safe storage of groundnuts. *Journal of Stored Products Research, 64*, 133-138. http://dx.doi.org/10.1016/j.jspr.2014.09.002

Tumusiime, Rhoda Peace, (2016). *Partnership for Aflatoxin Control in Africa (PACA), Press Release No 352/2016.*

Villers, P., (2013). Aflatoxins and Safe Storage. *Frontiers in Microbiology, 5,* 1-6.

Williams, J. H., (2011). Aflatoxin as a public health factor in developing countries and its influence on HIV and other diseases. *Peanut Collaborative Research Support Program, University of Georgia. World Bank Report #60371-AFR,* 1-95.

Williams, J. H., Phillips, T. D., Jolly, P. E., Stiles, J. K., Jolly, C. M., & Aggarwal, D. (2004). Human aflatoxicosis in developing countries: a review of toxicology, exposure, potential health consequences, and interventions. *Am.J.Clin.Nutr. 80,* 1106-1122.

Williams, S., Baributsa, D., & Woloshuk, C., (2014). Assessing Purdue Improved Crop Storage (PICS) bags to mitigate fungal growth and aflatoxin contamination. *Journal of Stored Products Research, 59,* 190-196. http://dx.doi.org/10.1016/j.jspr.2014.08.003

Zorya, S., Morgan, N., & Rios, L. D. (2011). Missing Food: The case of postharvest grain losses in Sub-Saharan Africa. *World Bank Report #60371-AFR.* Washington, DC.

19

Natural Health Products: Practices, Perceptions and Training Needs of Registered Dietitians

Valerie Dussault[1] & Marie Marquis[1]

[1]Department of Nutrition, University of Montreal, Canada

Correspondence: Valerie Dussault, Department of Nutrition, University of Montreal, Canada. E-mail: valerie.dussault20@hotmail.com

Abstract

Canadians have access to thousands of authorized natural health products (NHPs) and are also surrounded by many unauthorized NHPs, which may place them at risk of adverse effects. Consumers expect health professionals, including nutritionists, to be a source of information on NHPs. Current training programs suggest that registered dietitians may have little knowledge about NHPs. The *Ordre professionnel des diététistes du Quebec* (OPDQ) sent an electronic survey to registered dietitians who are members of the OPDQ to document their use, referral habits and sources of information for NHPs. The survey also explored respondents' perceptions of professional roles regarding NHPs as well as their perceptions of the effectiveness of specific NHPs. It measured their need for training on specific NHPs and on the health conditions that may be improved by NHPs. Data were analyzed with SPSS, through which descriptive statistics were obtained. A qualitative analysis was performed on the open-ended questions from the survey. A total of 295 questionnaires were analyzed. Among nutritionists, 93% have received requests for information about NHPs, 91% use or have used NHPs and 94% have recommended them. Also, 95% need training on NHPs and for various health considerations. Overall, they have a positive perception of their roles regarding this subject, with 77% indicating that the nutritionist should be a reliable source of information for NHPs. The major findings of our study are that the roles of Quebec nutritionists relating to NHPs are not clearly defined. Nutritionists need training on NHPs to provide sound nutritional advice for NHP users and, therefore, should develop a new area of nutritional practice.

Keywords: natural health product, supplement, alternative medicine, training, perception, nutritionist

1. Introduction

Growth has been impressive in the natural health product (NHP) market. Canadians have access to 55,000 authorized NHPs (Health Canada, 2013). In the United States, Americans spend $30.2 billion on complementary health approaches, most of which are NHPs. According to the National Center of Health Statistics, Americans are not the only ones doing this—the self-care movement is worldwide (Food and Drug Administration, 2014; National Institutes of Health, 2016).

Health Canada defines NHPs as naturally occurring substances that are used to restore or maintain good health. They are made from plants, animals, microorganisms or marine sources. They come in a wide variety of forms including tablets, capsules, tinctures, solutions, creams, ointments and drops. Often referred to as complementary or alternative medicines (CAMs), they include vitamins and minerals, herbal remedies, homeopathic medicines, traditional medicines (e.g., Chinese and East Indian medicines), probiotics, amino acids and essential fatty acids (Health Canada, 2016).

In 2010, over 70% of Canadians consumed an NHP, motivated mainly by health maintenance (85%), disease prevention (79%) and immune system strengthening (36%). To be legally sold in Canada, NHPs must be licensed and the Canadian sites that manufacture, package, label and import them must have site licenses (Health Canada, 2016). To obtain a license, labeling and packaging requirements must be met, good manufacturing practices must be followed, and proper safety and efficacy evidence must be provided to Health Canada. Still, Canadian consumers may choose, consciously or not, to use NHPs that have not been reviewed or authorized by Health Canada. The many differences in how countries regulate NHPs can put NHP consumers at risk. Interestingly, Health Canada states that a percentage of Canadians who use NHPs report adverse reactions after using NHPs. Those risks may be related to manufacturing problems, unproven claims leading consumers to use NHPs for serious conditions or to delay proper treatment, lack of information to make an informed choice, interaction with prescription drugs or other NHPs, or allergic reactions (Health Canada, 2016).

For these reasons, the Canadian Paediatric Society (Canadian Paediatric Society, 2005), Health Canada and the *Collège des Médecins du Quebec* (Goulet, 2006) advise Canadians to refer to a health professional before using any NHPs. However, most recommendations for NHP usage in Canada are still made by a friend or a family member (Goulet & Bell, 2005).

In the United States, the Academy of Nutrition and Dietetics has raised several concerns regarding the role of dietitian professionals with respect to complementary and alternative medicine and the skills recommended for dietetics professionals in the area of dietary supplements. These skills included knowing the risks and benefits for the most commonly used supplements, documenting clients use and clinical response to dietetic supplements and the ability to recommend NHP use if necessary (Decker-Touger & Thomson, 2003).

Although practices, attitudes and learning needs of nutritionists have been reported in North America (Gardiner et al.; Pomazak, 2009), no data are specific to registered dietitians (RDs) in Quebec. Moreover, in Quebec, professional training on CAMs is limited, suggesting the importance of emphasizing a continuum of professional training. Currently, RD professional roles are guided by a Code of Ethics that requires them to abstain from expressing opinions or giving nutritional advice without having full knowledge of the facts.

The current study aims to portray the practices, attitudes and training needs among RDs who are members of the OPDQ on the topic of NHPs. The objectives of the study are to document dietitians' use, referral habits and sources of information on NHPs. With regard to attitude, the study aims to document RD perceptions of their professional roles and on other professionals' roles regarding NHPs, as well as their perception of the effectiveness of specific NHPs. Finally, it aims to report on RD needs for training on specific NHPs and on health conditions that may benefit from NHP usage.

2. Materials and Methods

2.1 Instrument

Data were collected via a self-administered online questionnaire, available in French and English, integrated into Fluid Survey (http://fluidsurveys.com/), and pre-tested with seven RDs to assess clarity and the amount of time it took to complete. The data presented herein are specific to practices, attitudes and RD training needs.

Before the questionnaire component, a brief description of NHPs was provided so that respondents could properly identify NHPs in a Canadian context. This definition differentiated between NHPs, functional foods, nutraceuticals, medicines and food.

With regard to practices, respondents were asked whether they received questions about the NHPs in their professional practice. They were also asked about the NHPs they had recommended in the past year (choice of eight and open answer) and the health considerations for which the NHPs were recommended (choice of ten and open answer). To document their sources of information, the authors suggested ten possible answers and requested an example of the consulted sources.

With regard to attitudes, respondents were asked to identify two out of eight specialists whom they believed should be credible sources of information on NHPs. Respondents were also asked to select the role of nutritionists on NHPs among 12 options drawn from the Academy of Nutrition and Dietetics (Thomson et al., 2002). Their levels of agreement on the effectiveness of different NHPs (e.g., Omega-3 for cardiovascular disease) and on specific professional responsibilities (e.g., the importance of questioning the client about NHPs) were documented with eight statements, measured with a five-point Likert scale, and the option to select *do not know*.

Their needs for training on NHPs (eight choices), health conditions and NHP usage (11 choices and an open question) were collected.

Respondent profiles were created with information about training, areas of practice, professional affiliation and personal use of NHPs (choice of eight, plus *never consume*).

2.2 Recruitment

In 2014, the OPDQ sent all its registered dietitians (n=2,898) an email inviting them to take the survey. Respondents had one month to complete the questionnaire and a reminder was sent halfway to the deadline.

2.3 Statistical Analysis

Data were transferred from Fluid Survey to an Excel spreadsheet (2007, Microsoft Inc. version 14.0. Redmond,

WA) and imported into SPSS (version 19.0, SPSS Inc. Chicago, IL). Descriptive statistics were formulated. Data from open questions were compiled, synthesized by the first author and validated by the second author.

2.4 Ethical Approval

This project was approved by the Ethics Committee of Health Research of the Université de Montréal.

3. Results

3.1 Respondent Characteristics

In total, 295 completed questionnaires were obtained, representing 10.2% of OPDQ members. Characteristics of the respondents are presented in Table 1.

Table 1. Characteristics of RDs who participated in the Quebec consultation on NHPs (n=295)

	n	%
Sources of training on NHPs		
During undergraduate degree in nutrition	181	61.4%
Informal training	134	45.4%
Training obtained from continuum professional education	49	16.6%
Training offered at workplace	47	15.9%
None	24	8.1%
University attended		
Laval University, Quebec	115	39.0%
University of Montreal, Quebec	112	38.0%
McGill University, Quebec	35	11.9%
Other Canadian University	12	4.1%
American University	2	0.7%
Graduate studies in nutrition		
No	203	68.8%
Yes	73	24.7%
Area of practice		
Clinical nutrition	217	73.6%
Public nutrition/community	79	26.8%
Research	23	7.8%
Education	20	6.8%
Food service management	17	5.8%
Communications, public relations and marketing	11	3.7%
Food industry	6	2.0%
Other	6	2.0%
Pharmaceutical industry	3	1.0%
Naturopathic association affiliation		
No	260	88.1%
Yes	16	5.4%
Personal use of NHPs		
Vitamins	221	74.9%
Probiotics	181	61.4%
Minerals	140	47.5%
Fatty acid supplements	79	26.8%
Herbal remedies	54	18.3%
Homeopathic medicines	39	13.2%
Never consumed NHPs	26	8.8%
Amino acid supplements	14	4.7%
Traditional medicines	10	3.4%

3.2 Professional Practices

A total of 93% of RDs reported receiving questions on NHPs. The main NHPs about which they had made recommendations in the past year were vitamin supplements (87%), probiotics (78%) and mineral supplements (67%). The health concerns for which respondents recommended the use of NHPs were aging (46%), pregnancy

(41%) and cardiovascular diseases (30%). Among the 254 responses specifying other health problems, one third were about gastrointestinal disorders, such as ulcerative colitis, irritable bowel syndrome, clostridium difficile infections, constipation, diarrhea and celiac disease.

The main sources of information used by nutritionists with regard to NHPs are specific websites (n=238) (e.g., Health Canada Natural and Non-prescription Health Products Directorate, Passeport Santé, at www.passeportsante.net, Extenso, at www.extenso.org, and NHP manufacturers), Internet search engines (n=59) (e.g., Pubmed and Google), manufacturer brochures (n= 86) (e.g, Bio-K+, Probaclac and Centrum), reference materials (n= 70) (e.g., Dietary Reference Intakes published by Health Canada, at http://www.hc-sc.gc.ca/fn-an/nutrition/reference/table/index-eng.php), healthcare specialists (n=105) (e.g., pharmacists, nutritionists and doctors), science books (n=11) (e.g., on food and cancer or on interactions between drugs, nutrients and natural health products), scientific articles (n=68), training materials (n=73), newsletters from professional associations (n=34) and public interest articles (n=28).

3.3 Attitudes towards responsibilities and NHP effectiveness

According to RD, the two categories of specialists who should be a source of credible information on NHPs are pharmacists (85%) and nutritionists (77%). Among the other categories, physicians (16%), naturopaths (8%), homeopaths (1%) and osteopaths (1%) were proposed.

With regard to health professionals, a majority of RDs totally agreed that it is important to document client use of NHPs (78%), that all health professionals should acquire basic knowledge about NHPs (72%) and that health professionals should know the interactions, conditions of use, dosage, effectiveness and safety of NHPs in order to properly advise on using them (82%). Responses for each of the 12 statements on RD responsibilities with regard to NHPs are shown in Table 2.

Table 2. Respondents' perceived role of nutritionists with regard to NHPs (n=295)

	n	%
Educate the clients to consider "food first" before using an NHP	250	84.7
Assess the benefits NHPs could bring depending on the context and health condition of the consumer	210	71.2
Ensure collaboration with the interdisciplinary team about NHPs	193	65.4
Assess the safety of NHPs and the possible interaction with drugs, other NHPs and foods	163	55.3
Recommend the use of NHPs to nutritionally vulnerable consumers	159	53.9
Educate the public about NHPs and their regulation, labels and allegations	126	42.7
Identify the prevalence and reason for the use of NHPs	97	32.9
Critique, evaluate and conduct research on NHPs	89	30.2
Document the use of NHPs and client clinical response to report side effects	88	29.8
Closely monitor the health of a client who consumes an NHP	77	26.1
Monitor the laws and ethical considerations on NHPs	54	18.3
Monitor the quality of the NHPs according to good manufacturing practices	25	8.5

Respondents' level of agreement about the effectiveness of specific NHPs is shown in Table 3.

Table 3. Quebec registered dietitians' agreement regarding the efficacy of specific NHPs (n=295)

	Totally in disagreement % (n)	Somewhat in disagreement % (n)	In agreement % (n)	Somewhat in agreement % (n)	Totally in agreement % (n)	I don't know % (n)
Regarding the efficacy of an omega-3 supplement for cardiovascular disease risk	0.3% (1)	6.1% (18)	19.0% (56)	47.1% (139)	17.3% (51)	4.4% (13)
Regarding the efficacy of a mineral supplement for hyperactivity	5.4% (16)	16.6% (49)	11.5% (34)	2.4% (7)	11.5% (34)	58.3% (172)
Regarding the efficacy of a folic acid supplement to ensure normal pregnancy	0.7% (2)	0.3% (1)	2.0% (6)	7.8% (23)	82.4% (n=243)	1.0% (3)
Regarding the efficacy of a vitamin D and calcium supplement to decrease the risk of osteoporosis	0.3% (1)	1.7% (5)	6.1% (18)	33.9% (100)	50.5% (149)	1.7%(5)
Regarding the placebo effect of NHPs	0.7% (2)	1.7% (2)	6.8% (20)	29.8% (88)	43.7% (129)	11.5%(34)

Note. Dietitians' agreement was measured about specific NHPs.

3.4 Training Needs

The NHPs on which respondents request training are probiotics (73%), vitamins (63%), minerals (62%), fatty acid supplements (62%), medicinal plants (60%), amino acid supplements (56%), traditional medicines including Chinese medicines (37%) and homeopathic medicines (5%). Five percent reported no training needs.

Respondents cited a need for training on health conditions or topics linked to the use of NHPs such as aging (62%), cardiovascular diseases (57%) and regulatory and approval procedures of NHPs (57%). Also, of the 68 responses indicating other health concerns, pediatrics and digestive system were the most frequently reported, at 19%.

4. Discussion

Respondent profiles mainly comprised nutritionists from clinical and public health nutrition scopes of practice, which are representative of RD practices in Quebec. The Hirschkorn et al. study (Hirschkorn et al., 2013), conducted among 475 RDs from Ontario, also concluded that professionals from these areas of practice are more likely to have experience with NHPs and that they reported a demand for expertise on this subject.

Although 61% indicated having been trained on NHPs during their undergraduate studies in nutrition, 45% have learned independently about NHPs, suggesting that academic curriculums are incomplete. However, this learning method suggests that nutritionists are self-directed in their own professional development.

We should note that the naturopathy association of which 16% are members is neither regulated nor recognized by the *Office des professions du Québec*. This raises some issues, as a major motivation for nutritionists to become a member of this association may be so that insurers will consider them eligible for reimbursement of naturopath services; these same insurers do not to reimburse nutritionist services.

Regarding personal use of NHPs, the study conducted by Dickinson et al. (n=300) among nutritionists who consumed NHPs revealed that they were more likely to have a positive attitude towards the contribution of NHPs to health, which led to them recommending the products (Dickinson et al., 2012). More precisely, 96% consumed an NHP on a regular basis and 97% said that they had previously recommended an NHP to a client. Our results were similar; 91% of our respondents reported that they consume or have consumed an NHP, and 94% reported having recommended an NHP in the past year.

The review of the information sources recommended by the Academy of Nutrition and Dietetics relating to NHPs revealed that none of our respondents reported having consulted these sources of information. We note that most of these sources of information on NHPs must be purchased by Canadian nutritionists, while most of the information sources cited by our respondents are free. Furthermore, the use of websites published by NHP manufacturing companies as a source of information is a major concern, since the quality and objectivity of that information is questionable.

With respect to their professional responsibilities, nutritionists need to be vigilant about NHP regulations and approval mechanisms in and outside of Canada, considering the availability of unauthorized products for purchase. Lastly, nutritionists must also be aware of NHP interaction risks, either with medicines, other NHPs or food, in order to make safe recommendations (Government of Canada, 2012).

Respondents mentioned that pharmacists and nutritionists should be NHP specialists. According to *l'Ordre des pharmaciens du Quebec*, pharmacists have several responsibilities regarding the use of medications but are not responsible for the quality and the safety of over-the-counter medicines, including NHPs, which falls under the jurisdiction of Health Canada (Ordre des Pharmaciens du Québec, 2011). RDs referring to pharmacists for NHPs information is supported by the findings of Cashman et al., in which American nutritionists (n=160) said that pharmacists as well as naturopaths, doctors of traditional medicine or herbalists should be the pillars of information for NHPs. Moreover, Canadian nutritionists from the Atlantic provinces (Mihalynuk & Whiting, 2013), also considered pharmacists as being the most qualified health professionals on NHPs because of their presence at the point of sale. Interestingly, these same respondents discussed the importance of the interdisciplinary role of professionals to support collaboration, communication and knowledge about NHPs so that the messages conveyed to the public are consistent. Accordingly, two thirds of the respondents from our survey indicated that one of the three main responsibilities of nutritionists regarding NHPs was to cooperate with interdisciplinary teams.

The expected roles of nutritionists offered in the survey were drawn from the Academy of Nutrition and Dietetics. Four of the roles were perceived as expected roles by fewer than 30% of nutritionists. The roles refer to recommending high-quality NHPs, providing information about their use and side effects, monitoring the health conditions of clients who consume NHPs, and staying abreast of the legislative and ethical consideration regarding NHPs. These results deserve more attention since these responsibilities are certainly related to client safety.

The scientific literature on health and NHPs effectiveness supports better training on NHP use for aging (Rietsema, 2014), cardiovascular diseases (AbuMweis et al., 2014), oncology (Tutanc et al., 2013), endocrinology (Harbilas et al.), nephrology (Brazier, 2008), obesity (Government of Canada, 2012), menopause (Crawford et al., 2013), pregnancy (Health Canada, 2009) and for athletes (Frechette, 2009). Among other subjects reported by nutritionists, the pediatric sector deserves our attention, especially with regard to the effectiveness of NHPs for treating hyperactivity, a practice with which 14% of our respondents expressed agreement. However, for most complementary and alternative treatments, such as the use of an essential fatty acid supplement among children with attention deficits, there is not enough evidence to support recommendations (Bader & Adesman, 2012). On the other hand, professionals should be able to properly advise the parents of children who choose complementary and alternative approaches.

Findings from this study should not be compared to results obtained from the United States, since the regulation of NHPs differs between Canada and the United States. In the United States, the Food and Drug Administration monitors the NHPs after consumption and therefore is more reactive to product safety. Thus, a health professional's country of origin certainly affects his or her attitudes and use of the NHPs.

5. Conclusions

This research documented the experiences of Quebec registered dietitians' in terms of personal uses, information searches, attitudes and training needs with regard to NHPs. Quebec nutritionists often receive questions about NHPs. They have a positive attitude toward these products and are aware of their role in this matter. In addition to personally using NHPs, they also recommend them to clients for various health conditions. However, they express a need for training on different NHPs and on several health considerations or topics related to NHPs. The study also highlights a need for reliable sources of information and for interdisciplinary work.

It would be useful to inform professionals about the best manufacturing processes of NHPs currently in use (Chiu et al., 2014), expose them to the practices of alternative medicine practitioners and suggest protocols for recommending the use of NHPs. With proper training, the nutritionist should be able to assume the responsibilities suggested in this article, including assessing the benefits associated with the use of NHPs, considering the safety concerns, potential interactions and product quality, documenting their use and effects and staying abreast of the various regulations. The central role that food plays should also be emphasized (Academy of Nutrition and Dietetics, 2009).

The Internet would certainly be an effective tool for ensuring a continuum of training. The Dietitians of Canada association currently provides online training courses on vitamin and mineral supplements. A study has shown that these courses increase participants' knowledge by about 10% (Dietitians of Canada, 2013). Thus, it could be of interest to develop online courses on different NHPs, including medicinal plants, probiotics, homeopathic remedies, traditional remedies, amino acid supplements and essential fatty acids. These courses would enable RDs to offer better advice to the population and promote a complementary medicine approach with sound scientific evidence as a new area of nutritional practice.

Acknowledgments

We thank the *Ordre professionnel des diététistes du Québec,* which contributed to this study by sending an email invitation to participate to all RDs.

References

AbuMweis, S. S., Marinangeli, C. P., Frohlich, J., & Jones, P. J. (2014). Implementing phytosterols into medical practice as a cholesterol-lowering strategy: overview of efficacy, effectiveness, and safety. *Canadian Journal of Cardiology, 30*(10), 1225-1232. https://doi.org/10.1016/j.cjca.2014.04.022

Academy of Nutrition and Dietetics. (2009). Position of the American Dietetic Association: Nutrient Supplementation. *Journal of American Dietetic Association, 109*(12), 2073-2085. https://doi.org/10.1016/j.jada.2009.10.020

Bader, A., & Adesman, A. (2012). Complementary and alternative therapies for children and adolescents with ADHD. *Current Opinion in Pediatrics, 24*(6), 760-769. https://doi.org/10.1097/MOP.0b013e32835a1a5f

Brazier, J. L. (2008). *Quand les produits de santé naturels peuvent être dangereux pour le rein...Prévenir la néphropathie : un nouveau défi. Nutrition-Science en évolution. Rev OPDQ, 6*(1), 14-16.

Canadian Paediatric Society. (2005). Children and natural health products: What a clinician should know. *Paediatrics & Child Health, 10*(4), 236-241.

Cashman, L. S., Burns, J. T., Otieno, I. M., & Fung, T. (2003). Massachusetts Registered Dietitians' Knowledge, Attitudes, Opinions, Personal Use, and Recommendations to Clients About Herbal Supplements. *Journal of Alternative and Complementary Medicine, 9*(5), 735-46. https://doi.org/10.1089/107555303322524580

Chiu, Y. H., Lin, S. L., Tsai, J. J., & Lin, M. Y. (2014). Probiotic actions on diseases: implications for therapeutic treatments. *Food and Function, 5*(4), 625-634. https://doi.org/10.1039/c3fo60600g

Crawford, S. L., Jackson, E. A., Churchill, L., Lampe, J. W., Leung, K., & Ockene, J. K. (2013). Impact of dose, frequency of administration, and equal production on efficacy of isoflavones for menopausal hot flashes: a pilot randomized trial. *Menopause, 20*(9), 936-945. https://doi.org/10.1097/GME.0b013e3182829413

Decker-Touger, R., & Thomson, C. A. (2003). Complementary and alternative medicine: Competencies for dietetics professionals. *Journal of American Dietetic Association, 103*(11), 1465-1469. https://doi.org/10.1016/j.jada.2003.08.015

Dickinson, A., Bonci, L., Boyon, N., & Franco, J. C. (2012). Dietitians use and recommend dietary supplements: report of a survey. *Journal of Nutrition, 11*(14), 1-7. https://doi.org/10.1186/1475-2891-11-14

Dietitians of Canada. (2013). *Online Courses. Dietary Supplements - Vitamin and mineral supplements.* Retrieved   from http://www.dietitians.ca/Knowledge-Center/Learning-On-Demand/Learning-On-Demand-Store/lodStorePro duct.aspx?guid=516d04e7-c0cb-419e-bd5e-82bdb57675cf

Dorsch, K. D., & Bell, A. (2005). Dietary supplement use in adolescents. *Current opinion in pediatrics, 17,* 653-657. https://doi.org/10.1097/01.mop.0000172819.72013.5d

Food and Drug Administration. (2014). *Dietary Supplements.* Retrieved from http://www.fda.gov/food/dietarysupplements/default.htm_

Frechette, M. (2009). *Utilisation des suppléments alimentaires chez les athlètes d'élite québécois.* (Master's thesis, Université de Montréal, Québec, Canada).

Gardiner, P., Woods, C., & Kemper, K. J. (2006). Dietary supplement use among health-care professionals enrolled in an online curriculum on herbs and dietary supplements. *BMC Complementary and Alternative Medicine, 6*(21), 18. https://doi.org/10.1186/1472-6882-6-21

Goulet, F. (2006). Savez-vous si vos patients consomment des produits de santé naturels? *Bulletin LE COLLÈGE. Collège des médecins du Québec.*

Government of Canada. (2012). *Drug products that interact with warfarin.*Retrieved from http://healthycanadians.gc.ca/drugs-products-medicaments-produits/buying-using-achat-utilisation/products -canada-produits/drugs-devices-medicaments-instruments/warfarin-eng.php?

_ga=1.166613821.281827418.1467921823

Government of Canada. (2012). *The safe use of health products for weight lost. Healthy Canadians.* Retrieved from http://healthycanadians.gc.ca/drugs-products-medicaments-produits/buying-using-achat-utilisation/products-canada-produits/drugs-devices-medicaments-instruments/weight-loss-amaigrissants-eng.php?_ga=1.67138450.281827418.1467921823

Harbilas, D., Martineau, L. C., Harris, C. S., Adeyiwola-Spoor, D. C., Saleem, A., Lambert, J., Caves, D., Haddad, P. S. et al. Evaluation of the antidiabetic potential of selected medicinal plant extracts from the Canadian boreal forest used to treat symptoms of diabetes: part II. *Journal of Alternative and Complementary Medicine, 87*(6), 479-492. https://doi.org/10.1139/y09-029

Health Canada. (2009). *Prenatal nutrition guidelines for health professionals–background on Canada's food guide.*

Health Canada. (2013). *The approach to natural health products.* Retrieved from http://www.hc-sc.gc.ca/dhp-mps/prodnatur/nhp-new-nouvelle-psn-eng.php

Health Canada. (2016). *About natural health products regulation in Canada.* Retrieved from http://www.hc-sc.gc.ca/dhp-mps/prodnatur/about-apropos/index-eng.php

Hirschkorn, K., Rishma, W., & Heather, B. (2013). The role of natural health products (NHPs) in dietetic practice: results from a survey of Canadian dietitians. BMC *Complementary and Alternative Medicine, 13*(156), 1-8. https://doi.org/10.1186/1472-6882-13-156

Mihalynuk, T., & Whiting, S. (2013). The role of dietitians in providing guidance on the use of natural health products. *Canadian Journal of Dietetic Practice and Research, 74*(2), 58-62. https://doi.org/10.3148/74.2.2013.58

National Institutes of Health. (2016, June 22). *Americans spent $30,2 billion out-of-pocket on complementary health approaches.* Retrieved June 22, 2016, from https://www.nih.gov/news-events/americans-spent-302-billion-out-pocket-complementary-health-approaches_

Ordre des Pharmaciens du Québec. (2011). *Natural health products.* Retrieved from http://www.opq.org/fr-CA/grand-public/comportement-responsable/produits-de-sante-naturels/_

Pomazak, E. (2009). *The diet-s survey (dietetic interns' early thoughts about supplements): the relationship of dietetic interns' knowledge, attitudes and confidence of dietary supplements.* (Master's thesis, Northern Illinois University, Illinois, United States).

Rietsema, W. (2014). Unexpected recovery of moderate cognitive impairment on treatment with oral methylcobalamin. *Journal of the American Geriatrics Society, 62*(8), 1611-1612. https://doi.org/10.1111/jgs.12966

Thomson, C., Diekman, C., Fragakis, A. S., Meerschaert, C., Holler, H., Devlin, C., & American, Dietetic, Association. (2002). Guidelines regarding the recommendation and sale of dietary supplements. *Journal of American Dietetic Association, 102*(8), 1158-1164. https://doi.org/10.1016/S0002-8223(02)90257-9

Tutanc, O. D., Aydogan, A., Akkucuk, S., Sunbul, A. T., Zincircioglu, S. B., Alpagat, G., & Erden, E. S. (2013). The efficacy of oral glutamine in prevention of acute radiotherapy-induced esophagitis in patients with lung cancer. *Contemporary Oncology, 17*(6), 520-524. https://doi.org/10.5114/wo.2013.38912

20

Domestic Cooking Effects of Bambara Groundnuts and Common Beans in the Antioxidant Properties and Polyphenol Profiles

V. Nyau[1], S. Prakash[2], J. Rodrigues[3] & J. Farrant[3]

[1]Department of Food Science and Nutrition, University of Zambia, Lusaka, Zambia

[2]Department of Chemistry, University of Zambia, Lusaka, Zambia

[3]Department of Molecular and Cell Biology, University of Cape Town, 7701 Rondebosch, South Africa

Correspondence: V. Nyau, Department of Food Science and Nutrition, University of Zambia, Lusaka, Zambia. E-mail: vincentnyau@yahoo.co.uk

Abstract

Processing of legumes before consumption has several effects on micronutrients, macronutrients and phytonutrients. This study was undertaken to investigate the effect of domestic processing on antioxidant activities and phenolic phytochemicals of the red bambara groundnuts and red beans. The study employed *in vitro* antioxidant assays (DPPH and FRAP) to screen for antioxidant properties, HPLC-PDA-ESI-MS and Folin Ciocalteu assay to screen for phenolic phytochemical profiles. Domestic cooking displayed positive effects on the antioxidant activity and phenolic phytochemical profiles of the two legumes. The free radical scavenging speed increased 10-fold in the methanolic extract from cooked red bambara groundnuts compared to uncooked. By contrast, the free radical scavenging speed increased 20-fold in the methanolic extract from cooked red beans compared to uncooked. HPLC-PDA-ESI-MS profiles of the cooked red bambara groundnuts and red beans revealed a number of emergent phenolic compounds, mainly flavonoids. These data indicate that cooking appear to enhance the nutraceutical profiles of the legumes investigated.

Keywords: domestic cooking, polyphenols, antioxidant, bambara groundnuts, common beans

1. Introduction

Generally most food legumes have to be processed before consumption. Different processing methods are applied depending on the intended use of the final product and the availability of the processing facilities. The most common method of processing is domestic cooking and involves boiling the seed legumes until soft using fire wood or electricity as heating sources.

Concentration of plant secondary metabolites having antioxidant activity is affected by a number of factors, including genetics and growing conditions (Kalt, 2005). Processing is another important factor that can impact total antioxidant activity (Papas, 1996). Knowledge about the fate of total antioxidant activity as a result of home processing may have a significant impact on consumers' food selection and processing (Danesi, 2009). Few studies have been done on seed legumes to investigate the effect of domestic processing on the total antioxidant activity and the phytochemical profiles. Although common beans are widely consumed all over the world, very little information is available in the literature regarding the changes in total phenols, total flavonoids and antioxidant activities following food preparation methods (Akillioglu and Karakaya, 2010).

In this study, the effect of domestic cooking on the antioxidant activities and phenolic phytochemical profiles of the red Zambian market classes of common beans and bambara groundnuts was investigated. Previous studies have shown that red beans possess excellent antioxidant activities in the raw form (Nyau et al., 2016) and therefore need to be evaluated further to determine hydrothermal effects on these activities. On the other hand, HPLC-PDA-ESI-MS profiling of the red bambara groundnuts revealed more phenolic compounds than the brown bambara groundnuts investigated (Nyau et al., 2015). This prompted further investigations in the current study to assess the changes in the polyphenolic phytochemical profiles that occur due to hydrothermal cooking. The methanolic extract was used because most likely it contained both hydrophilic and hydrophobic compounds. Antioxidant properties of cooked seeds were measured based on their free radical scavenging activity and ferric reducing power. The free radical scavenging ability of the methanolic extracts is reported on the basis of kinetic behaviour of the DPPH free radicals with antioxidants in the two legumes. The ferric reducing power is reported on the basis of the number of mmoles Fe^{2+} produced in the reduction of Fe^{3+} by the antioxidants in the extracts.

2. Materials and Methods

2.1 Sample Collection

The red common beans (*Phaseolus vulgaris* L) and red bambara groundnuts (*Vigna subterranea* L. Verdc) were obtained directly from the farmers immediately after harvest. The two legumes are shown in Figure 1. In order to make the samples representative, an attempt was made to collect each seed type from 15 farmers in the growing areas with not less than 0.5 kg per farmer and a total of 15 samples per seed was collected.

| Red bambara groundnuts | Red beans |

Figure 1. Red bambara groundnuts and red beans investigated

2.2 Cooking Treatment

Approximately 400 g of the seeds in 1.5 L water of each legume were cooked using the traditional cooking method commonly used in Zambia. Seeds were boiled in tap water on a hot plate at the temperature of 100 ± 5 °C, until they felt soft using the finger compression test. The finger compression test is a sensory based approach in which texture is treated as a perception or how a food material feels with the fingers. It is a very rapid and useful method employed to determine firmness or the degree of softness by consumers (Mitcham et al., 1996). In this experiment, the beans were considered soft if they were able to deform under moderate pressure when compressed between the index finger and the thumb. It took 410 and 430 minutes for the red beans and Bambara groundnuts to cook respectively.

2.3 Preparation of Extracts of Cooked Beans and Bambara Groundnuts

The cooked seeds together with the water that remained after cooking were immediately frozen at – 80 °C and freeze dried to obtain the dried material that was later ground to a powder. Approximately 15 g of seed powder in 150 ml of 70% methanol was sonicated for 30 minutes at 25°C using the Eumax UD500SH 40 kHz ultrasonic bath. After extraction, the mixture was centrifuged at a speed of 10,000 rpm for 15 minute using Beckman Coulter JE centrifuge. The resulting supernatant was first concentrated to 30 ml by evaporation under reduced pressure in a rotary evaporator (Buchi R-210 model, Switzerland) to remove methanol. The extract was then frozen at -80 °C and freeze dried to obtain a powdered methanolic extract using the Telstar LyoQuest -85 freeze dryer. The freeze dried extracts were stored at -4°C until further analysis

2.4 Preparation of Extracts from the Raw Beans and Bambara Groundnuts

Approximately 15 g of seed powder in 150 ml of 70% methanol was sonicated for 30 minutes at 25°C using the Eumax UD500SH 40 kHz ultrasonic bath. After extraction, the mixture was centrifuged at a speed of 10,000 rpm for 15 minute in Beckman Coulter JE centrifuge. The supernatant was first concentrated to 30 ml by evaporation under reduced pressure in a rotary evaporator (Buchi R-210 model, Switzerland) to remove methanol. The extract was then frozen at -80 °C and freeze dried to obtain a powdered methanolic extract using the Telstar LyoQuest -85 freeze dryer. The freeze dried aqueous and methanolic extracts were stored at -4°C until further analysis.

2.5 Determination of Total Polyphenols

Total polyphenols were determined by the Folin Ciocalteu assay according to the method of Makkar et al., (2000). To 100 μl of sample extract, 400 μl of distilled water was added followed by the addition of 250 μl Folin Ciocalteu reagent. 20 % Sodium carbonate (1.25 ml) was then added and the mixture was incubated for 40 min. Absorbancies were read at 725 nm after 40 minutes using a spectrophotometer (Ultrospec 1000 model, England) against the blank (70% methanol or water) depending on whether it was the water or 70 % methanol extract. The amount of total polyphenols was calculated as gallic acid equilvalents from the calibration curve of gallic acid standard solution and expressed as mg gallic acid equivalents/ 100 g DW. The experiment was conducted three times and all measurements were performed in triplicate.

2.6 HPLC-DAD-ESI-MS Instrumentation and Chromatographic Conditions

The freeze dried 70% methanolic extract powder of common beans were analysed using a Waters ZMD 4000 system that was equipped with a Waters 2690 HPLC, Waters 996 photodiode array, ZMD mass spectrophotometer, 717 Plus autosampler, and a quaternary pump (Waters Corp, Milford, MA, USA). Separations were carried out on a 300 x 3.9 mm, 4 μm reversed phase Nova-Pak C18 (Waters) column that was maintained at 40°C. The photodiode array detector (PDA) was linked directly to a sprayer needle where ions were generated by electrospray ionisation (ESI) in a negative mode. The mobile phase A consisted of 5% (v/v) acetonitrile/water, containing 0.1% (v/v) formic acid and mobile phase B consisted of 100% acetonitrile containing 0.1% (v/v) formic acid. The sample was injected at a volume of 25 μl. The elution profile consisted of a stepwise linear gradient from 0% to 28% solvent B for 22 minutes with a flow rate of 0.3 ml/min. The PDA detector was set to a scanning range of 200 to 700 nm and the UV-Vis absorption spectra were recorded online during the HPLC analysis. Phenolic acids and flavonols were detected at 280 and 360 nm, respectively. Continuous mass spectra data were recorded on a full scan negative ionisation mode for a mass range of m/z 85 to 1000. The capillary voltage was set at 2.5 kV, the cone at 20 V and the extractor at 5 V. Nitrogen gas was used for nebulising and drying at different fragmentation voltages. Data acquisition was controlled using MassLynx 4.1 (Micromass, Waters Corp., Beverly, MA, USA). The gradient solvent system used for the analysis of phenolic compounds is summarised in Table 1.

Table 1. Gradient solvent system for analysis of phenolic compounds by HPLC -PDA -ESI-MS

Time (minutes)	Composition of the mobile phase (%)	
	*Mobile phase (A)	*Mobile phase (B)
1	100	0
22	72	28
22.50	60	40
23	0	100
24.50	0	100
25	100	0
26	100	0

Mobile phase A consisted of 5% (v/v) acetonitrile/water, containing 0.1% (v/v) formic; mobile phase B consisted of 100% acetonitrile containing 0.1% (v/v) formic acid.

2.7 Preparation of the Samples for HPLC-DAD-ESI-MS Analysis

Preparation of the test solution for HPLC-DAD-ESI-MS was done according to the procedure by Gülçin et al., (2010) with slight modifications. One hundred mg of the freeze dried 70% methanolic extract was dissolved in 5 ml of ethanol-water (50:50 v/v). One hundred μl of the prepared extract was transferred into a 5 ml volumetric flask and diluted to the volume with ethanol-water (50:50). From the final solution, an aliquot of 1.5 ml was transferred into a capped autosampler vial and 25 μl of the sample was injected into the HPLC-DAD-ESI-MS system. Identification of phenolic compounds was accomplished using UV spectra and ESI-MS spectral data and by comparison with published data reported in the literature. Authentic standards were also used where available by comparing their chromatograms with those of the samples. The available standards were *t*-ferrulic acid, gallic acid, salicylic acid, *p*-coumaric acid, epicatechin and catechin.

2.8 HPLC-PDA-ESI-MS Quantification of Phenolic Compounds

Quantification of individual phenolic compounds could only be done where authentic standards were available. The available standards were *t*-ferrulic acid, gallic acid, salicylic acid, *p*-coumaric acid, epicatechin and catechin. The concentrations of ferrulic acid, gallic acid, salicylic acid, *p*-coumaric acid, epicatechin and catechin were obtained from the linear regression equations of the standard curves (Gülçin et al., 2010). The experiment was conducted three times and all measurements were done in duplicate.

2.9 Determination of Free Radical Scavenging Activity of the Bambara Groundnuts

DPPH stable free radicals are reduced to DPPH-H leading to discoloration from purple to yellow and consequently a decrease in absorbance. The degree of discoloration indicates the scavenging potential of the antioxidant compounds (Pal et al., 2008). This assay therefore involves the measurement of hydrogen atom transfer or electron donation from a potential antioxidant to free radical molecules (Becker et al., 2004). First, it was important to study the kinetic behaviour of the extracts towards DPPH free radicals when the freeze dried extracts from each legume were added at the same concentration. The knowledge of the kinetics of atom transfer is important because free radicals in the organism are short-lived species, implying that the impact of a substance as an antioxidant depends on its fast reactivity towards free radicals (Villaño et al., 2007). The free radical scavenging kinetic determinations were adapted from (Villaño et al., 2007). Under the experimental conditions used, the DPPH concentration was in large excess with respect to that of the extracts in order to follow pseudo first-order kinetics. This was done to exhaust the hydrogen donating capacity of the extracts. The excess concentration of DPPH (200 mM) was determined to be the optimum concentration after performing a number of runs with the extracts. This was the only way the excess DPPH concentration could be determined since it was not possible to work it out based on the DPPH: antioxidant molar ratios as the antioxidants in the extracts were not pure compounds. In the assessment of the kinetic behaviour, 2 ml of the extracts were added at the same concentration (400 µg / ml) to 2 ml of DPPH radical solution (200 mM) prepared in 95% methanol. The reaction was run at room temperature within a time period of 80 minutes. The absorbances of the mixture were automatically measured every 10 seconds using the spectrophotometer at 517 nm connected to a computer and the output was displayed using SWIFT 1000 software (Ultraspec 1000 model, England). From the reaction between an antioxidant and DPPH;

(DPPH) + (Y-H) → DPPH-H + (Y), it can be deduced that:

$$-\frac{d[DPPH]}{dt} = k[DPPH][Y-H] \qquad (1)$$

Considering that DPPH was in excess and therefore the experiment was under pseudo first-order conditions, one can say:

$$InA = InA_o - kt \qquad (2)$$

Where A_0 is the absorbance of the reaction mixture (DPPH and the extract) at $t = 0$; A is the absorbance of the reaction mixture (DPPH and extract) at time t.

The pseudo first order rate constant 'k' for the reaction of the antioxidants in the extracts and DPPH in the first seconds of the reaction was calculated from the slopes of InA versus time plots.

The percentage of DPPH remaining at any time t can be determined as:

$$\%DPPH_{remaining} = \frac{A_t}{A_0} X100 \qquad (3)$$

(Villaño et al., 2007)

Where A_0 is the initial absorbance and A_t is the absorbance at time = t, both measured at 517 nm respectively. Plots of percentage DPPH versus time were constructed to show the disappearance pattern of the DPPH with time in the presence of each extract.

2.10 Determination of Ferric Reducing Antioxidant Power (FRAP)

The FRAP assay was used to determine the ferric reducing antioxidant power of the two legumes (Benzie, 1996). The method measures the ferric reducing ability of the antioxidants compounds in the extracts. At low pH, ferric-2,4,6-tri-2-pyridyl-s-triazine (TPTZ) complex (Fe^{3+} TPTZ) is reduced to the ferrous form Fe^{2+} in the presence of the antioxidant producing an intense blue colour with an absorption maximum at 593 nm. Powdered sample of bambara groundnuts or beans (5 g) in 50 ml of 70 % methanol or water was sonicated for 30 minutes at 25 °C followed by centrifugation at 10,000 rpm for 15 minutes at 4 °C to obtain a clear supernatant. Working FRAP reagent was prepared by mixing 25 ml of acetate buffer (300 mM, pH 3.6); 2.5 ml ferric chloride solution (prepared by dissolving 54 mg ferric chloride in 10 ml distilled water) and 2.5 ml TPTZ solution (prepared by dissolving 31 mg TPTZ in 40 mM HCl at 50 °C). The mixture was placed in a water bath at 37 °C for 10 minutes. The assay was performed as follows: 1 ml of water and 80 μl of the test sample were pipetted into a cuvette. About 600 μl of the incubated FRAP reagent was added to the cuvette and mixed by inversion. A reagent blank was prepared as above with 80 μl water added instead of the test sample. The change in absorbance was recorded at 593 nm using a spectrophotometer after exactly 4 minutes (Ultrospec 1000 model, England). The amount of Fe^{2+} produced from the reduction of Fe^{3+} by the extract was calculated from the standard curve prepared from ferrous sulphate solution and results were expressed as mg Fe^{2+} / 100 g dry sample. The experiment was conducted three times and all measurements were performed in triplicate.

2.11 Statistical Analysis

Statistical analysis was performed using S-PLUS 6 Windows Professional 2001. Experimental results were expressed as mean values ± standard error. Data was analysed using two-sample t-test. Values at $p < 0.05$ were considered statistically significant.

3. Results

3.1 Free Radical Scavenging Activity of Cooked Red Bambara Groundnuts and Red Beans

Figure 2 presents the disapearance of the DPPH free radicals in the presence of methanolic extract of cooked and uncooked red bambara groundnuts. The initial decay of DPPH was faster in the presence of methanolic extracts from the cooked red bambara groundnuts than in the uncooked. The pseudo first-order rate constant was 0.43 min^{-1} in the cooked and 0.042 min^{-1} in the uncooked bambara groundnuts (Table 2). This is a very desirable property for the antioxidants because free radicals are very fast reactive species that equally need fast- acting scavengers. There was a significant difference in the amount of DPPH free radicals scavenged at the end of the incubation time by the extracts from the uncooked and cooked red bambara groundnuts. It was intriguing to note that the amount of the DPPH free radicals scavenged by the extracts from the cooked red bambara groundnuts was similar to that scavenged by the positive reference standard trolox towards the end of the assay period chosen (Table 2). This observation implies that the cooked nuts most likely, contained more antioxidant compounds than the uncooked nuts.

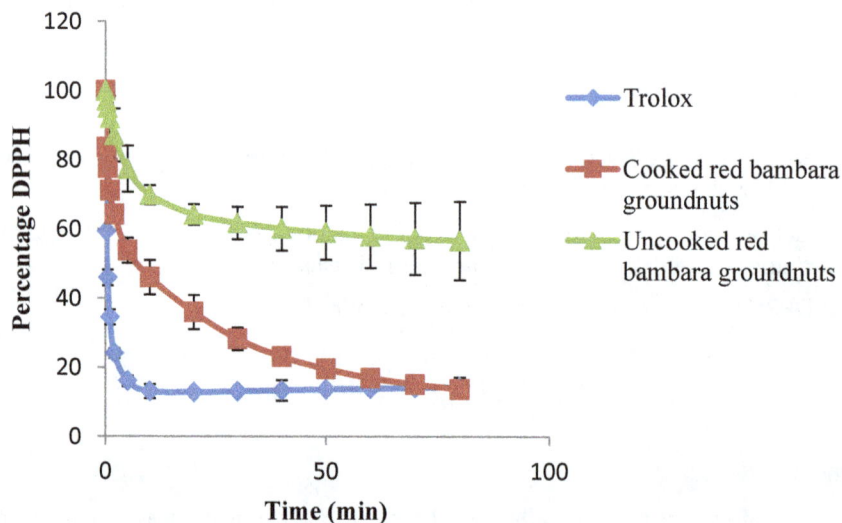

Figure 2. Disappearance pattern of DPPH free radicals with time in the presence of 70% methanol extracts of cooked and uncooked red bambara groundnuts

Table 2. Pseudo-first order rate constant of antiradical (Y-H) in cooked red bambara groundnuts and red beans and the amount of DPPH scavenged after 80 minutes of incubation

Market classes of bambara groundnuts and common beans		Pseudo-first order rate constant (K) [min^{-1}]	Amount DPPH quenched [%] after 80 minutes incubation
Red bambara groundnuts	Control (uncooked)	0.042 [a]	44.01 ± 2.96 [a]
	Cooked	0.430 [b]	86.38 ± 1.36 [b]
Red beans	Control (uncooked)	0.050 [a]	62.01 ± 2.21 [a]
	Cooked	1.120 [b]	65.22 ± 2.36 [b]

Trolox K = 1.55, amount quenched in 80 minutes = 85.5%. Means in the same column for each legume type with different superscripts were significantly (p <0.05) different

The free radical scavenging pattern by antioxidants in the cooked and uncooked red beans is presented in Figure 3. The disappearance pattern of the DPPH in the presence of the extract from the cooked red beans was much faster than that of the uncooked. There was a very sharp initial decay with a pseudo first-order rate constant (K) of 1.12 (min^{-1}) for the cooked red beans compared to 0.05 (min^{-1}) for the uncooked beans (see Table 2 above). The cooked extract had a better pseudo first-order rate kinetic than the extract from the uncooked beans. As discussed above with respect to red bambara groundnuts, this is a very desirable attribute for the antioxidants because free radicals react extremely quickly and require fast-acting antioxidants.

The results from our study are in agreement with Akillioglu and Karakaya (2009) who reported the increase in the amount of DPPH free radicals scavenged when common beans were cooked by first soaking in water for 3 hours prior to cooking. However, the study by Akillioglu and Karakaya did not report the free radical scavenging pattern by elaborating on the kinetics of the DPPH free radicals and the antioxidants in common beans in the course of the reaction.

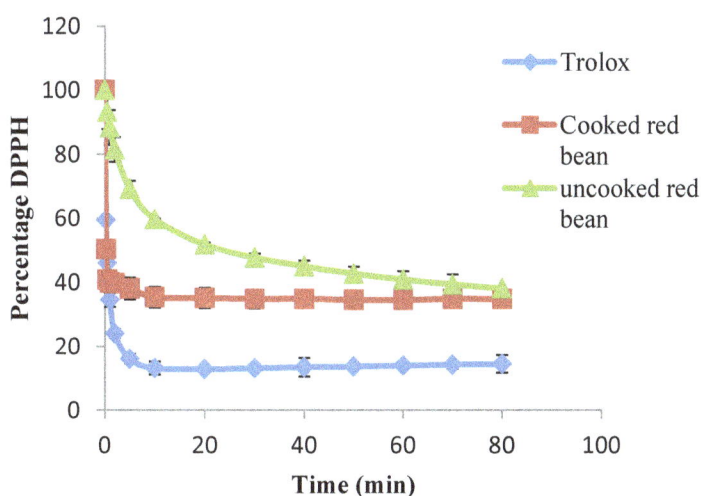

Figure 3. Disappearance pattern of DPPH free radicals with time in the presence of 70% methanolic extracts of cooked and uncooked red beans

3.2 Ferric Reducing Antioxidant Power of the Cooked Red Bambara Groundnuts and Red Beans

The FRAP values of the cooked red bambara groundnuts and red beans are presented in Table 3. The FRAP values of the uncooked and cooked red bambara groundnuts were not significantly different. However, the extract from the cooked red beans had greater reducing power than the extract from the uncooked material, suggesting that the cooked sample contained more antioxidant compounds or more reactive antioxidants than the uncooked.

Table 3. FRAP values for the cooked red bambara groundnuts and red beans

Market classes of bambara groundnuts and common beans		FRAP values (mmole Fe^{2+}/ 100 g DW)
Red Bambara groundnuts	Control (uncooked)	8.01 ± 0.41 [a]
	Cooked	8.55 ± 0.02 [a]
Red beans	Control (uncooked	4.81 ± 0.01 [b]
	Cooked	5.05 ± 0.04 [b]

Means in the same column for each legume type with different superscripts were significantly (p <0.05) different.

3.3 Total Polyphenol Levels in Cooked Red Bambara Groundnuts and Red Beans

Total polyphenol levels in cooked red bambara groundnuts and red beans are presented in Table 4. There was an increase in total polyphenol content in both the red bambara groundnuts and red beans after cooking. Total polyphenol content increased by 6% in red bambara groundnuts and 41% in red beans respectively.

Table 4. Total polyphenol concentration of cooked red bambara groundnuts and red beans

Market classes of bambara groundnuts and common beans		Total polyphenols (mg GAE / 100 g DW)
Red Bambara groundnuts	Control (uncooked)	109.12 ± 2.35 [a]
	Cooked	116.03 ± 3.27 [b]
Red beans	Control (uncooked)	84.50 ± 2.55 [a]
	Cooked	119.31 ± 3.04 [b]

Means in the same column for each legume type with different superscripts were significantly (p <0.05) different.

3.4 Levels of Individual Phenolic Compounds in Cooked Red Bambara Groundnuts and Red Beans

Quantitative contents of t-ferrulic acid, p-coumaric acid, catechin, epicatechin and salicylic acid of cooked red bambara groundnuts and red beans are presented in Figures 4 and 5. Generally, there was an increase in all phenolic compounds investigated in red bambara groundnuts, except for salicylic acid which showed a 59% decrease after cooking. Epicatechin increased by 92%, followed by catechin (54%), t-ferrulic acid (39%) and p-coumaric acid (30%) respectively. Similarly, there was an increase in the concentration of all the phenolic compounds investigated in red beans after cooking, except for salicylic acid which was completely missing. Phenolics in the flavonol category recorded higher increase than phenolic acids. Epicatechin increased by 96%, followed by catechin (80%), p-coumaric (41%) and t-ferrulic acid (40%) respectively.

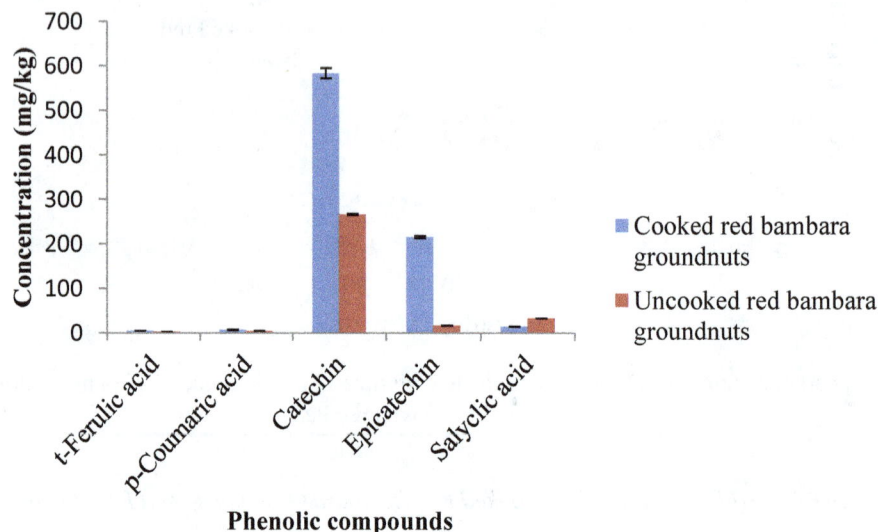

Figure 4. Concentration of individual phenolic compounds of 70% methanol extracts of cooked and uncooked red bambara groundnuts

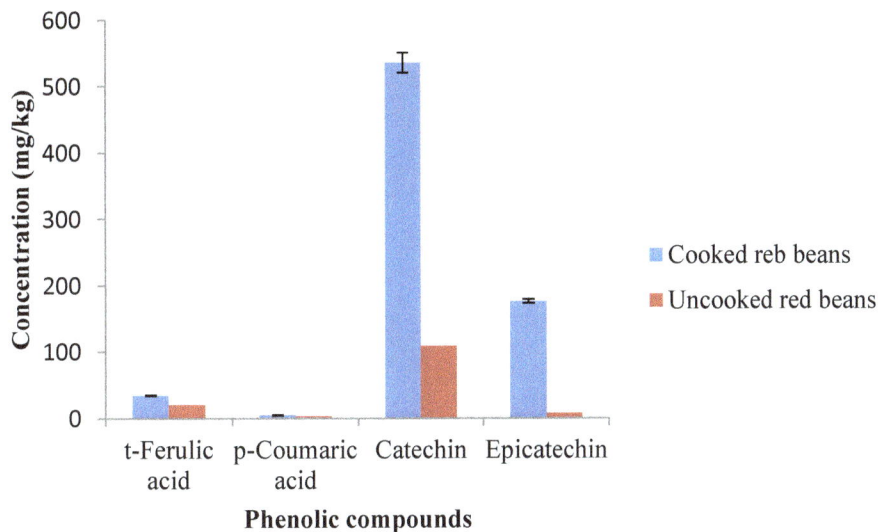

Figure 5. Concentration of individual phenolic compounds of 70% methanol extracts of cooked and uncooked red beans

3.5 HPLC-PDA-ESI-MS Profiles of the Cooked Red Bambara Groundnuts and Red Beans

The HPLC-PDA-ESI-MS chromatograms of methanolic extracts from the uncooked and cooked red bambara groundnuts and red beans are compared in Figures 6 & 7 (for red bambara groundnuts) and 8 & 9 (for red beans) respectively. A slight drift in the retention times of peaks was observed in these runs. This may be attributed to minor temperature changes or the increase in the back pressure in the column. Shifts in laboratory temperature and a slight increase of the back pressure in the column may cause drifting retention times in long automated operations (Waters Corporation, 2002).

Significant differences were observed in the HPLC-PDA-ESI-MS profiles of uncooked and cooked seeds for both legumes. In the profile of cooked red bambara groundnuts (Figure 6), there were ten emergent deprotonated molecules [M – H⁻] of m/z 341, 639, 305, 495, 577, 451, 323, 389, 625 and 463 respectively. In the profile of the cooked red beans (Figure 8), there were five new deprotonated molecules [M – H⁻] of m/z 164, 451, 608, 463 and 447 that emerged respectively. Differences in the number of emergent compounds between bambara groundnuts and common beans confirm that there are variations in the amount, types and distribution of phytochemicals in legumes.

For both red bambara groundnuts and red beans, there were variations in the peak heights (concentrations) for the compounds that were detected in the uncooked and cooked seeds. Peak heights were higher in the extracts of the cooked than the uncooked seeds, confirming that more phytochemicals are extracted when the seeds are cooked. These results suggested that there are changes in the phytochemical profiles of bambara groundnuts and common beans after the cooking process and validates our earlier observation on the changes in total polyphenol contents and concentrations of individual phenolic compounds. The observed changes are positive and dispel the generally held concerns that nutrients and phytonutrients are significantly lost during the cooking process.

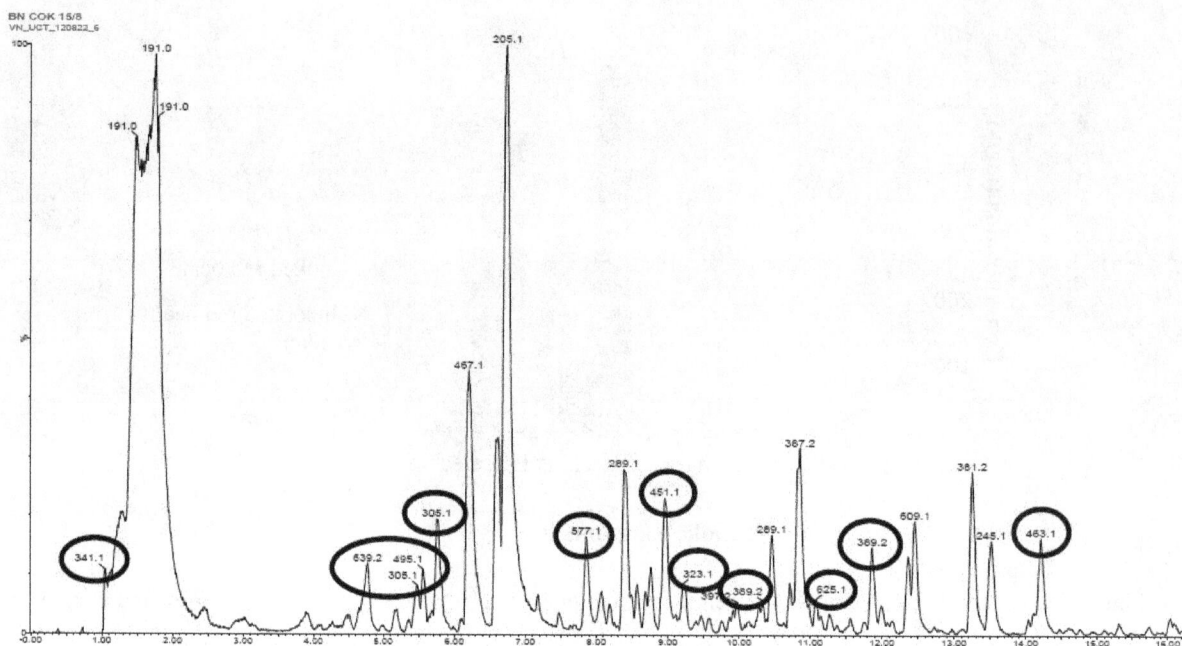

Figure 6. HPLC-PDA-ESI-MS chromatogram of 70% methanol extract of cooked red bambara groundnuts. represents emergent compounds after processing. Number on top of each peak is the m/z of the deprotonated molecule M – H] ¯ of each compound

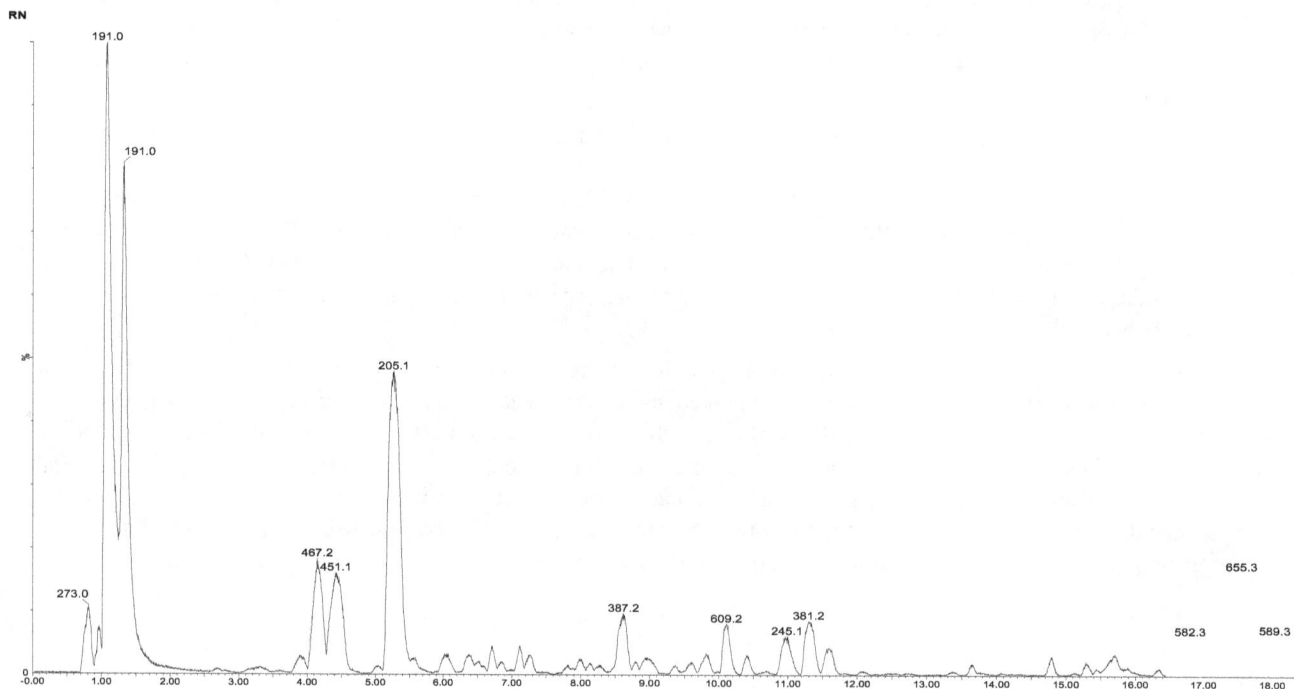

Figure 7. HPLC-PDA-ESI-MS chromatogram of 70% methanol extract of the uncooked red bambara groundnuts. Number on top of each peak is the m/z of the deprotonated molecule [M – H] – of each compound

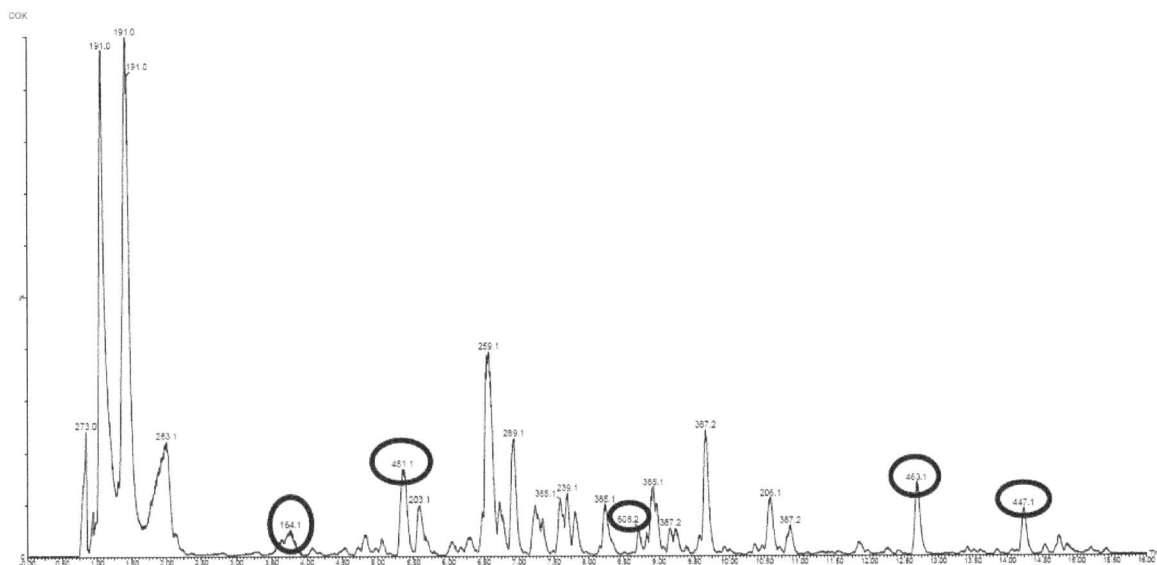

Figure 8. HPLC-PDA-ESI-MS chromatogram of 70% methanol extract of cooked red beans. represents emergent compounds after processing. Number on top of each peak is the m/z of the deprotonated molecule [M – H]⁻ of each compound

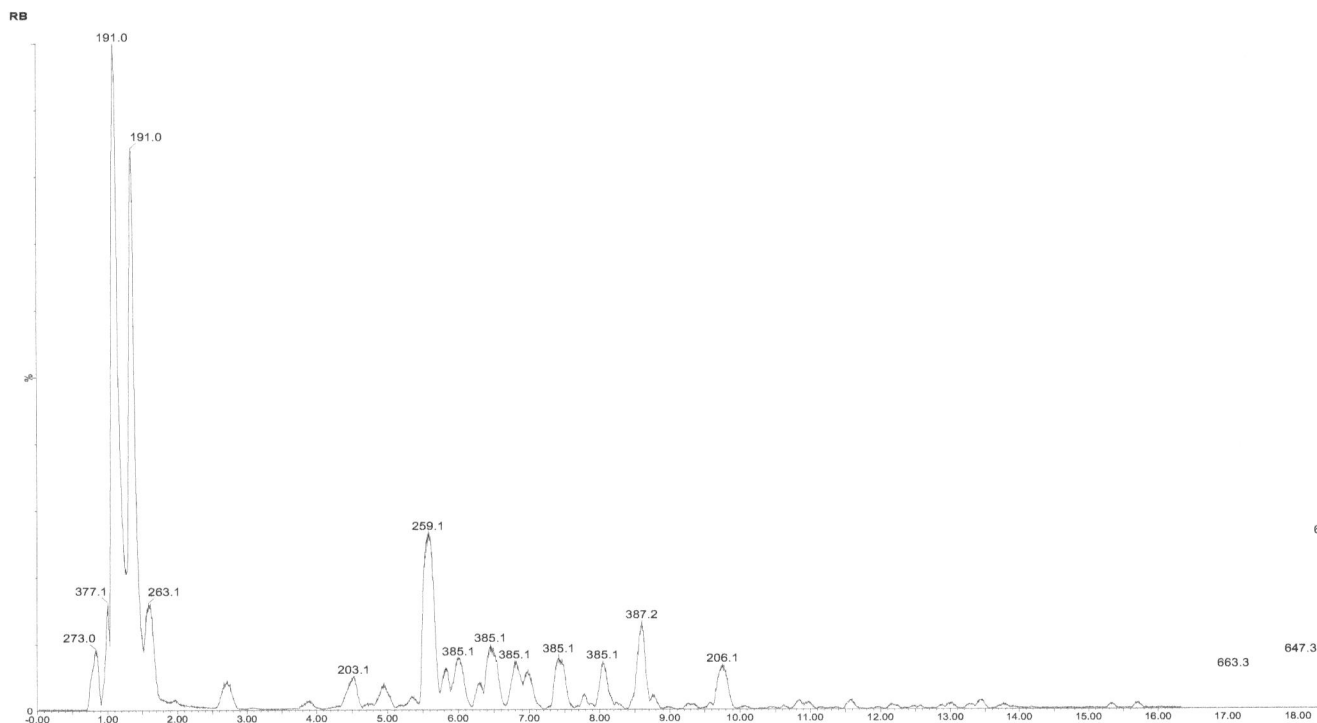

Figure 9. HPLC-PDA-ESI-MS chromatogram of 70% methanol extract of the uncooked red beans. Number on top of each peak is the m/z of the deprotonated molecule [M – H]⁻ of each compound

Based on the literature data and fragmentation pattern, attempts were made to identify the new emerging compounds in both the cooked red beans and red bambara groundnuts. Some emergent compounds that were tentatively identified are presented in Tables 5 and 6 respectively.

Table 5. Emerging phenolic compound tentatively identified in cooked red bambara groundnuts

Parent ion [m/z]	Fragment ions [m/z]	Tentative identification
341	179, 161	Caffeic acid hexoside [v]
305	289,263, 247,219	Gallocatechin [w]
389	185,157,143	Resveratrol glucoside [w]
625	239,179, 164	Caffeic acid derivative [x]

[v] Hassain et al., 2010, [w] Amandeep et al., 2010, [x] Rabaneda et al., 2003, [x] Gouveia and Castilho 2011

Table 6. Emerging phenolic compound tentatively identified in cooked red beans

Parent ion [m/z]	Fragment ions [m/z]	Tentative identification
451	289	Catechin glucoside [a]
608	301	Quercetin conjugate [b]
463	301	Quercetin-3-O-glucoside [b]
447	287	Kaempferol glucoside [b]

[a] Estrella et al., 2011, [b] Lin et al., 2008

4. Discussion

Cooking has displayed positive changes in the antioxidant activity and the polyphenol phytochemical profiles of both red beans and red bambara groundnuts. The increase in polyphenol concentration may be attributed to the release of bound polyphenols as a result of cooking process. Generally, complex polypenols have the ability to bind with other natural compounds within the cell. According to Harbaum (2007), cell wall bound phenolics are present as monomeric, dimeric, or oligomeric compounds which are esterified to the cell components (carbohydrates, lignin, pectin and proteins). These cell components may undergo various changes during the cooking process. For instance, proteins may undergo thermal denaturation in which a number of bonds are weakened and broken, resulting in the loss of the tertiary structure. The loss of protein structure affects the polyphenol-protein interaction and bound phenolic compounds may be released. On the other hand, long chain carbohydrates tend to break down into simple sugars when cooked while simple sugars may form syrups. These changes may affect carbohydrate-polyphenol interaction resulting in the release of bound phenolic compounds. The study by Akillioglu and Karakaya (2010) on the total polyphenol content of cooked common beans that were first soaked in water for 3 hours reported a 78% increase in total polyphenol content, which was higher compared to the increase that has been observed in the present study. This could be due to the differences in the methodology and the samples used. In our study, pretreatment by soaking the samples prior to cooking was not done. According to Akillioglu and Karakaya (2009), differences between total polyphenol content of the beans cooked without soaking and cooked after soaking in water may be due to the increase in the efficacy of the heat process to extract phenolic compounds from the food matrix.

Changes in the concentration of invidual phenolic compounds after cooking can be explained in two ways: firstly, the bound phenolics present are released during the heating process due to changes that take place to the components to which they bind. Secondly, phenolic compounds which exist as oligomers and polymers (condensed tannins or proanthocyanidins) may disintegrate to release different constitutive units when heated. According to Cheynier (2005), phenolic compounds are highly unstable and are rapidly transformed into various products when the plant cells are damaged (for instance, during processing). The large increase in the concentration of flavanols (catechin and epicatechin) after cooking as observed in this study supports the second assumption. Catechin, epicatechin, epigallocatechin and gallocatechin are the main monomeric units present in condensed tannins (Shadkami et al., 2009). Proanthocyanidins may occur as polymers containing as much as 50 catechin units (Wahle et al., 2010). HPLC-PDA-ESI-MS profiling of the extracts from the cooked seeds revealed ten and five new compounds in the cooked red bambara groundnuts and red beans respectively. In cooked red bambara groundnuts, new compounds tentatively identified include caffeic acid hexoside, gallocatechin, resveratrol glucoside and a caffeic acid derivative. New compounds tentatively identified in red beans include catechin glucoside, quercetin-3-O-glucoside, kaemferol glucoside and quercetin conjugate.

Thermal processes have a large influence on the availability of phenolic compounds in food regardless of the method used. Changes in the polyphenol content have been reported in other foods as a result of heating using other methods other than domestic boiling. Roasting of peanuts at 130 °C for 33min caused an increase in the total polyphenol content (Yu et al., 2005); similar results are observed for cashew nuts when they were roasted using the same processing conditions (Chandrasekara and Shahidi, 2011). In apple juice processing, an increase in temperature from 40 °C to 70 °C caused increase of flavonoid content by 50% (Gerard and Roberts, 2004). According to the study by Fuleki and Ricardo-DaSilva (2003), pasteurization of grape juice increased the concentration of catechin and procyanidins in cold pressed juices. In these situations, an increase of temperature improves the extraction of phenolic compounds from foods. In other foods, thermal processes have been reported to decrease the content of phenolic compounds. Significant losses are noticed in tomato sauce pasteurized at 115 °C for 5 min (Valverdú-Queralt et al., 2011). A loss of 40% in total phenolic compounds was observed in strawberries pasteurized at 85C for 5 min (Hartman et al., 2008). It seems reasonable to assume that each type of food responds differently with regards to polyphenol stability when subjected to different thermal processes. This may be ascribed to the difference in the food matrix. The food matrix can act as a barrier to heat effect or induce the degradation polyphenols that exist as polymers and oligomers. It is very difficult to dissociate the thermal processing effect from the food matrix effects when discussing changes that occur due to thermal processing (Irina and Mohamed, 2012).

The antioxidant activities of both the cooked red beans and red bambara groundnuts were higher compared to the uncooked. The free radical scavenging speed increased 10-fold in the presence of methanolic extract from cooked red bambara groundnuts compared to uncooked. By contrast, there was a 20-fold increase in the presence of the methanolic extract from cooked red beans compared to uncooked. This finding is of great significance because free radicals react very quickly and require equally fast-acting scavengers. Yet again, extracts from the cooked red beans had greater FRAP derived total antioxidant power compared to the uncooked. However, the FRAP values for the uncooked and cooked red bambara groundnuts were not significantly different. Changes in the antioxidant activities are positive and may be attributed to the increase in the concentrations of total polyphenols and individual phenolic compounds. Further, the increase in the speed of DPPH free radical scavenging reaction by the antioxidants in the cooked legumes may be attributed to the new emerging compounds. There is a possibility that the emergent compounds have good free radical scavenging abilities. An increase in the antioxidant activity of various foods subjected to thermal processes has been reported by many workers (Chandrasekara and Shahidi, 2011; Freeman et al., 2010; Hartman et al., 2008; Sharma and Gujral 2011). Emergent compounds can have antioxidant activity sometimes higher than the initial phenolic compounds (Buchner et al., 2006). Due to thermal processing, synergies between antioxidant compounds and the food matrix can occur resulting into enhanced antioxidant activity of polyphenolic compounds (Wang et al., 2011). Increase in the antioxidant acitivity of flavonoids has been reported in thermally treated food matrix (Freeman et al., 2010). It seems that products of degradation, which are assumed to be more reactive than the initial phenolic compound and the synergistic interaction between the antioxidant compounds and the food matrix are key factors that may be ascribed to the enhanced antioxidant activity of thermally processed foods in some cases.

5. Conclusion

The study has demonstrated that cooking has positive effects on the antioxidant activities and phenolic phytochemical profiles of common beans and bambara groundnuts. In both common beans and bambara groundnuts, antioxidant activities are higher in the cooked samples compared to the uncooked. Cooking favoured the release of phenolic compounds and subsequently resulted in high concentrations of total polyphenols and individual phenolic compounds. Additionally, HPLC-PDA-ESI-MS advanced analytical technique revealed that new phenolic compounds emerge after cooking, and these may have an additive effect on the antioxidant activities. New compounds tentatively identified in cooked red bambara groundnuts include caffeic acid hexoside, gallocatechin, resveratrol glucoside and a caffeic acid derivative. In cooked red beans, new compounds tentatively identified include catechin glucoside, quercetin-3-O-glucoside, kaempferol glucoside and a quercetin conjugate. Cooking therefore enhances the nutraceutical profiles in both common beans and bambara groundnuts.

References

Akıllıoğlu, G. H., & Karakaya, S. (2009). Effect of Some Domestic Cooking Methods on Antioxidant Activity, Total Phenols and Total Flavonoid Content of Common Beans. *Akademil Gida, 7*(6), 6-12.

Akillioglu, G. H., & Karakaya, S. (2010). Changes in Total Phenols, Total Flavonoids, and Antioxidant Activities of Common Beans and Pinto Beans after Soaking, Cooking, and in vitro Digestion Process. *Food Sci. Biotechnol., 19*(3), 633-639. http://dx.doi.org/10.1007/s10068-010-0089-8

Amandeep, K. sandhu, & Liwei, G. (2010). Antioxidant Capacity, Phenolic Content, and Profiling of Phenolic Compounds in the Seeds, Skin, and Pulp of *Vitis rotundifolia* (Muscadine Grapes) As Determined by HPLC-DAD-ESI-MSn. *J. Agric. Food Chem, 58*, 4681-4692 4681. http://dx.doi.org/10.1021/jf904211q

Becker, E. M., Nissen, L. R., & Skibsted, L. H. (2004). Antioxidant evaluation protocols: Food quality or health effects. *Eur. Food Res. Technol, 19*, 561-571. http://dx.doi.org/10.1007/s00217-004-1012-4

Benzie, I. F. F. (1996). *Clin. Biochem. 29*, 111-116.

Buchner, N., Krumbein, A., Rhon, S., & Kroth, L.W. (2006). Effect of thermal processing on the flavonol rutin and quercetin. *Rapid Comm. in Mass Spectro., 20*, 3229-3235. http://dx.doi.org/10.1002/rcm.2720

Chandrasekara, N., & Shahidi, F. (2011). Effect of roasting on phenolic content and antioxidant activities of whole cashew nuts, kernels and testa. *J. Agric. Food Chem, 59*, 5006-5014. http://dx.doi.org/10.1021/jf2000772

Cheynier, V. (2005). Polyphenols in foods are more complex than often thought. *Am. J. of Clin. Nutr.* 81(suppl), 223S-9S.

Danesi (2009). PhD thesis Biological Effects of Bioactive Components and Extracts derived from Edible plants Commonly used in Human Nutrition. PhD thesis, University of Bologna

Estrella, I., Aguilera, Y. Benitez, V., Rosa, M. Esteban,. Martín-Cabrejas (2011). Bioactive phenolic compounds and functional properties of dehydrated beans flours. *Food Res. Int. 44*, 774-780. http://dx.doi.org/10.1016/j.foodres.2011.01.004

Freeman, B. L., Eggett, D. L., & Parker, T. L. (2010). Synergistic and antagonistic interactions of phenolic compounds foung in navel oranges. *J. Food Scie., 75*(6), C570-C576. http://dx.doi.org/10.1111/j.1750-3841.2010.01717.x

Fuleki, T., & Ricardo-Da-Silva, J. M. (2003). Effect of cultival and processing methods on the contents of catechin and procyanidins in grape juice. *J. Agric. Food Chem, 51*, 640-648. http://dx.doi.org/10.1021/jf020689m

Gerard, K. A., & Roberts, J. S. (2004). Microwave heating of apple mash to improve juice yield and quality. *Food Scie. and Techno., 37*, 551-557. http://dx.doi.org/10.1016/j.lwt.2003.12.006

Gouveia S., & Castilho P. C. (2011). Characterisation of phenolic acid derivatives and flavonoids from different morphological parts of *Helichrysum obconicum* by a RP-HPLC–DAD–ESI-MS^n method. *Food Chem, 129*, 333-344. http://dx.doi.org/10.1016/j.foodchem.2011.04.078

Gülçin, E., Bursal, H. M., Sehitoglu, M., Bilsel, & Gören, A. C. (2010). Polyphenol contents and antioxidant activity of lyophilized aqueous extract of propolis from Erzurum, Turkey. *Food Chem. Toxicol, 48*, 2227-2238. http://dx.doi.org/10.1016/j.fct.2010.05.053

Harbaum, B. (2007). Characterization of Free and Cell-Wall-Bound Phenolic Compounds in Chinese *Brassica* Vegetables. PhD thesis, Christian-Albrechts-Universität zu Kiel.

Hartmann, A., Patz, C. D. Andlauer, W., Dietrich, H., & Ludwig, M. (2008). Influence of processing on quality parameters of strawberries. *J.Agric. Food Chem*, 56(20), 9484- 9489. http://dx.doi.org/10.1021/jf801555q

Hassain, M., Dilip K., Brunton, Nigel, Martin-Diana, A., & Barry-Ryan, C. (2010). Characterization of Phenolic Composition in Lamiaceae Spices by LC-ESI-MS/MS. *J. Agric. Food Chem, 58*(19), 10576-10581. http://dx.doi.org/10.1021/jf102042g

Irina, I., & Mohamed, G. (2012). Biological activitities and effects of food processing on flavonoids as phenolic antioxidants. Adv. in Appl. Biotechnol., ISBN: 978-953-307-820-5, In Tech. http://dx.doi.org/10.5772/30690

Kalt, W. (2005). Effects of Production and Processing Factors on Major Fruit and Vegetable Antioxidants. *J Food Sci* 70(1), R11-R19. http://dx.doi.org/10.1111/j.1365-2621.2005.tb09053.x

Lin, L. Z., James, M. Harnly, Matcial, S. Pastor-Corrales, & Devanand, L. (2008). The Polyphenolic profiles of common beans (*Phaseolus vulgaris* L.). *Food Chem, 107*, 399-410. http://dx.doi.org/10.1016/j.foodchem.2007.08.038

Makkar, H. P. S., Hagerman, A., & Mueller-Harvey, I. (2000). Quantification of tannins in tree foliage – a laboratory manual, FAO/IAEA Working Document. IAEA, Vienna, 23-24.

Mitcham, B., Cantwell, M., & Kader A. (1996). Methods for Determining Quality of Fresh Commoditites. *Perishables Handling Newsletter*, February Issue, 85, 1.

Nyau, V., Prakash, S., Rodrigues, J., Farrant, J. (2015). Antioxidant Activities of Bambara Groundnuts as Assessed by FRAP and DPPH Assays. *Am. J. of Food and Nutr., 3*(1), 7-11.

http://dx.doi.org/10.12691/ajfn-3-1-2

Nyau, V., Prakash, S., Rodrigues, J., Farrant, J. (2016). Screening Different Zambian Market Classes of Common Beans (*Phaseolus vulgaris)* for Antioxidant Properties and Total Phenolic Profiles. *J. of Food and Nutr. Res.,* *4*(4), 230-236. http://dx.doi.org/10.12691/jfnr-4-4-6

Pal, D. K., Kumar, S., Chakraborty, P., & Kumar, M. A. (2008). A study on the antioxidant activity of *Semecarpus anacardium* L. f. Nuts. *J. of Nat. Rem., 8*, 160-163.

Papas, A. (1996) Determinants of antioxidant status in humans. *Lipids, 31*(1) S77-S82. http://dx.doi.org/10.1007/BF02637055

Rabaneda, F. S., Olga, Ja´uregui, Rosa Maria Lamuela-Ravento Jaume Bastida, Francesc Viladomat & Carles Codina. (2003). Identification of phenolic compounds in artichoke waste by high performance liquid chromatography–tandem mass spectrometry. *J. of Chrom. A* 1008, 57-72. http://dx.doi.org/10.1016/S0021-9673(03)00964-6

Shadkami, F., Estevez, S., & Helleur, R. (2009). Analysis of catechins and condensed tannins by thermally assisted hydrolysis/methylation-GC/MS and by a novel two step methylation. J. *of Analy. and Appl. Pyroly, 85*, 54-65. http://dx.doi.org/10.1016/j.jaap.2008.09.001

Sharma, P., & Gujral, H. S. (2011). Effect of sand roasting and microwave cooking on antioxidant activity of barley. *Food Res. Int., 44*, 235-240 http://dx.doi.org/10.1016/j.foodres.2010.10.030

Valverdú-Queralt, A., Medina-Remón, A., Andres-Lacueva, C., & Lamuela-Raventos, R. M. (2011). Changes in phenolic profile and antioxidant activity during production of diced tomatoes. *Food Chem., 126*, 1700-1707. http://dx.doi.org/10.1016/j.foodchem.2010.12.061

Villaño, D., M. S. Fern´andez-Pach´on, M. S., M. L. Moy´a, M. L., Troncoso, A. M., & Garc´ıa-Parrilla, M. C. (2007). Radical scavenging ability of polyphenolic compounds towards free radicals. *Talanta, 71*, 230-235.

Wahle, K. W., Brown, I., Rotondo, D., & Heys, S. D. (2010). Plant Phenolics in the Prevention and Treatment of Cancers. *Adv. Exptl. Med. Biol. 698*, 36-51. http://dx.doi.org/10.1007/978-1-4419-7347-4_4

Wang, S., Meckling, K. A., Marcone, M. F., Kakuda, Y., & Tsao, R. (2011). Synergistic, additive, and antagonistic effects of food mixtures on total antioxidant capacities', *J. Agric. Food Chem., 59*, 960-968. http://dx.doi.org/10.1021/jf1040977

Waters Corporation (2002). HPLC troubleshooting guide. American Laboratory and Waters Corporation, 720000181EN, 08/02.

Yu, J., Ahmedna, M., & Goktepe, I. (2005). Effects of processing methods and extraction solvents on concentration and antioxidant activity of peanut skin phenolic. *Food Chem., 90*, 199-206. http://dx.doi.org/10.1016/j.foodchem.2004.03.048

PERMISSIONS

LIST OF CONTRIBUTORS

Sara Najdi Hejazi and Valérie Orsat
Bioresource Engineering Department, McGill University, Ste-Anne-de-Bellevue, Canada

Amegovu K. Andrew
Department of Food Science & Technology, College of Applied and Industrial Sciences, University of Juba. P. O. BOX 83, Juba, South Sudan

Anuradha Vegi and Clifford A. Hall III
Department of Plant Sciences, North Dakota State University, North Dakota, USA

Charlene E. Wolf-Hall
Department of Veterinary and Microbiological Sciences, North Dakota State University, North Dakota, USA
Office of the Provost, North Dakota State University, North Dakota, USA

Wendy L. Lizárraga-Mata, Celia O. García-Sifuentes, Susana M. Scheuren-Acevedo, María E. Lugo-Sánchez, Libertad Zamorano-García, Juan C. Ramirez-Suárez and Marcel Martinez-Porchas
Centro de Investigación en Alimentación y Desarrollo, A. C. Carretera a la Victoria, Km 0.6. Hermosillo, Sonora,C.P. 83304,México

Ferouz Ayadi, K. Muthukumarappan and S. Kannadhason
Department of Agricultural and Biosystems Engineering, South Dakota State University, USA

Kurt A. Rosentrater
Department of Agricultural and Biosystems Engineering, Iowa State University, USA

Kouadio N. Joseph, Akoa E. Edwige, Kra K. A. Séverin and Niamké L. Sébastien
Laboratoire de Biotechnologies, UFR Biosciences, Département de Biochimie, Université Félix Houphouët-Boigny, 22 BP 582 Abidjan22, Côted'Ivoire

Elisa A. S. F. Boin
Estoril Higher Institute for Tourism and Hotel Studies, Av. Condes de Barcelona, 801, Estoril, 2769-510, Portugal
BLC3 Association –Technology and Innovation Campus, Rua Nossa Senhora da Conceição, 2, Lagares, Oliveira do Hospital, 3405-155, Portugal

Cláudia M. A. M. Azevedo and Manuela M. Guerra
Estoril Higher Institute for Tourism and Hotel Studies, Av. Condes de Barcelona, 801, Estoril, 2769-510, Portugal

João M. S. A. Nunes
BLC3 Association –Technology and Innovation Campus, Rua Nossa Senhora da Conceição, 2, Lagares, Oliveira do Hospital, 3405-155, Portugal

Karola R. Wendler
Agriculture & Agri-Food Canada, Lacombe Research Centre, Canada , Canada , Canada
Delacon Biotechnik GmbH, Weissenwollfstr. Steyregg, Austria

Francis M. Nattress, Jordan C. Roberts, Ivy L. Larsen and Jennifer L. Aalhus
Agriculture & Agri-Food Canada, Lacombe Research Centre, Canada , Canada , Canada

Jordan J. Rich and Kurt A. Rosentrater
Department of Agricultural and Biosystems Engineering, Iowa State University, 3327 Elings Hall, Ames, IA 50011, USA

Areli H. Peredo-Luna, Aurelio Lopez-Malo, Enrique Palou and María Teresa Jiménez-Munguía
School of Engineering, Department of Chemical, Environmental and Food Engineering, Universidad de las Américas Puebla, Puebla, Mexico

Riin Karu and Ingrid Sumeri
Competence Centerof Food and Fermentation Technologies, Akadeemia tee 15A, 12618 Tallinn, Estonia

Adepoju, Oladejo Thomas
Department of Human Nutrition, Faculty of Public Health, College of Medicine, University of Ibadan, Ibadan, Nigeria

Ajayi, Kayode
Department of Human Nutrition and Dietetics, College of Medicine and Health Sciences, Afe Babalola University, Ado-Ekiti, Ekiti State, Nigeria

Fabrice F. D. Dongho, Inocent Gouado, Lambert M. Sameza, Adélaïde M. Demasse and Annie R. N. Ngono
Department of Biochemistry, Faculty of Science, University of Douala, P.O. Box 24157 Douala, Cameroon

Raymond S. Mouokeu
Institute of Fisheries and Aquatic Sciences, University of Douala, P.O. Box 2701, Douala, Cameroon

Florian J. Schweigert
Institute of Nutritional Science, University of Potsdam, Arthur-Scheuert-Allee 114-116, 14558 Bergholz-Rehbrücke, Germany

José Fernando Haro-Maza and José Ángel Guerrero-Beltrán
Departamento de Ingeniería Química, Alimentos y Ambiental. Universidad de las Américas Puebla. Ex Hda. Santa Catarina Mártir, San Andrés Cholula, Puebla 72810, Mexico

Jason R. Croat and William R. Gibbons
Biology & Microbiology Department, South Dakota State University, Brookings, SD 57007, USA

Mark Berhow
USDA, Agricultural Research Service, National Center for Agricultural Utilization Rearch; Peoria, IL 61604, USA

Bishnu Karki and Kasiviswanathan Muthukumarappan
Agricultural & Biosystems Engineering Department, South Dakota State University, Brookings, SD 57007, USA

Djedjro C. Akmel, Arsène. L. I Nogbou, Ibrahima Cissé, Kouassi E. Kakou, Kisselmina Y. Koné, Nogbou E Assidjo and Benjamin Yao
Laboratoire des Procédés Industriels, de Synthèses, de l'Environnement et des Energies Nouvelles, Institut National Polytechnique Félix Houphouet Boigny (LAPISEN/ INP-HB), Yamoussoukro, Côte d'Ivoire

Abimbola Rashidat Ezeh
University of Lagos, Nigeria
University of Texas Medical School, Houston, Texas, USA

Olusimbo Olugbo Aboaba
University of Lagos, Nigeria

Barbara E Murray
University of Texas Medical School, Houston, Texas, USA

Ben Davies Tall
Center for Food Safety and Applied Nutrition, US Food and Drug Administration, USA

Stella Ifeanyi Smith
Nigerian Institute of Medical Research, Nigeria

Mesfin W. Tenagashaw
Department of Food Science and Technology, Jomo Kenyatta University of Agricultureand Technology, Nairobi, Kenya
Faculty of Chemical and Food Engineering, Bahir Dar Institute of Technology, Bahir Dar University, Bahir Dar, Ethiopia

Glaston M. Kenji and John N. Kinyuru
Department of Food Science and Technology, Jomo Kenyatta University of Agricultureand Technology, Nairobi, Kenya

Eneyew T. Melaku
Department of Food Science and Applied Nutrition, Addis Ababa Science and Technology University, Addis Ababa, Ethiopia

Susanne Huyskens-Keil
Division Urban Plant Ecophysiology, Research Group Quality Dynamics/Postharvest Physiology, Faculty of Life Sciences, Humboldt-Universitätzu Berlin, Berlin, Germany

Philippe Villers
President, GrainPro, Inc., Concord MA,USA

Valerie Dussault and Marie Marquis
Department of Nutrition, University of Montreal, Canada

V. Nyau
Department of Food Science and Nutrition, University of Zambia, Lusaka, Zambia

S. Prakash
Department of Chemistry, University of Zambia, Lusaka, Zambia

J. Rodrigues and J. Farrant
Department of Molecular and Cell Biology, University of Cape Town, 7701 Rondebosch, South Africa

Index